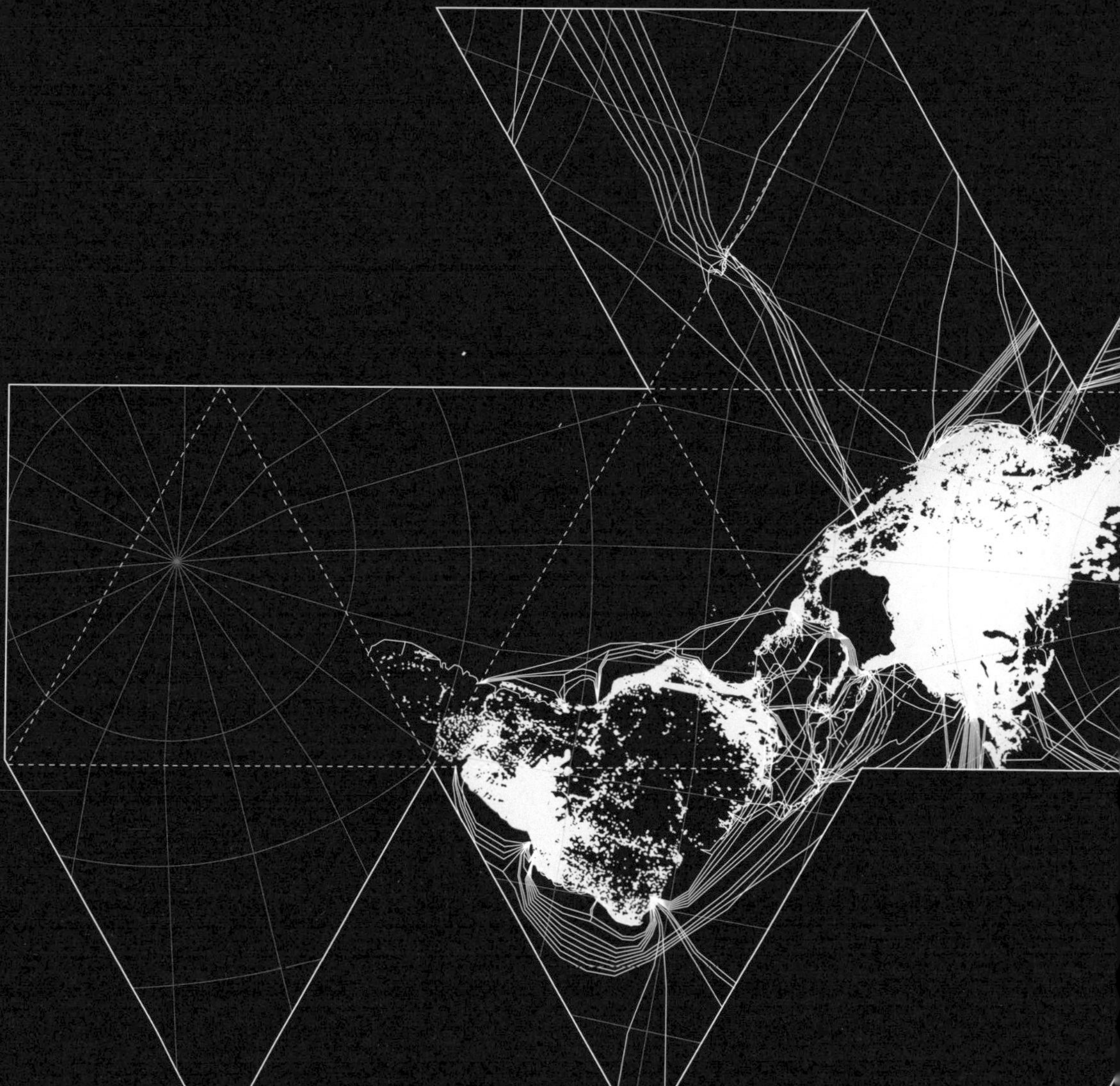

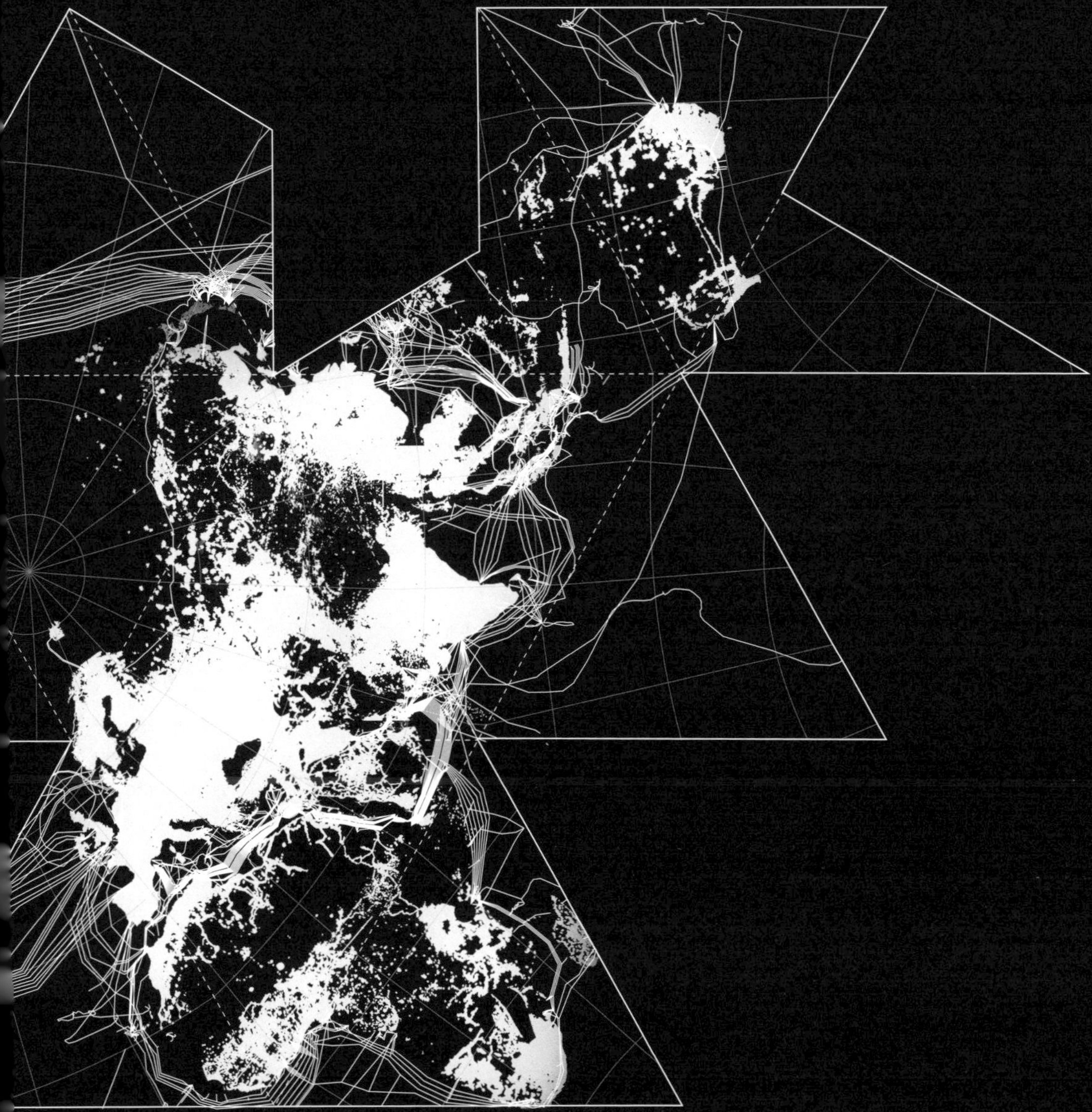

As of 2022, there are over 1.35 million kilometers of submarine internet cables in service. Their disparate distribution across the globe reveals territories of (dis)connection. The areas represented in black are places without cell phone reception or internet Wi-Fi connection. Sources: Buckminster Fuller Institute, The White Spots Project, TeleGeography. Drawing by Léa Alapini.

# DATAPOLIS

EXPLORING THE FOOTPRINT OF DATA ON OUR PLANET AND BEYOND

Paul Cournet, Negar Sanaan Bensi

nai010 publishers

# INTRODUCT

## INTRODUCTION: ON DATAPOLIS

Negar Sanaan Bensi, Paul Cournet

### To Be-Come Invisible

Driving in the woods near Eindhoven, in the south of the Netherlands, we are trying to navigate our route on Google Maps as the network gets weaker. This area is one of the very few zones in the country that has a low density of 5G towers. After a few hours of driving through the Dutch landscape, defined by its indefinite flatness, we finally arrive at our destination, a big cottage surrounded by trees and covered in plants. Birds are singing in the background. As we approach the cottage, we are requested to put our phones in flight mode and leave them in the car, which is parked outside the plot at a short distance. We are here to meet a community of people who suffer from EHS, or Electromagnetic hypersensitivity, generally defined as 'a claimed sensitivity to electromagnetic fields, to which negative symptoms are attributed'.[1] We sit in the garden and chat with them. 'I used to have a normal job, but over the years I've become more and more sensitive to electromagnetic fields with all of these wireless devices. I used to have intense headaches at home, at work, or even going to the city. My life became impossible. So I moved here to the forest, far from communication antennas and wireless networks,' one of the ladies explains. 'It's almost impossible to be invisible,' another states, 'this creates this constant state of anxiety.' This is the moment we realize just how much our bodies and environment have become entangled with the 'infosphere' that is a 'technical network, consisting of telegraphy, telephony, television, radio, radar, satellites, and the Internet, which covers the globe and enables global exchange of data as well as the organization of transport for people and goods',[2] but also ideas and histories. As Peter Wiebel states in the preface to the 'Infosphere' exhibition in ZKM: 'Today, people live in a globally interconnected world in which the biosphere and the infosphere are interfused and interdependent.'[3] This infosphere, however, is never smooth or homogeneous, as it is often described or anticipated within the context of globalization. Rather, it is striated and heterogeneous. It is a geopolitical sphere.

A few months after our visit to the EHS community, another very different context highlighted this entanglement of the bio- and the technosphere. The protest movement with the slogan of 'woman, life, freedom' erupted across Iran. During this time, the Internet was shut down inside the country by the government in an attempt to stop news of police violence against citizens from leaking out. Even plans for establishing a national Internet network or a new Internet-classed society to limit the access of the general population to the Internet were suggested. The aim of all such actions was to keep people in the dark and invisible.[4]

Perhaps what can be considered as a shared problem in these seemingly opposite situations resides in what

# ON: ON DAT

Bernard Stiegler sees in the Internet as a 'pharmakon'[5] that can become 'a technique for hyper-control', leading to the state of 'social dis-integration' by submitting further to automatism. A state that results from 'the unprecedented aspect of the digital unification that it allows articulations between all these automatisms: technological, social, psychic and biological- and this is the main point of neuro-marketing and neuro-economics'[6]—and one might add neuro-politics. The logic of such processes stems from the calculative power and speed of algorithms devised in this technology to capture and exploit any traces of behavioural expressions of its users.

Hence, according to Stiegler, there exists the necessity of an 'ars of control', as a therapeutic 'invention'—and not purely as a mode of resistance as Deleuze suggested in the context of his hypothesis on 'The Societies of Control'. Through such an art—which should become ars again[7]—a new knowledge is produced, constituting new forms of existence and new ways of living, thinking, doing things, and reasoning to do things, and eventually 'new conditions of subsistence'.[8]

But that sounds like bickering with a many-headed demon.

## The Many-Headed Demon

Data and its infrastructure is a many-headed demon. In the words of Benjamin Bratton, it might resemble an 'amorphous Amoeba-like creature'[9] that moves and alters its shape by extending and retracting its pseudopods to cover the Earth and its beyond. Within that amorphousness, it contains opposing forces and narratives. Something like Ravana the Indian demon, who has several heads and hands. While holding many qualities that make him a learned scholar, he is widely portrayed as an evil character who can change into any form he wishes.

Leading to an epoch that is often referred to as the information age, this many-headed demon has changed the ways we relate to our world. There have been various tendencies to describe the inherent contractions within this technosphere. Some consider this networked infrastructure, distributed in time and space, as 'the driving connective forces of much-debated processes of globalisation',[10] which stem from the powerful ideology of constructing a smooth world made of inter-networks with no hierarchy.[11] Others call it 'Surveillance Capitalism',[12] which 'claims the human experience as free raw material for its advanced manufacturing processes'[13] for a market surplus, creating giants like Google, Facebook, Apple, Microsoft and Amazon, whose control of data gives them enormous power in a so-called 'data economy'.

Today we have increasingly gotten used to inhabiting this bio-infosphere,[14] where data, algorithms, barcodes and 'the cloud' are ubiquitous to our bodies and daily activities.[15] As we slowly dissolve ourselves in living and working in the world of connectivity, regardless of time zones, geographies and political borders, we intend to forget about it. This disappearance as Timothy Morton suggests is a quality of things when they are functioning nicely in our world.[16]

Despite this ubiquity, we hardly comprehend the mechanisms of this infosphere, a complex world that can be glimpsed through tangible and intangible evidence: from the global Internet cable networks and so-called non-human architecture of data centres and IXPs to automated landscapes of factories and distribution centres; from the effect and footprint of data on the Earth to climate monitoring and the formation of global and planetary institutions; from the ownership system, political mapping and legal complexity and ambiguity of data to the appearance of digital identities; from the changing conditions of work and labour to expanded logistical systems. All these entangled complex territories, spaces, organizations, systems and entities constitute what we call Datapolis, which in our understanding requires proper exploration, analysis, conceptualization, projection and representation.

## Datapolis

Datapolis was originally a research/design studio and a project that we initiated in 2019 at the Faculty of Architecture at Delft University of Technology. The Datapolis studio explored the physical and spatial aspects of data, its effect on the environment, and its relevance to architectural research and design. When we started this project, the first question simply was: What is the cloud? What is its materiality and how do we account for both the heavy weight and ephemerality of an entity such as the cloud? As we went on and the topic unfolded, we faced different angles and questions. Are data and its infrastructure merely another technological advancement for us to utilize? Is it a worldview and a mode of believing as, for example, Harari puts it in the emergence of 'Dataism'?[17] Or is it a conceptual and philosophical development in the way that, for example, Alexander Galloway discusses the concept of the digital?[18]

Datapolis is a term coined by Francis Pisani in the book, *A Journey through Smart Cities,* to describe a 'city completely managed using data collected by technological infrastructure'.[19] It sits in extreme opposition to 'Participolis', a city 'in which citizens participate in the design and management of the space in which they live'.[20] While such a binary framework is arguably problematic, it is also a reminder of Antoine Picon's statement that 'a whole array of intermediaries exist between technocratically inspired control and freely consented cooperation'.[21] He writes:

> In the domain of the smart city, we are indeed witnessing the emergence of projects inspired by neocybernetics, based on an ideal of a hierarchical control of information and of the behaviour that results from it. In contrast to this interpretation of the new urban ideal, which can easily be criticised for its technocratic tendency, is a more participatory vision founded on individuals cooperating freely, following the example of businesses such as Wikipedia ... The future intelligence of cities will reside in the coexistence of these two approaches, and above all in the interactions that such a cohabitation is sure to provoke.[22]

In this book, we follow Antoine Picon in not viewing Datapolis as the opposite of Participolis. Datapolis does not epitomize a technocratic tendency that necessarily sits in opposition to participatory visions. Instead, it is an attempt to 'give form' to the contradictory desires inherent to the idea of a city. As Massimo Cacciari explains, we want the city to work like a machine with extreme effectivity and efficiency, containing a complex of flows and functions and we also demand it to give us 'peace', to be our 'sojourn', like a 'womb' holding us safe.[23]

Datapolis is about cities in transition that due to their unstable nature cannot be reduced to any of these tendencies. Moreover, it is not 'a' city, it is always cities, territories, cultures, histories and so on. Datapolis is an invitation to explore these territories, their inhabitants, their ways of life, relations, cultures, infrastructures, organizational systems and so on. It is an invitation to join us in reading it, mapping it and conceptualizing it, as well as projecting and imagining it.

### Between Thing and Data: A New Rapport

Certainly data and its infrastructure have had great technical and social impacts. Opening up questions of ontology, thinkers like Gilbert Simondon and Bruno Latour have suggested a different 'rapport' between man and 'technical object'[24] as a coupling between living and non-living. Away from its object-like instrumentalization, they refer to this rapport as a networked and complex assemblage between object-subject and object-object. In this 'open system' the relationship between human and non-human cannot be reduced to the master-slave dialectic as it was understood in traditional theories of the machine. Other important writers, like Donna Haraway, have also rejected the rigid boundaries separating 'human' from 'animal' and 'human' from 'machine'.[25] The important writings of Bernard Stiegler go one step further, stating that humans and technics are constitutive of each other. He decentralizes the concept of the human by approaching it not only as a historical or social construct, but a technical one through a process of exteriorisation—that is, by extending beyond human boundaries (bodily, cognitive and temporal) into technics (technic here is no longer opposed to culture and nature, rather it involves all sorts of arts and artefacts, including architecture).[26] In this sense, both Simondon and Stiegler are critical of [pure] automatism. For Simondon this is because of automatism's intention to exclude man and indeterminacy and hence its closed system; which is, in fact, contrary to what 'technicity' is—that is, an open system that perpetuates the co-existence and co-relation of living and non-living and is an ensemble of man and machine.[27] For Stiegler, automation's crisis stems from the analytical power of algorithms that 'are conveyed outside of any intuition in the Kantian sense—that is, outside of any experience'.[28]

The decentralization of the human in contemporary Western philosophy and its reunion with the world of objects and other beings has occurred at the same time as an increasing technological integration of our buildings, spaces and territories. With the drive towards full automatism, fundamental ontological and epistemic questions are being raised in our discipline. Currently, within the realm of architecture and urbanism the topic of data—as ambiguous as it is—is very relevant and highly debated. Many existing scholarly works have attempted to unravel the importance and impact of data on the [trans]formation of our cities, territories and spaces[29] and modes of understanding them.[30] Yet, speculations on how this can be integrated into real spatial and territorial thinking require much further work.

Fundamental questions such as: Do we need a different understanding of architecture itself? Or: Have we acquired new ways to get to know the world?[31] are now the core of many discussions in architecture, landscape and urban design. Authors like Keller Easterling remark that: 'Buildings are often no longer singularly crafted enclosures, uniquely imagined by an architect, but reproducible products set within similar urban arrangements.'[32] More than being static objects, they have become 'active forms'. A kind of thinking that is finding its way into the discussions on architecture as an affordance that can 'operate, affect, and interact as environments, entities, and beings',[33] and also considers architecture as a mediatory and manipulative procedure that articulates the configuration of information systems of spatial arrangements.[34]

Writers like Picon, on the other hand, while recognizing the shifts occurring in architecture in the aftermath of digitization and Big Data, have different ideas: to go back to the problem of meaning for the sake of reinventing or reinterpreting the fundamentals of architecture.[35] At the same time authors like Martin Pawley question totally the monumental ambitions of architecture in resolving our current technological and infrastructural mess. According to him, architecture instead can find its relevance in bringing 'together the anonymity of the "hot site", the efficiency of the "Big Shed", the ephemeralization of the petrol station, and the limitless horizons of virtual reality, and exposes all of them to the methodology of automated industrial production'.[36]

Despite the massive evolution of technology in the digital age, it is often stated that compared with other epochs where transformations were caused by technological and infrastructural developments, physically, cities have not been modified much. While Picon assumes that this morphological change is yet to happen,[37] Shannon Mattern suggests how new technologies are often being grafted onto existing infrastructures, a horticultural analogy that could provide an alternative to reinstating the city as a tree, but only if 'we learn how to recognize our urban grafts' layered and entangled manifestations, to discern the stories behind each cut and fusion, and to recognize the ethics and politics of grafting technique'.[38]

Such a reading can lead us to the expanded territories formed through these grafted technologies, taking us beyond the city itself to the massive spatial transformations that have occurred in the countryside, deep in the soil, in the oceans, and as far as outer space. These territories are where we find the physical footprints of our infosphere, laid out or submerged or floating. They suggest a need to host other inhabitants—human and non-human—within our architectural spaces. One can argue that we are witnessing a great reterritorialization of built environments, a great spatial shift that has suspended the scale of architecture somewhere between the microscopic scale of bacteria and the geologic scale of the Earth.

## The Book

In this book, we intended to go beyond the mere instrumentality of digital technology and the existing discourse around digital architecture and smart cities. Instead, we are interested in data's own infrastructure, its tectonics and materiality, its inherent ethical and environmental contradictions, as well as the related political and cultural productions. For us, the challenge has been to go beyond the pure description of these systems and infrastructures as only services and backgrounds, in order to surpass the technological problem-solving mandates by placing them within wider political, economic, social, environmental and aesthetic assemblages. These assemblages have become an inherent part of the book, allowing different questions to intervene in the text through various material representations such as maps, different forms of drawings, photographs and texts.

In the book, we follow two main trajectories: the first trajectory is an ontological investigation per se. It attempts to define what 'the cloud' is and how it operates. From the systems and infrastructures behind the Internet to the apparatus, gizmos and buildings that can transcend scales and temporal dimensions. Not only do we map, read and analyse, but we also look into alternative conceptualizations and projections. The second trajectory tries to explore how data penetrates our existence, not only by affecting the ways we live and work, or design and make cities, but by offering distinct ways of life and organization that otherwise would not have been possible. On the one hand, we investigate data as an ethical and political phenomenon. On the other, we perceive how data through its apparatus and sensors gives agency to things that previously had no voice, gives 'visibility' to invisible things, and forms new imaginaries.

The book is structured like a barcode, it represents the combination of the different textual and visual materials that we produced and gathered throughout the process. It contains four main chapters that can be read or looked at autonomously, and yet are complementary. Many intersections and interrelations are floating thematically between different chapters, and we hope that the reader will enjoy exploring them.

The first chapter is an elongated introduction, including maps that question the European-centric representation of data infrastructures, a timeline that calls for multiplicity and non-linearity in history and reveals the gaps in the existing historiographies of technology. It narrates alternative stories of infrastructural spaces. The emergence of digital technology is rooted in changes in theories and inventions, software and hardware, episteme and technical objects, modes of learning and doing, perceiving and projecting and hence it cannot be perceived linearly. It is a series of entangled events that form the infosphere we live in today. While paradoxically the notion of the timeline suggests a linearity of events, we state that an oligopticon[39] reading of it allows for the multiplicity of narratives to emerge and intersect. Hence, the presented retroactive timeline illustrates a possibility for transversal rather than linear readings of events through several different narratives, superposing the apparent linearity of the background. As the number of narratives increases, the constellation of lines and points will take over the linear organization of events. The timeline is accompanied by a text by Georg Vrachliotis, 'Notes on an Archaeology of Design Data Literacy', which adds another narrative to such a constellation.

The chapter ends with the Catalogue of Gizmos, which suggests alternative relationships between technical objects, other beings and the human body. As Kees Kaan suggests in the glossary of his text 'On Gizmos, Gadgets, Objects and Data', 'the gizmo is a weird gadget whose real name and use has been forgotten'. In this sense these objects almost need to be revisited and their places reallocated in our world.

The second chapter focuses on a grounded investigation of data infrastructure and the automated landscapes that are heavily dependent on it. Through a photographic journey in the Netherlands, Paul Swagerman explores the layered boundaries, materiality, aesthetics, spatial and organizational complexity of data [infra]structures and highly automated landscapes that are sometimes devoid of human presence, and are often hidden within vast ubiquitous boxes or under the ground in the bunkers inherited from the World Wars. Through a rather personal gaze, this collection documents the observations of the photographer while visiting nine different locations in the country: the point where the undersea Internet cable reaches the land from the North Sea, the buildings storing data online or offline, the landscapes of highly automated industries and logistical landscapes, all the way to a counter-narrative for the Datapolis, where the struggle for invisibility is practiced daily by communities in the least connected parts of the country, in the forests or deserted dune landscapes.

The third chapter contains all the textual contributions to the book. The first part gathers three important pieces of already published works by three female writers on the topic, which together build a trilogy about technology on Earth, Grafts and Race. This part includes excerpts by the well-established authors Kate Crawford, Shannon Mattern and Ruha Benjamin. In a way, all three texts explore the scenes behind the clean face and the structural and social hierarchies traceable within AI and data technology. They unravel tales and stories that urge us to more carefully localize the tectonic of data technology, to (re)consider justice and (re)incorporate others' intelligence when utilizing these technologies.

The second part, however, involves a younger generation of authors, artists and designers who reflect on data, its infrastructure and agency. From different angles, the papers in this chapter present alternative conceptions and readings of data and its infrastructure. In her paper 'From Data Centre to Data Forest', Marina Otero Verzier delves into new architectural paradigms for storing data, attuned to social and ecological challenges, such as plant-DNA storage. She then encourages us to imagine the forest as the potential paradigm of a non-Cartesian, situated and open-ended data-storing architecture. In her paper 'Vibrant Data: The Sewer as Information Infrastructure', Sabrina Chou foregrounds the sewer as an information infrastructure, by tracing flows of faecal data and digital information to gesture towards an embodied metaphysical conception of data as a vibrant, relational and social substance. Following Chou, Ali Fard, in his paper 'On the Sociotechnical Production of Planetary Platforms', unravels technology platforms as well-designed sociotechnical entities that combine ideology, myth and marketing to construct amorphous and all-encompassing narratives, while actively hiding their dependence on the infrastructural landscapes, material geographies and operational extensions. Departing from the metaphor of the digital cloud, in her paper 'The Digital Cloud: Between Geological Extraction and Meteorological Hyper-Responsiveness', Natalie P. Koerner explores the spatialities, materialities and temporalities of data centres. Describing and unpacking the cloud through planetary analogies, she questions the innate neo-colonial injustices embedded in the material reliance of 'the cloud' on earth minerals, extraction and climate politics.

One of the most apparent and important changes facilitated by data flows and digital technology has happened in our abilities to see, remodel, represent and navigate. These new technologies have raised fundamental questions not only about how we perceive, relate to and imagine the space we inhabit, but also about the way we monitor, survey and surveil the environment and people.[40] Departing from these questions, Gökçe Önal, in her paper '"Here Be Dragons": The Liminal Topographies of Statistical Imaging', problematizes the all-seeing capacity associated with the accuracy of digital earth engines, arguing for an unseen ecology of energetic gaps, numeric approximations and rejected probability distributions growing together with the visible end of our cartographic screens. In the following contribution, Joost Grootens, in his paper 'This Way: Expressive Self-Tracking as Creative Practice', addresses the surveillance issues surrounding satellite navigation and fitness apps, as well as the economic models employed by the technology companies running the latter. Through a post-representational reading of GPS drawing, he emphasizes the ambiguity of authorship and the degree of completeness of digital production.

Extending on the topic of remodelling our world via digital technologies, in her paper 'Blockchain and the Real-Estate-Media Complex', Marija Marić shows how the emergence of fractional ownership and entire markets for developing and trading digital land has prompted the real-estate market to be organized around the uses and promises of blockchain technologies. She rightfully argues for a deeper understanding of such technologies, not only as technical innovations but above all as new imaginaries of social coordination and trust, based on their capacity to redefine our property relationship with the built environment and reappropriate the urban commons. Following Marić, in his artistic work 'Life Needs Internet: Handwritten Letters on Digital Culture', Jeroen van Loon archives the global influences of the Internet on our lives through local and personal testimonials, and intimate stories. Through these individual stories, he brings situated views that sometimes exceed their local and temporal coordinates and find commonalities.

The last chapter of the book includes the works produced by the students of Datapolis Studio. This chapter envisions other possibilities to, as Rania Ghosn and El Hadi Jazairy suggest 'speculate on ways of living with'[41] the legacy of Datapolis, as a way for architecture to intervene and project imaginaries that can assemble again the technical, political, social, environmental and aesthetic, and give them form. In these visions, architecture brings together the various scales, systems, things and spaces that often remain unrelated. It is exactly in the making of these relations and alternative scenarios that architecture finds its relevance, which is mostly visible in the projection and organization of the final drawings. The architectural drawing is a strong speculative medium to investigate and represent possible scenarios for the gatherings of technology and environment. The final drawings of each project were originally made at a large scale, 841 × 1189 mm in size, with a high level of detail. The nature of these drawings has meant that they are difficult to fit within the format of this book, but we have shown excerpts and details to illustrate particular points. This series of projects touch upon various themes, some of which have also been addressed in texts and visual forms of photographs in other chapters.

To conclude, the ambition of this book was to become a platform to explore different spectrums, perspectives, opportunities and challenges that digitization poses to the thinking and making of our environment, and in that sense, we do not claim to have reached solutions. We have tried to go beyond the opposing narratives on technology— that is, emancipation or oppression—and have instead explored the complex assembly of layered immanent forces within Datapolis. Bringing such an assembly to spatial thinking, we have often reminded ourselves to not lean towards one form of intelligence over another. Instead, we have asked whether it is possible, to integrate and assemble various forms of intelligence in our imaginaries, worlds, spaces and cities—that is, in our Datapolis.

Notes

1 Wikipedia, Electromagnetic Hyper Sensitivity.
2 Peter Wiebel, preface, Infosphere: The Transformation of Things into Data, ZKM, https://zkm.de/en/exhibition/2017/05/infosphere.
3 Ibid.
4 In recent years, we have seen around the world the increasing role of data and Internet on various sociopolitical events. For example, during the 2019 Hong-Kong protests, one of the largest series of demonstrations in the history of Hong Kong, the data infrastructure was employed both by the police and the protestors: to identify and control the protests in the first case, and to organize and create new tactics of fluidly moving the protestors, in the second case. Or the escalation of Internet censorships during the Arab Spring on the one hand and the role of Social media as enabling and facilitating the voice of the protestors on the other. Or the rise of radical groups on social media platforms and their impact on election poles in the United States—and so on.
5 According to Stiegler, because Internet is originally technological it can, like every other technics, be ambivalent, and hence be a pharmakon in its designated Greek meaning: 'At once the poison, the cure and the scapegoat.' Bernard Stiegler, Colette Tron and Daniel Ross, 'Ars and Organological Inventions in Societies of Hyper-Control', *Leonardo* 49/5 (2016), 480–484.
6 Ibid., 482.
7 *Ars* is the Latin equivalent of the Greek *techne*. 'The diminutive of "ars" is "articulum"—i.e. "a little art".' When Stiegler suggests that art becomes 'ars', he means that art meets 'techne', to literally be skillful again, standing both for knowledge and know-how. So for 'art of hyper-control' to be a 'therapeutic invention', it would 'need to be inseparable from a juridical, philosophical, scientific, political and economic inventiveness'. Ibid., 483.; Vilém Flusser and John Cullars, 'On the Word Design: An Etymological Essay', *Design Issues* 11/3 (1995), 50–53.
8 Ibid., 481.
9 Benjamin Bratton, *The Terraforming* (Moscow: Strelka Press, 2019).
10 Stephen Graham and Simon Marvin, *Splintering Urbanism: Networked Infrastructures, Technological Mobilities and the Urban Condition* (London: Routledge & Kegan Paul, 2001), 8.
11 Stephen Graham and Simon Marvin explain that in the Western world, since the Second World War, a powerful ideology has been built of how we perceive the infrastructural networks as services that 'add cohesion to territory' and 'bind cities, regions and nations into functioning geographical or political wholes' and are 'freely available to all individual at equal costs within a particular administrative area'. Ibid.
12 Shoshana Zuboff, *The Age of Surveillance Capitalism: The Fight for the Future at the New Frontier of Power* (London: Profile Books, 2019).
13 Ibid., 14.
14 This bio-infosphere is a new assemblage, a multiplicity that is made up of heterogeneous spheres constituting a new world.
15 It goes without saying that this state of living is not equally smooth everywhere, but certainly its presence and effects are extended everywhere on Earth and beyond.
16 Timothy Morton, *Being Ecological* (Cambridge, MA: MIT Press, 2018), 45.
17 Yuval N. Harari, *Homo Deus: A Brief History of Tomorrow* (New York: Harper Collins, 2017).
18 Alexander Galloway, *Laruelle: Against the Digital* (Minneapolis: University of Minnesota Press, 2014); Alexander Galloway, 'Golden Age of Analog', *Critical Inquiry* 48/2 (2022), 211–232.
19 Francis Pisani, *A Journey through Smart Cities: Between Datapolis and Participolis* (UNESCO/Netexplo Observatory, 2015), 9.
20 Ibid.
21 Antoine Picon, *Smart Cities: A Spatialised Intelligence* (Chichester: Wiley, 2015), 16.
22 Ibid.
23 Massimo Cacciari, 'Nomads in Prison', *Casabella* 705 (2002), 106–108.
24 Gilbert Simondon, *On the Mode of Existence of Technical Objects*, translated by Cécile Malaspina and John Rogove (Minneapolis: Univocal Publishing, 2017).
25 Donna Haraway, 'A Cyborg Manifesto: Science, Technology, and Socialist-Feminism in the Late Twentieth Century', in: Donna Haraway, *Simians, Cyborgs and Women: The Reinvention of Nature* (New York; Routledge, 1991), 149–181.
26 Bernard Stiegler, *Technics and Time, 1: The Fault of Epimetheus*, translated by Richard Beardsworth and George Collins (Stanford: Stanford University Press, 1989 [1994]); Robert Alexander Gorny and Andrej Radman, 'From Epiphylogenesis to General Organology', *Footprint* 16/1 (2022), 3–20; Andrés Vaccari and Belinda Barnet, 'Prolegomena to a Future Robot History: Stiegler, Epiphylogenesis and Technical Evolution', *TRANSFORMATIONS Journal of Media & Culture* 17, Bernard Stiegler and the Question of Technics (2009).
27 Simondon, *On the Mode of Existence*, op. cit (note 24).
28 Bernard Stiegler et al., 'Ars and Organological Inventions', op. cit. (note 5), 480.
29 See, for example: Anthony M. Townsend, *Smart Cities: Big Data, Civic Hackers, and the Quest for a New Utopia* (New York: W.W. Norton & Company, 2013), or several books by William J. Mitchell or the writings of Carlo Ratti on the impact of Big Data and digital technologies on city and architecture.
30 See, for example: Michael Batty, *The New Science of Cities* (Cambridge, MA: MIT Press, 2013).
31 Weaving Worlds: speculations between affect & evidence, https://www.weaving.world
32 Keller Easterling, *Extrastatecraft: The Power of Infrastructure Space* (London: Verso, 2014).
33 Jenny Sabin, 'Transformative Research Practice: Architectural Affordances and Crisis', *Journal of Architectural Education* 69/1 (2015), 63–71.
34 Michael Barton, 'Synphronesis', in: Amir Ameri and Rebecca O'Neal Dagg (eds.), *The Ethical Imperative:106th Annual Meeting* (New York: ACSA, 2018), 443–445.
35 Antoine Picon, *Digital Culture in Architecture: An Introduction for the Design Professions* (Basel: Birkhäuser/Springer Verlag, 2010).
36 Martin Pawley, *Terminal Architecture* (London: Reaktion Books,1998).
37 Picon, *Smart Cities*, op. cit. (note 21).
38 Shannon Mattern, *The City Is Not a Computer: Other Urban Intelligences* (Princeton, NJ: Princeton University Press, 2021), 11.
39 Bruno Latour proposed the oligoptican (and oligoptica) in confrontation with Michel Foucault's use of 'panoptican', as an ideal of a total surveillance, and 'panorama' as the representation of a total view. Instead, he suggests to allow for the multitude of more localized and heterogeneous relations, views, realities and actors to intertwine. Oligoptica was a medium through which Latour could practice and visualize his ideas of actor-network theory 'to trace more sturdy relations and discover more revealing patterns by finding a way to register the links between unstable and shifting frames of reference rather than by trying to keep one frame stable'. Bruno Latour, *Reassembling the Social: An Introduction to Actor-Network-Theory* (Oxford: Oxford University Press, 2005); Bruno Latour, Emilie Hermant and Susanna Shannon, *Paris ville invisible* (Paris: La Découverte, 1998), translated from the French by Liz Carey-Libbrecht as *Paris: Invisible City*, http://www.bruno-latour.fr/sites/default/files/downloads/viii_paris-city-gb.pdf.
40 Laura Kurgan, *Close Up at a Distance: Mapping Technology and Politics* (New York: Zone Books, 2013).
41 Ghosn Rania, El Hadi Jazairy and Design Earth (Firm), *Geostories : Another Architecture for the Environment* (New York NY USA: Actar, 2018).

# RETROACTI

RETROACTIVE TIMELINE OF DATA INFRASTRUCTURE

Patrycja Raszka, Theodor Reinhardt, Fabio Sala, Yiyin Yu

TIMELINE
extended version

# EDGE OF TOMORROW

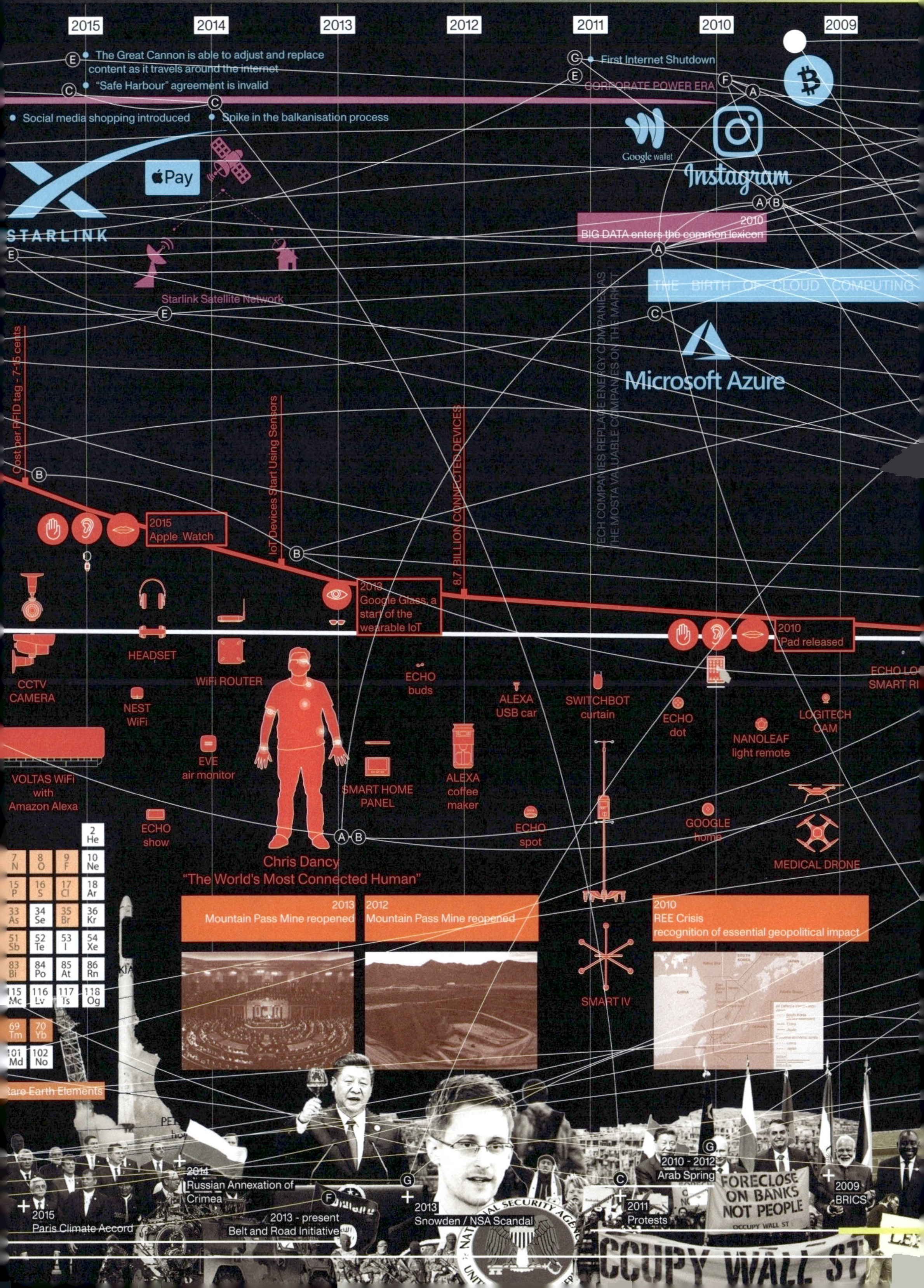
2015
2014
2013
2012
2011
2010
2009
The Great Cannon is able to adjust and replace content as it travels around the internet
"Safe Harbour" agreement is invalid
Social media shopping introduced
Spike in the balkanisation process
First Internet Shutdown
CORPORATE POWER ERA
Google wallet
Instagram
Pay
STARLINK
Starlink Satellite Network
2010
BIG DATA enters the common lexicon
THE BIRTH OF CLOUD COMPUTING
Microsoft Azure
TECH COMPANIES REPLACE ENERGY COMPANIES AS THE MOST VALUABLE COMPANIES ON THE MARKET
Cost per RFID tag - 7-15 cents
IoT Devices Start Using Sensors
8.7 BILLION CONNECTED DEVICES
2015
Apple Watch
2013
Google Glass, a start of the wearable IoT
2010
iPad released
CCTV CAMERA
HEADSET
WiFi ROUTER
NEST WiFi
EVE air monitor
VOLTAS WiFi with Amazon Alexa
ECHO show
ECHO buds
SMART HOME PANEL
ALEXA USB car
ALEXA coffee maker
ECHO spot
SWITCHBOT curtain
ECHO dot
NANOLEAF light remote
LOGITECH CAM
GOOGLE home
MEDICAL DRONE
SMART IV
ECHO LO
SMART RI
Chris Dancy
"The World's Most Connected Human"
2 He
7 N
8 O
9 F
10 Ne
15 P
16 S
17 Cl
18 Ar
33 As
34 Se
35 Br
36 Kr
51 Sb
52 Te
53 I
54 Xe
83 Bi
84 Po
85 At
86 Rn
115 Mc
116 Lv
117 Ts
118 Og
69 Tm
70 Yb
101 Md
102 No
Rare Earth Elements
2013
Mountain Pass Mine reopened
2012
Mountain Pass Mine reopened
2010
REE Crisis
recognition of essential geopolitical impact
2015
Paris Climate Accord
2014
Russian Annexation of Crimea
2013 - present
Belt and Road Initiative
2013
Snowden / NSA Scandal
NATIONAL SECURITY AGENCY
2011
Protests
2010 - 2012
Arab Spring
FORECLOSE ON BANKS NOT PEOPLE
OCCUPY WALL ST
OCCUPY WALL ST
2009
BRICS

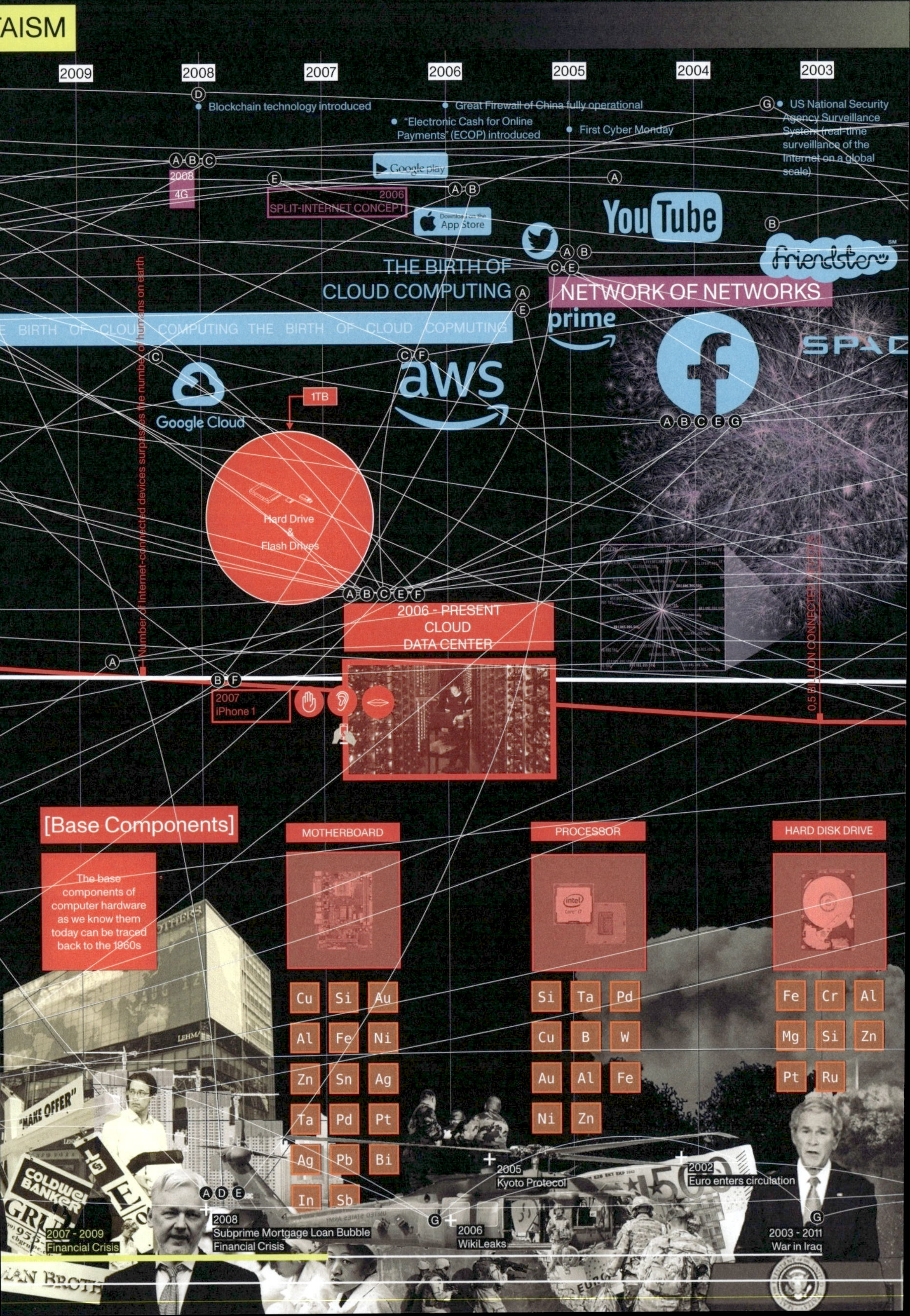

TAISM
2009
2008
2007
2006
2005
2004
2003
Blockchain technology introduced
Great Firewall of China fully operational
"Electronic Cash for Online Payments" (ECOP) introduced
First Cyber Monday
US National Security Agency Surveillance System (real-time surveillance of the Internet on a global scale)
2008 4G
2006 SPLIT-INTERNET CONCEPT
Google play
App Store
YouTube
friendster
THE BIRTH OF CLOUD COMPUTING
NETWORK OF NETWORKS
THE BIRTH OF CLOUD COMPUTING THE BIRTH OF CLOUD COPMUTING
prime
aws
Google Cloud
Number of Internet-connected devices surpasses the number of humans on earth
1TB
Hard Drive & Flash Drives
2006 - PRESENT CLOUD DATA CENTER
2007 iPhone 1
0.5 BILLION CONNECTED DEVICES
[Base Components]
The base components of computer hardware as we know them today can be traced back to the 1960s
MOTHERBOARD
Cu Si Au Al Fe Ni Zn Sn Ag Ta Pd Pt Ag Pb Bi In Sb
PROCESSOR
Si Ta Pd Cu B W Au Al Fe Ni Zn
HARD DISK DRIVE
Fe Cr Al Mg Si Zn Pt Ru
2005 Kyoto Protocol
2002 Euro enters circulation
2008 Subprime Mortgage Loan Bubble Financial Crisis
2007 - 2009 Financial Crisis
2006 WikiLeaks
2003 - 2011 War in Iraq

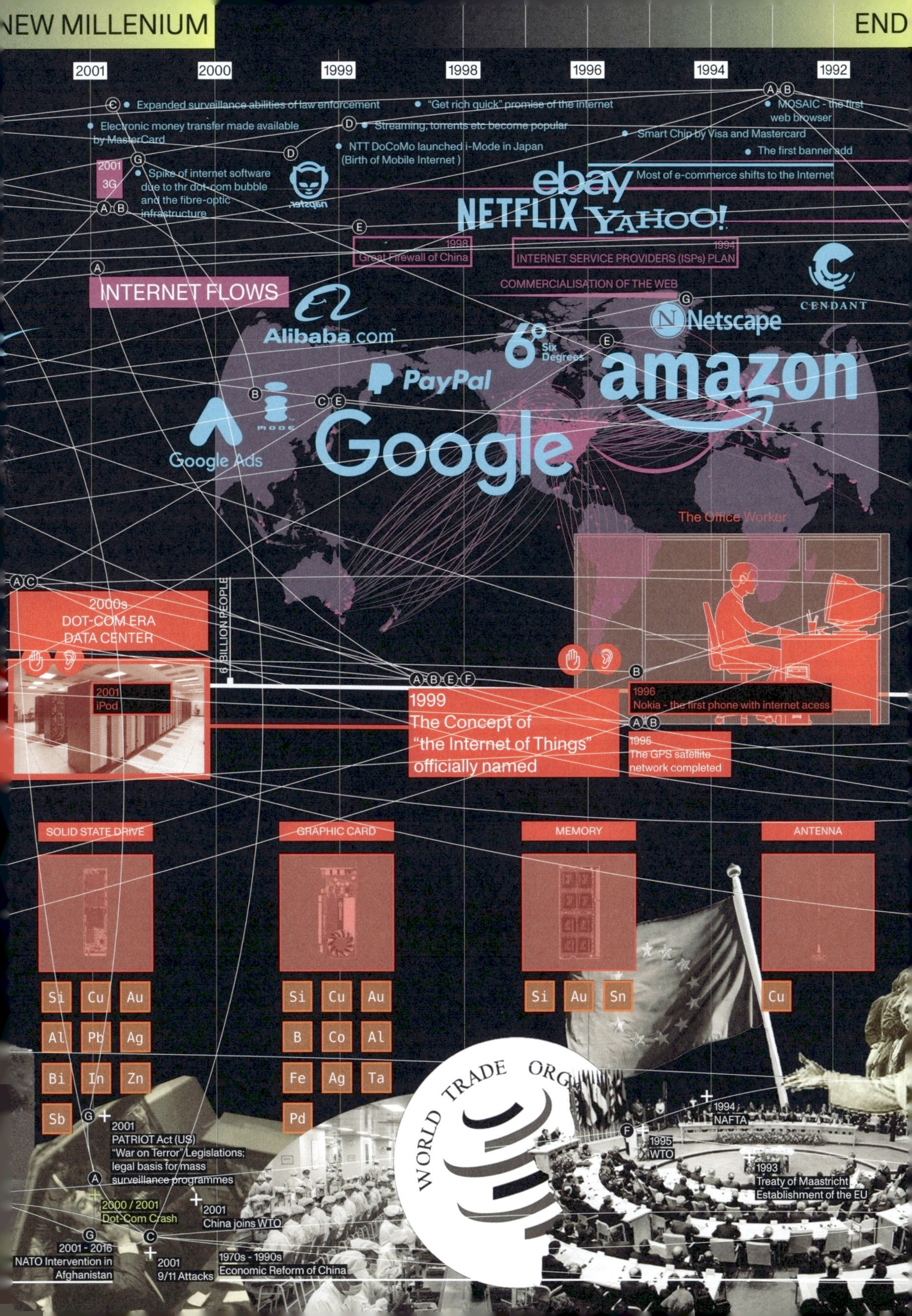
NEW MILLENIUM
END
2001
2000
1999
1998
1996
1994
1992
Expanded surveillance abilities of law enforcement
"Get rich quick" promise of the internet
MOSAIC - the first web browser
Electronic money transfer made available by MasterCard
Streaming, torrents etc become popular
Smart Chip by Visa and Mastercard
NTT DoCoMo launched i-Mode in Japan (Birth of Mobile Internet )
The first banner add
2001
3G
Spike of internet software due to thr dot-com bubble and the fibre-optic infrastructure
napster
ebay
Most of e-commerce shifts to the Internet
NETFLIX
YAHOO!
1998
Great Firewall of China
1994
INTERNET SERVICE PROVIDERS (ISPs) PLAN
CENDANT
INTERNET FLOWS
COMMERCIALISATION OF THE WEB
Alibaba.com
Netscape
6° Six Degrees
amazon
PayPal
i MODE
Google Ads
Google
The Office Worker
2000s
DOT-COM ERA
DATA CENTER
6 BILLION PEOPLE
2001
iPod
1999
The Concept of "the Internet of Things" officially named
1996
Nokia - the first phone with internet acess
1995
The GPS satellite network completed
SOLID STATE DRIVE
GRAPHIC CARD
MEMORY
ANTENNA
Si Cu Au
Al Pb Ag
Bi In Zn
Sb
Si Cu Au
B Co Al
Fe Ag Ta
Pd
Si Au Sn
Cu
2001
PATRIOT Act (US)
"War on Terror" Legislations; legal basis for mass surveillance programmes
2000 / 2001
Dot-Com Crash
2001
China joins WTO
2001 - 2016
NATO Intervention in Afghanistan
2001
9/11 Attacks
1970s - 1990s
Economic Reform of China
WORLD TRADE ORG
1994
NAFTA
1995
WTO
1993
Treaty of Maastricht
Establishment of the EU

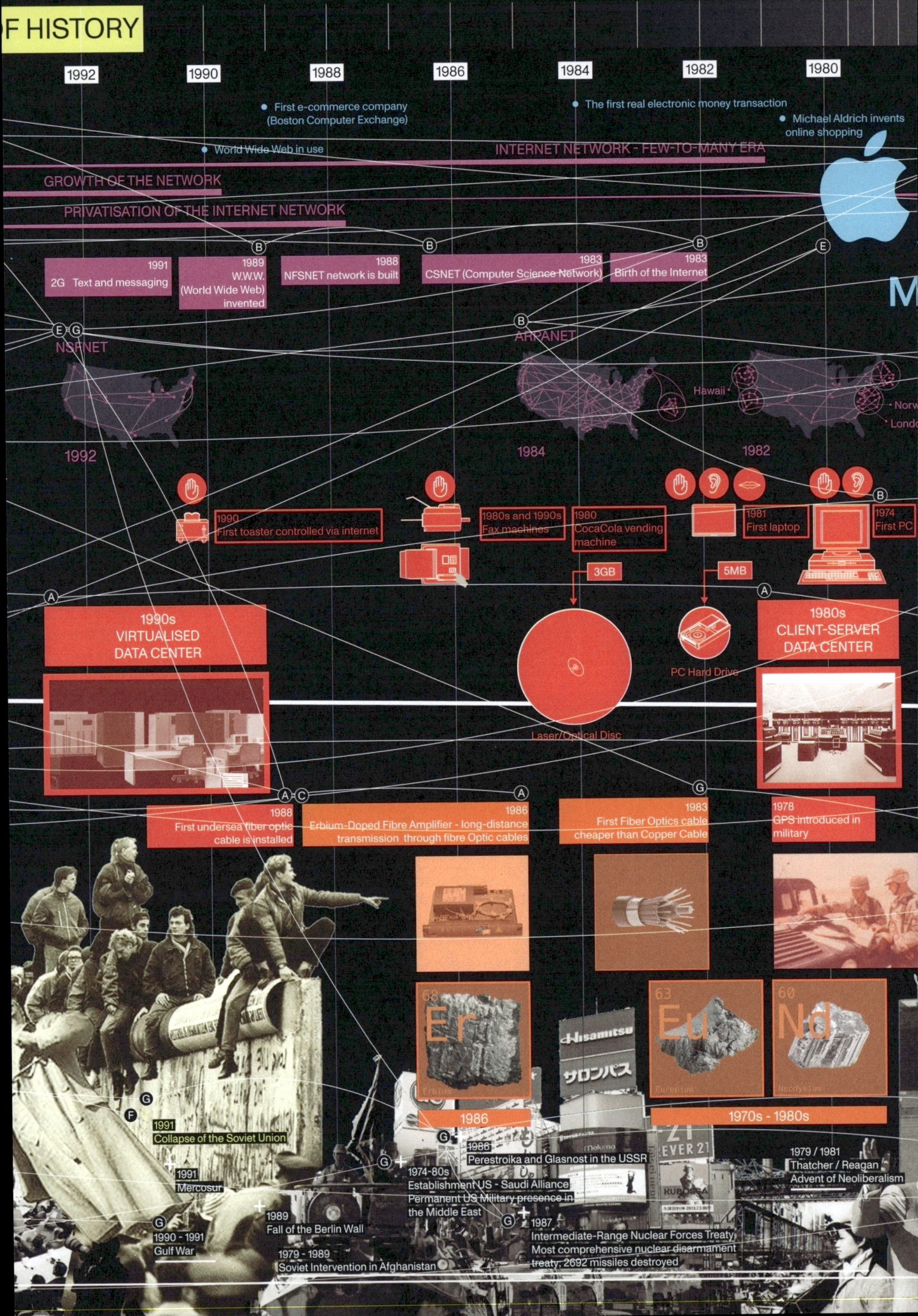

F HISTORY
1992
1990
1988
1986
1984
1982
1980
First e-commerce company (Boston Computer Exchange)
The first real electronic money transaction
Michael Aldrich invents online shopping
World Wide Web in use
INTERNET NETWORK - FEW-TO-MANY ERA
GROWTH OF THE NETWORK
PRIVATISATION OF THE INTERNET NETWORK
1991 2G Text and messaging
1989 W.W.W. (World Wide Web) invented
1988 NFSNET network is built
1983 CSNET (Computer Science Network)
1983 Birth of the Internet
NSFNET
ARPANET
Hawaii
1992
1984
1982
1990 First toaster controlled via internet
1980s and 1990s Fax machines
1980 CocaCola vending machine
1981 First laptop
1974 First PC
3GB
5MB
1990s VIRTUALISED DATA CENTER
Laser/Optical Disc
PC Hard Drive
1980s CLIENT-SERVER DATA CENTER
1988 First undersea fiber optic cable is installed
1986 Erbium-Doped Fibre Amplifier - long-distance transmission through fibre Optic cables
1983 First Fiber Optics cable cheaper than Copper Cable
1978 GPS introduced in military
68 Er Erbium
63 Eu Europium
60 Nd Neodymium
1986
1970s - 1980s
1991 Collapse of the Soviet Union
1991 Mercosur
1990 - 1991 Gulf War
1989 Fall of the Berlin Wall
1979 - 1989 Soviet Intervention in Afghanistan
1974-80s Establishment US - Saudi Alliance Permanent US Military presence in the Middle East
1986 Perestroika and Glasnost in the USSR
1987 Intermediate-Range Nuclear Forces Treaty Most comprehensive nuclear disarmament treaty, 2692 missiles destroyed
1979 / 1981 Thatcher / Reagan Advent of Neoliberalism

# NUCLEAR-DIGIT

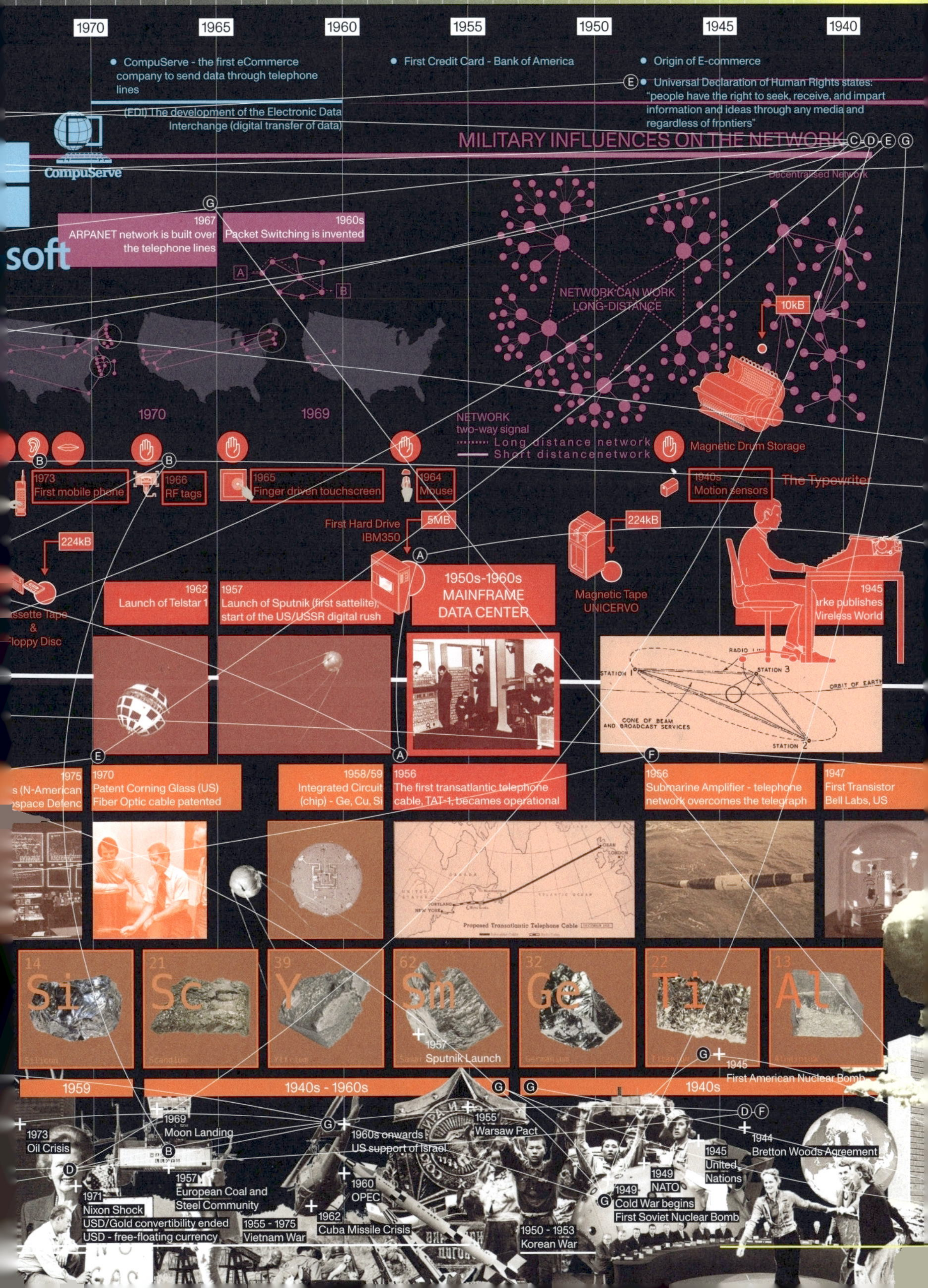

L AGE
1940
1930
1920
1910
1900
1890
1880
Western Union launches metal money (early form of credit card)
NETWORK
two-way signal
Long distance network
Short distancenetwork
NO LONG-DISTANCE SIGNALS
DIGITAL NETWORK - TELEPHONE
AT&T
1877
First telephone line
1878
First commercial telephone switchboard
1878
AT&T founded
1879
1G
YPE OF DATA SENT: AUDITORY SIGNALS / VOICE
ANALOG-TO-DIGITAL CONVERSION PROCESS
ANALOG
DIGITAL
1940s
SUPERCOMPUTER
DATA CENTER
1920s
Teleprinters
2 BILLION PEOPLE
1890
TABULATION
MACHINE
1876
Telephone
1920
Launch of KDKA; first commercial radio station
1914-1918
Wireless telegraph used during the WWI
1914
First test of a transcontinental telephone line
1877
First telephone line laid in Boston
74
W
Tungsten
51
Sb
Antimony
1930s - 1940s
1917
October Revolution
1933
Hitler comes into power
1929 - 1933
The Great Depression
1939 - 1945
WWII
1912
RMS Titanic sinks
1891
Hong Kong Stock Exchange
1891
Shanghai Stock Exchange
1878
Tokyo Stock Exchange
1776
US Independence

AGE OF THE EMPIRE
1860
1850
1840
1800
1700
1600
...
First telegraph money transfer
NETWORK
one-way signal
Long distance network
Short distancenetwork
ANALOG NETWORK - TELEGRAPH
Distributed Network
Centralised Network
1825 Electromagnetism invention
1828 First telegraph (USA)
1835 Samuel Morse proved that signals could be transmitted by wire
Invention of the Morse Code
TYPE OF DATA SENT: TEXT AND FACSIMILE
ENCODING/DECODING
ANALOG
80B
Punch Card
Paper Storage
The Chalcolithic Man
1868 Typewriter
1830s / 1840s Telegraph
1 BILLION PEOPLE
0.6 BILLION PEOPLE
0.5 BILLION PEOPLE
1866 First transatlantic telegraph cable
1844 First telegraph line is build in the USA
1816 First telegraph signal through a copper wire
2000BC - 1960s Horse carriers as main form of sending informaion
3000BC Ancient methods of comunication
1814 Advent of Pax Britannica
Britain attains global dominance after France is disadvantaged in the Vienna Congress
1698 London Stock Exchange
1602 Dutch East India Company First ever traded stock
1792 New York Stock Exchange
1683 Bank of Amsterdam issues first fiat money
cheaply manufactured goods
semi-processed materials (rum, cotton...)
raw maerials, slaves
29 Cu Copper
10.000BC
Colonialism
Economic Exploitation Through Colonial Chain of Added Value
[1] African Colonies - Appropriation by Disposession
[2] American Colonies - Semi-Processing of Imported and Extracted Resources through Slave Labour
[3] Metropolitan State - Cheap, Industrial Manufacturing; Administration; Finance;
1492 Discovery of America
1522 First circumnavigation of the Earth
1494 Treaty of Tordesillas First significant colonial division
1839 - 1860 Opium Wars
Western agression against China for access to Chinese market

# NOTES ON A

## NOTES ON AN ARCHAEOLOGY OF DESIGN DATA LITERACY

Georg Vrachliotis

'In recent years, "data" has become one of the most important aspects of architectural design. With the advent of new technologies, architects are now able to collect and analyse data about everything from building materials to weather patterns. This data can be used to create more efficient and sustainable buildings. In some cases, data can even be used to create entire new neighbourhoods or cityscapes.' The sentences sound like a quote, one that could have come from an interview about the history of digital culture or the interconnections between architecture and technology. But as authentic as this statement sounds, it was never said. Anyone who thinks they hear the voice of an architect, computer scientist or historian is mistaken. Strictly speaking, the words do not come from a human being, but from the language software Generative Pre-Trained Transformer, GPT-3 for short. The program is the third generation of artificial intelligence, considered one of the most impressive products our digital culture has produced to date. On a timeline of disruptive technologies, it currently occupies the most current position. GPT-3 is a text generator that can write poems and dramas, provide answers to complex questions on topics such as love or trust, discuss the weather or international climate policy with us—and that has also written the short statement on the subject of 'data and architectural design' mentioned above. The *New York Times* calls the program 'amazing, spooky, humbling and more than a little terrifying'.[1]

Design is more than just generating texts and images using intelligent machines, as is made possible, for example, by the increasingly powerful text-to-image software, such as DALL-E and Midjourney.[2] But such examples make it clear where the journey could also go for architecture and urban design. While physical models are still regarded as an essential medium of knowledge production, there has been an ever-widening spectrum of machine learning models for discussion for some time. A glimpse at the architecture history of digital culture makes us realize that data technologies are far too social to be understood only technically, just as societies are far too driven by data technologies to be understood only socially.[3]

The architecture history of design data literacy is a long search to find a proper culture form for the structural form of the computer. It is fundamentally about how we, as humans, want to behave towards operational technologies, and how to interact and learn with tools and machines to design buildings, explore cities and think about societies. In the twentieth century, we were primarily concerned with learning about machines; instead, we are now concerned with how to teach machines to design. This paradigm shift is accompanied by recalibrating the fundamental cultural technique of learning and interacting with tools to produce new imaginaries about the living environment and, thus, of the architect's role as a creative communicator and explainer linking design, data and society. This change also brings new cultural, ethical, and political questions about how to rethink authorship, creativity, diversity or the digital gab in societies from a data perspective. The debate about biased training data, for example, has repeatedly shown in recent years that the humanistic ideal of learning has become a complex discourse about the power of technical intelligence on an industrial scale.[4]

The sphere of the digital has emancipated itself from being a pure history of tools machines and interfaces to a planetary narration of information networks, feedback, digital ecosystems and data economies based on machine intelligence. It is as if we need to read Marshall McLuhan's *The Invisible Environments*, published in 1967, in the opposite direction to understand that it is now the algorithmic environments that learn something about us.

> As we move into the world of integral, computerized knowledge, mere classification becomes secondary and inadequate to the speeds with which data can now be processed. As data can be processed very rapidly we move literally into the world of pattern recognition, out of the world of mere data classification. One way of putting this is to say that our children today live in a world in which the environment itself is a teaching machine made of electric information.[5]

McLuhan suggests that digital media have the power to challenge and recalibrate existing social and cultural structures in societies. Or, as architecture historian Antoine Picon put it, our world today is 'an heiress of the universe opened up by cybernetics and electronic art in the 1950s and 1960s'. 'But who would have thought,' Picon continues, 'that digital life would get under our skin to this degree?'[6] It doesn't take long to realize that this is a rhetorical question that is completely in step with our times. The radical nature of the question is already evident in the fact that even the idea of digitization creeping 'under our skin' is no longer a metaphor. For some time now, we have been experiencing new cultural forms of a recoding, transcoding and converting of spaces, objects, forms, surfaces, materials and the human body. We have, according to the argument behind the following reflections, become actors in a data-based world of operationality that will lead to new cultural, aesthetic, social and political design paradigms within the natural and built environment. So what does it mean to design for a data-driven society that is seeking a balance between an ecological sense of resources and artificial intelligence?

## Operational Thinking

Since the mid-twentieth century, architects debated whether machines would cause severe damage to the supposedly humanistic core of design.[7] Almost unnoticed, a second intellectual battleground of entirely different dimensions developed on the peripheries of this debate, which was burgeoning, especially in the 1970s, almost ten years before the World Conference on Computers and Education in Lausanne would talk about 'computer literacy' for the first time.[8] Instead of *computer-aided design*, it was about *computer-aided instructions*, so not about the digitization of design but of learning. But this should not obscure how closely the one was coupled to the other—and still is today. What connects the datafication of design and learning is the technologization of intellectual work itself. The automation wave of the 1970s was, therefore, nothing less than a large-scale attempt to externalize and operationalize tacit knowledge and make it accessible and valuable for the societies. Digitization has not only brought about new tools, but a fundamental change in 'intellectual work' is up for debate in the future, changing design and learning. With a view to the history of digitization, two different but essentially related forms of knowledge production are thus at issue: processes of learning are, to a certain extent, also processes of designing and vice versa, whether drawing machines or learning automatons. It was about the technologization of thinking itself.

The roots of the so-called educational technology are crucial for repositioning design in the age of data. They go back to the field of 'human engineering', an interdisciplinary branch of research established at the time of the Second World War. Its goal was to optimize the interaction between humans and machines. The knowledge gained in the context of the US military was transferred to various areas: this included not only Buckminster Fuller's Dymaxion Houses[9] and Konrad Wachsmann's US Aircraft Hangar,[10] but also the automation of learning itself. Learning automata were an attempt to integrate the school as a social institution into the system competition of the Cold War, which meant not only system building and comprehensive schools, but also learning automata and cybernetic pedagogy. The catalyst for this development was the 'Sputnik shock', which grew from the awareness that the Soviet Union had gained a technological lead over the capitalist West with its successful launch of an earth-orbiting satellite in 1957.[11] The reaction of the American government was not long in coming: in 1958, only one year later, President Dwight D. Eisenhower established not only the defence institution DARPA and the space agency NASA, but also the National Defense Education Act, which entailed a large-scale reform of the American school curriculum and led to the upgrading of the technical intelligence in the country. The focus was on greater individualization of instruction and developing new kinds of interdisciplinary knowledge production. This included so-called creativity research[12] and brainstorming as well as the development of special thinktanks and 'auto-instructional programmes'. The individualization of learning led to the founding of Educational Technology, or 'edtech' for short, a new branch of research in the education industry, based in particular on the behaviourist theories of the influential psychologist and behavioural scientist B.F. Skinner.[13] Edtech promised efficiency in every respect, knowledge production and ideological protection from communist ideas. In fact, the 1960s were the golden age of the so-called teaching machines. The star of this movement was called PLATO—short for Programmed Logic for Automated Teaching Operations—and was probably the most influential teaching machine system of the post-war period. Assembled by engineers, educationalists, mathematicians and psychologists at the University of Illinois from parts of an old radar set, the new machine required the decomposition of all content into so-called 'teaching quanta' according to behaviourist theories. Content and topics were no longer to be thought out of complex contexts but sequentially and additively, as if on an assembly line. If you like, the automation of teaching went hand in hand with the automation of learning. The goal was to build a smooth control loop of learning consisting of theory, hardware, software, information and user. The focus was on creating devices that could be operated with simple buttons. In the operation of the early teaching machines, the hand was regarded as the central tool of sensory perception and haptic feedback. It is probably hardly known that Norbert Wiener, for example, also developed a special device for the deaf in 1948, one year after the publication of his epoch-making book on cybernetics. This device translated spoken language into vibration patterns that the deaf could feel with the help of tiny sensors on their fingertips and thus understand. Wiener did not develop this so-called correlator any further, but the project illustrates that the development of the early teaching machines was predominantly a question of hardware and the design of haptic human-machine communication, and that physical space played hardly any role in this. This changed with the beginning of new network technologies. Instead of just individual places, it was now about global systems, networked thinking and communication environments. Architects rediscovered the physical space, not to design it, but to overcome it.

## From Places to Networks

McLuhan's vision about the disruptive potential of electronic media on education developed into an equivalent debate about space and new communication technologies. In the late 1960s several architects, including Toyo Ito, Buckminster Fuller, Yona Friedman and Constantin Doxiadis, began to interpret the building not as a single object, but as a global network of virtual nodes. Swiss architect Fritz Haller, for instance, had realized a series of innovative school and university buildings since the 1950s and was also known for his USM Haller building systems and the USM Haller furniture system.[14] For the competition for a new university campus for the École Polytechnique Fédérale de Lausanne (EPFL), he submitted a project that attempted to leap from an architectural building system to a technological communication system. According to Haller, a university is no longer an autonomous institution but a node in a worldwide knowledge network linked by telephone, television, data interconnection and high-speed trains. Added to this is the idea that universities are no longer limited to academic discourse but radiate into society as 'cultural centres'. This network thinking was the highlight of the project. Haller, who had visited the recently completed university buildings in West Germany for the competition, developed a speculative scheme of how the spaces of learning would develop in the future. According to Haller, the 'general transformation trend' of school models will develop in four stages: from the traditional hierarchizing model to an ever-flatter hierarchy. First, the head teachers will disappear and then the teachers; then classical frontal teaching will be replaced by dynamic forms of community and team teaching. In the fourth and final stage, the school is a network of interdisciplinary learning made up of nodes. Specific types of space are assigned to the different levels under the heading 'Building Structure'. The first three levels can be brought into line with examples of Haller's school buildings, such as the Höhere Technische Lehranstalt Brugg-Windisch (1961–1966). In the last level, he proposes an open-plan school as a system open on all sides, with separable rooms, depending on the visual, acoustic and climatic requirements of teaching. But by the end of the 1970s, the time for visionary school-building projects seemed to have finally run out. And so Haller's network university was not awarded a prize. The competition jury found the vision too radical and not sufficiently application-oriented. Haller expressed his disappointment to the Studiengruppe für Systemforschung (Study Group for Systems Research), co-founded by Horst Rittel, a mathematician and former teacher at Ulm School of Design: 'The experts called the work interesting, but too progressive—too inhuman—too un-architectural. Maybe later on, you can build something like that.'[15] Haller held on to the idea of expanding spaces of learning through communication technologies. In 1996, only a few years after the foundations of the modern World Wide Web were laid at the European Organisation for Nuclear Research (CERN), Haller was also sure: 'Schools will no longer be schools. They will be communication nodes of a global network of relationships and data, in which people of all levels of knowledge and rank will act within their possibilities and create new values.'[16]

What Haller describes here as a vision for education would later become the basis of a platform-based data economy. His idea of open networks and knowledge societies builds on concepts developed since the 1970s. In particular, Daniel Bell's *The Coming of Post-Industrial Society*,[17] published in 1974, laid the foundation for our current discussion on the spatial impact of platforms and logistics on the built environment.

## From Instructions to Discovery

In 1970, the same year Haller submitted his entry for the EPFL competition, Marshall McLuhan published an essay titled 'Education in the Electronic Age', with which he arguably created one of the most influential blueprints for the digital culture of learning. The division into chairs, according to McLuhan, also belonged to the dissecting and specializing age of printing, a time still without feedback. McLuhan said that anyone could learn about anything on their own if they were given encouragement and tools. And this is precisely where the utopian potential of the early digital tool culture lies. So, it was educational issues that decided the future of computers and societies, from the earlier teaching machines to Fritz Haller's global university networks. McLuhan was convinced that you could transform learning if you transformed the spaces of learning. Modern communication technologies could even liberate societies from the rigid thought patterns of school institutions so that, in the end, they would practically no longer be needed. Technology's reform potential was no longer located in schools, but in the outside space, cities and the environment. It is no coincidence that McLuhan spoke of 'classrooms without walls', a wonderful metaphor that could also be used as a proxy for the countless experimental places of learning in the digital age.

The changes have gone on outside, not inside the school. The outside environment, perhaps for the first time in history, is, in terms of information, many times more heavily laden than the inside environment of the school. What is going on inside the school is puny and nourished compared to what goes on the moment the child steps outside ... What goes on inside the school is an interruption of education, of the education available in the current environment. In the electric age, people make their world in an entirely new way; the whole environment is created.[18]

McLuhan tried to overcome schools as institutional and built spaces, which may have been because he saw in the mass media, and especially in communication technologies, a new kind of global feedback infrastructure for unfolding independent and individual learning. Education in the electronic age was a matter of discovery and exploration.

### From Miniaturization to Ecosystems

Haller's network architecture and McLuhan's vision of education must be seen in the context of another debate. In the 1960s, much revolved around digitizing basic cultural techniques, such as writing, arithmetic or reading, which meant reading, writing or calculating and drawing. Thin graphic lines flickered on architects' screens initially introduced by the military industry (ARPA), and the traditional coupling of drawing and seeing, as well as seeing, was scrutinized by a new kind of computational knowledge. The intuitive dialogue between the hand and the creative eye was thus severely disrupted. Drawing is linked to design and, accordingly, to thinking, and many architects feared the automation of one also meant the automation of the other. Machine intelligence, therefore, seemed to be about a double cultural devaluation: that of the architect as the sole decisionmaker and that of the design process as a creative genius technique. This was prompted by new digital infrastructures as well as machine storage, processing and communication.

The early computer avant-garde understood that you had to focus on people's behaviour and therefore also on their mental schematic. Influential figures such as Steven Coons laid the conceptual foundation for a machine world in which the mathematical control of abstract input and output variables rather than the mechanics of the physical object could be regarded as the characteristics of a machine. The boundaries between man and machine, nature and culture were to be overcome in order to arrive at a new kind of behaviourist machine thinking and finally at a superordinate method of algorithmic world analysis.

Exemplary for such a behaviourist view was Augmenting Human Intellect, a human–machine theory developed by electrical engineer Douglas Engelbart. He believed computers could expand the cognitive abilities of humans, and designed a remarkable vision of the future for the architect. With the sentence 'Let us consider an augmented architect at work',[19] he began to describe the working process of a computer-aided architect. This 'augmented architect'[20] was to have a screen and a small keyboard at his workplace, which he could use to communicate with the machine. 'With a "pointer", he [the architect] indicates two points of interest, moves his left hand rapidly over the keyboard, and the distance and elevation between the points indicated appear on the right-hand third of the screen.'[21] It is possible to rotate a drawing constructed in this way. Using the keyboard, the architect can also enter metric data. After several steps, the first outlines of the building would emerge. At the same time, the mechanical architecture assistant calculated the possible effects of the designed building and tested them under different parameters. All the data produced in the course of such a working process—which, interestingly enough, Engelbart understood to mean not only 'the building design' but also 'its associated thought structure'—could ultimately be saved on a 'tape' and retrieved at any time.

With Marshall McLuhan, we could say that Engelbart understood computers as a creative prosthesis, that is, an extension by means of which the architect's cognitive and physical abilities can be extended and technically enhanced.

The term 'augmented architect' makes it clear that the architect should be woven into an information technology milieu and that the digital should be wrapped around him like a second skin. The architect's drawing table was transformed into a digital ecosystem of interfaces, tools and databases.

## Towards Design in Open Data Societies

The image of the future is by no means just a question of data—although in many ways it is. Today, almost 50 years after McLuhans *invisible environments*, Bell's *post-industrial society* and the dazzling wave of cyberneticization of learning, and designing, modern forms of societies have emerged that could perhaps best be described as 'delivery societies' or 'cloud economies', linking people and buildings to platforms, interfaces, data centres and machine intelligence. Data is no longer just traded like an important raw material. It also forms the economic basis of perhaps the most powerful promise of a predictable future. French philosopher Cornelius Castoriadis claimed that 'no category apart from the *imaginary* allows us to reflect on the idea of society'.[22] This is an appealing and clever assertion, as it gets to the heart of something that seldom attracts much attention, that is that we are constantly searching for new models of community and living together. This is no longer about the conscious fusion of aesthetics and life, however, as was so vehemently demanded by the artistic avant-garde movements of the last century. Rather, it seems, a new kind of data-based reading of the entire environment is up for debate. Whether bodies, buildings, cities, landscapes, oceans or climate, we are at the beginning of a large-scale operationalization project in which, in terms of media technology, there is no longer an inside or an outside and whose goal is nothing less than a planetary rereading. In other words: the computerization of the world makes us realize what it means to be part of a globally operating industrial complex. This is especially true for architectural production. Materials, objects and capital circulate in an infrastructural matrix whose scale and impact we are only gradually beginning to understand. Thanks to more precise simulation models of material and substance cycles, we are increasingly able to document and research the consequences of an environment that has been completely transformed by humans. Digital mappings are now emerging that will lead to new insights and simultaneously question traditional supply and production systems. We are only beginning to grasp, for example, that the *way* buildings are manufactured and constructed around the world is a process that no longer takes place only on the Earth's surface, but also leaves traces deep inside it. Copper and lithium are mined and processed for the global construction industry. Even the ocean has changed from a once mythical place into a vast geopolitical infrastructure project, whose story can no longer be told without oil platforms, undersea cables, floating server farms and forensic oceanography.

How can construction, raw materials and digitization be brought together in the future? And what is actually the geological footprint of future data-based digital architectural production? These are tricky but nonetheless important questions that are not only about innovation, but also about responsibility, and in the coming decades they will certainly have a greater impact on construction than we may have expected. The twenty-first century will require us to design in an open-data-based and ecological way, which is nothing less than developing design data literacy for the age of machine intelligence. The question of what it means to design in such societies, or even how we can produce meaning at all from the vast amounts of environmental data, is the other side. In the future, we will not be able to avoid computer vision and machine learning, because human intuition will eventually reach its natural limits when faced with so much data. We are collecting more and more data about materials, the environment and the climate, but the faster this data is collected, the more likely we are to be confronted with the limitations of our own ability to make judgements. We will have to admit to ourselves that we will neither be able to evaluate nor judge the complex structure of open-data societies without automated analytical capabilities.

Instead of purely technical aspects, today's focus is increasingly on social, ecological and ethical aspects of platform technologies, such as the digital divide, privacy, ownership, accessibility, sustainability and public trust in clean and qualitative architectural datasets. These issues form perhaps one of the most significant architectural challenges in the coming decade, aiming to sovereignly position and critically examine the creative, social and political potential of operational design thinking in open data societies.

Notes

1 Farhad Manjoo, 'How Do You Know a Human Wrote This?', *New York Times*, 29 July 2020, https://www.nytimes.com/2020/07/29/opinion/gpt-3-ai-automation.html, accessed 26 November 2022.
2 See: Matthew Mullane, 'Seeing History with Computer Vision: The Architecture of Text-to-Image Synthesis', and Erik Hermann, 'Mambo Dogface to the Banana Patch? Talking Wrong with Deep Learning', both papers presented at the 9th Jaap Bakema Study Centre conference on 'Building Data. Architecture, Memory, and Imaginaries', 23 November 2022.
3 See: Georg Vrachliotis, *The New Technological Condition: Architecture and Design in the Age of Cybernetic* (Basel: Birkhäuser, 2022).
4 See: Safiya Umoja Noble: *Algorithms of Oppression: How Search Engines Reinforce Racism* (New York: NYU Press, 2018).
5 Marshall McLuhan, 'The Invisible Environment: The Future of an Erosion', *Perspecta* 11 (1967), 161–167.
6 Antoine Picon, 'Architektur und Wissenschaft: Wissenschaftliche Exaktheit oder produktives Missverständnis?', in: Ákos Morávanszky and Ole W. Fischer (eds.), *Precisions: Architektur zwischen Wissenschaft und Kunst* (Berlin: JOVIS, 2008), 78.
7 See: Georg Vrachliotis, 'Architecture, Computer and Technological Unrest: Towards an Architecture History of Anxiety', in: Teresa Fankhänel and Andres Lepik, *The Architecture Machine: The Role of Computers in Architecture* (Basel: Birkhäuser, 2020), 28–34.
8 Computers in Education: Proceedings of the IFIP TC-3 3rd World Conference on Computers in Education, Lausanne, 27–31 July 1981.
9 See: Federico Neder, Elsa Lam and Mark Wigley, *Fuller Houses: R. Buckminster Fuller's Dymaxion Dwellings and Other Domestic Adventures* (Baden: Müller, 2008).
10 See: Konrad Wachsmann, *The Turning Point of Building: Structure and Design* (New York: Reinhold, 1961); and Georg Vrachliotis, 'Konrad Wachsmann's Educational Concepts', in: Beatriz Colomina et al. (eds.), *Radical Pedagogies* (Cambridge, MA: MIT Press, 2022).
11 See: Barbara Barksdale Clowse: *Brainpower For The Cold War: The Sputnik Crisis and National Defense Education Act of 1958* (Westport: Greenwood Press, 1981).
12 See: Bregje F. van Eekelen, 'Uncle Sam Needs Your Ideas: A Brief History of Embodied Knowledge in American World War II Posters', *Public Culture* 30/1 (2018), 113, 1527–8018.
13 B.F. Skinner, *The Technology of Teaching* (New York: Appleton-Century-Crofts, 1968).
14 See: Laurent Stalder and Georg Vrachliotis (eds.), *Fritz Haller: Architekt und Forscher* (Zurich: gta Verlag, 2015).
15 Fritz Haller, Letter to Thorbjörn Mann, 16 July 1970; gta Archive, ETH Zurich: Fritz Haller Archive, file 189–0316.
16 Fritz Haller, 1996; gta Archive, ETH Zürich: Fritz Haller Archive,
17 Daniel Bell, *The Coming of Post-Industrial Society* (New York: Harper Colophon Books, 1974).
18 Marshall McLuhan: "Education in the Electronic Age", in: *Interchange* 1/4 (1970), p 1
19 Douglas C. Engelbart, 'Augmenting Human Intellect: A Conceptual Framework', Summary Report AFOSR-3223 prepared for the Air Force Office of Scientific Research, Stanford Research Institute, Menlo Park, CA, October 1962, 1.
20 Ibid., 4.
21 Ibid.
22 Cornelius Castoriadis, *Gesellschaft als imaginäre Institution: Entwurf einer politischen Philosophie* (Frankfurt am Main: Suhrkamp, 1990), 274.

1 : 2

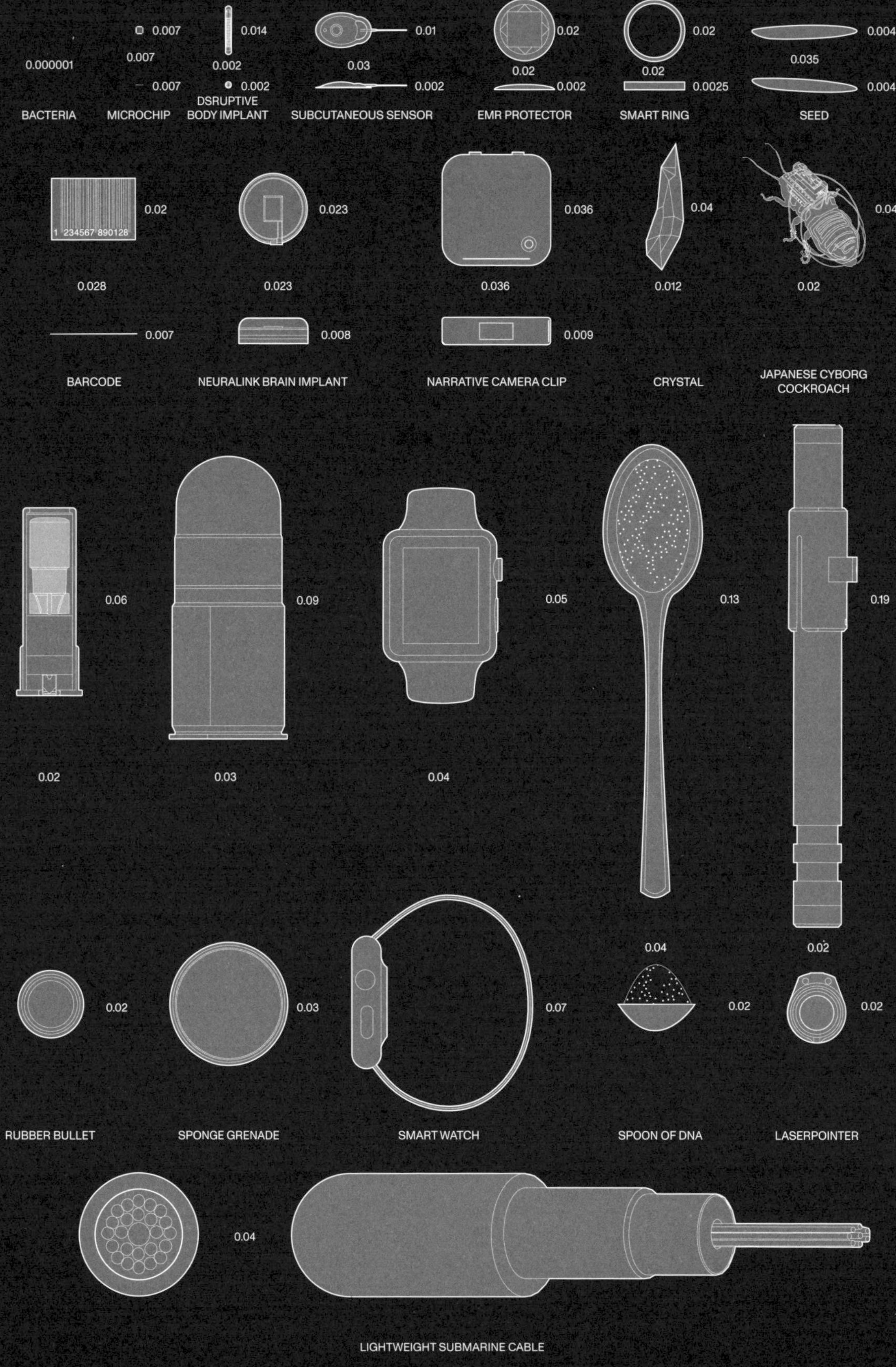

0.000001
BACTERIA
0.007
0.007
0.007
MICROCHIP
0.014
0.002
0.002
DSRUPTIVE BODY IMPLANT
0.01
0.03
0.002
SUBCUTANEOUS SENSOR
0.02
0.02
0.002
EMR PROTECTOR
0.02
0.02
0.0025
SMART RING
0.0045
0.035
0.0045
SEED
1 234567 890128
0.02
0.028
0.007
BARCODE
0.023
0.023
0.008
NEURALINK BRAIN IMPLANT
0.036
0.036
0.009
NARRATIVE CAMERA CLIP
0.04
0.012
CRYSTAL
0.045
0.02
JAPANESE CYBORG COCKROACH
0.06
0.02
0.02
RUBBER BULLET
0.09
0.03
0.03
SPONGE GRENADE
0.05
0.04
0.07
SMART WATCH
0.13
0.04
0.02
SPOON OF DNA
0.19
0.02
0.02
LASERPOINTER
0.04
LIGHTWEIGHT SUBMARINE CABLE

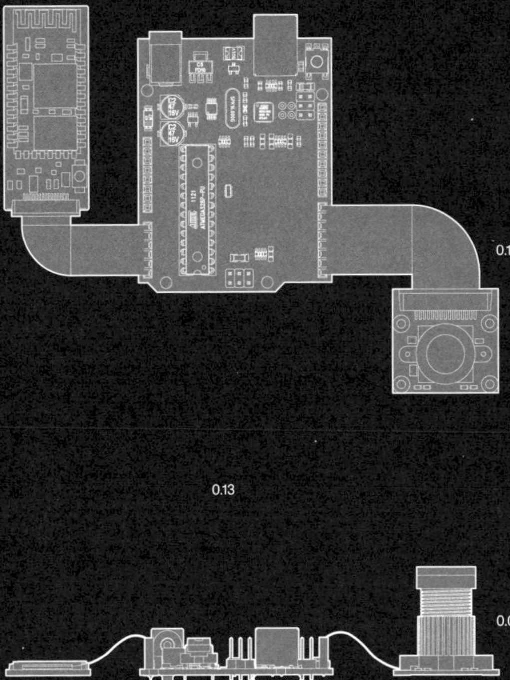


ARDUINO WITH BLUETOOTH AND CAMERA MODULE

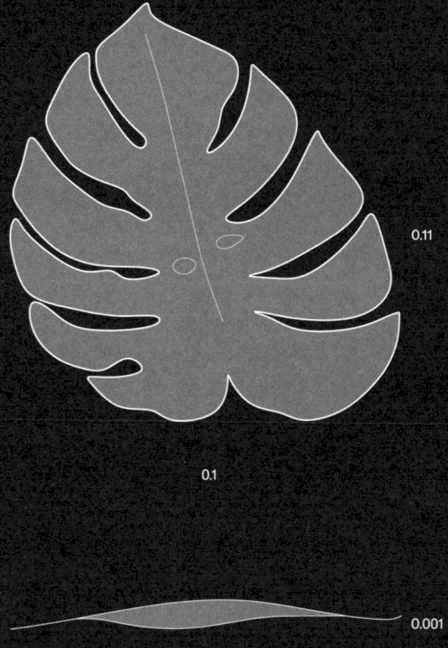


LEAF

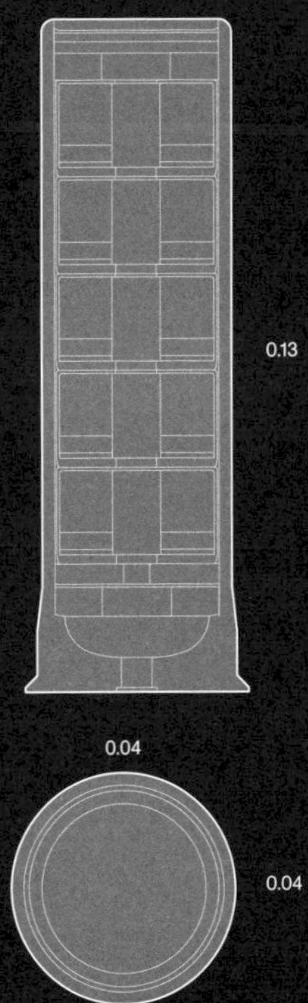


TEARGAS ROUND

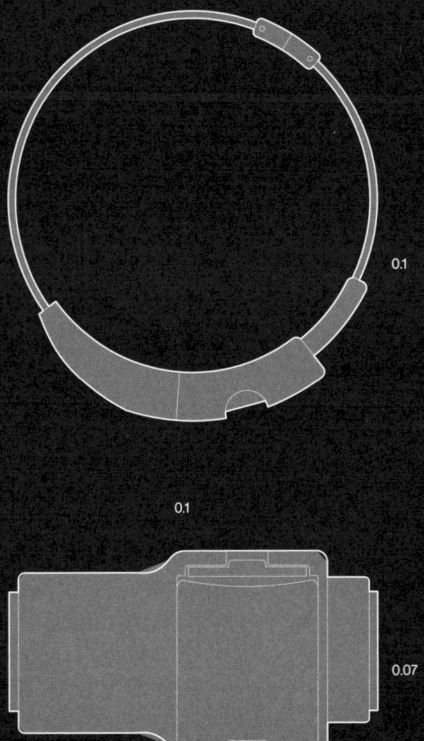


ANKLE BRACELET

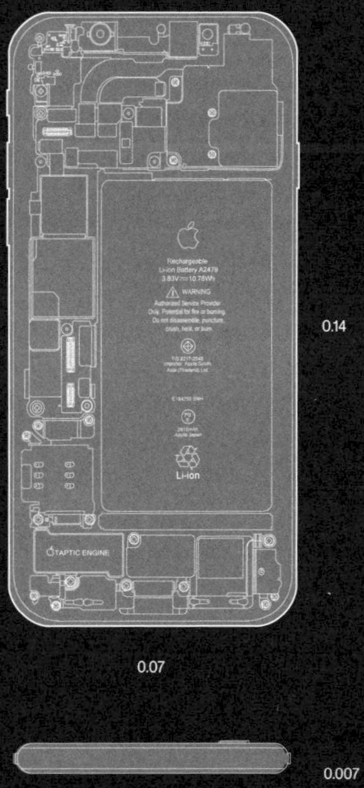


SMARTPHONE

1:20

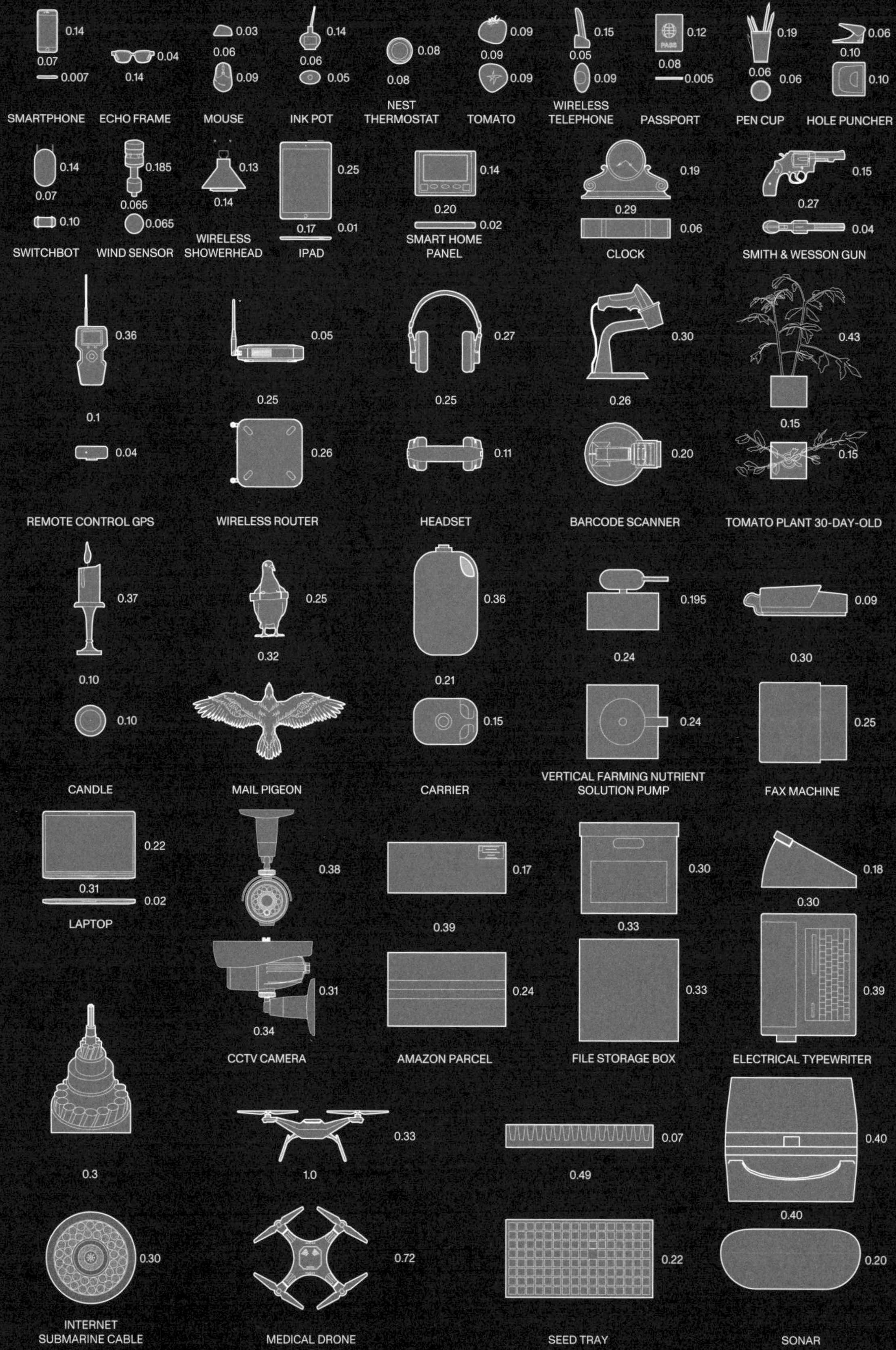
0.14
0.07
0.007
SMARTPHONE
0.04
0.14
ECHO FRAME
0.03
0.06
0.09
MOUSE
0.14
0.06
0.05
INK POT
0.08
0.08
NEST THERMOSTAT
0.09
0.09
0.09
TOMATO
0.15
0.05
0.09
WIRELESS TELEPHONE
0.12
0.08
0.005
PASSPORT
0.19
0.06
0.06
PEN CUP
0.06
0.10
0.10
HOLE PUNCHER
0.14
0.07
0.10
SWITCHBOT
0.185
0.065
0.065
WIND SENSOR
0.13
0.14
WIRELESS SHOWERHEAD
0.25
0.17
0.01
IPAD
0.14
0.20
0.02
SMART HOME PANEL
0.19
0.29
0.06
CLOCK
0.15
0.27
0.04
SMITH & WESSON GUN
0.36
0.1
0.04
REMOTE CONTROL GPS
0.05
0.25
0.26
WIRELESS ROUTER
0.27
0.25
0.11
HEADSET
0.30
0.26
0.20
BARCODE SCANNER
0.43
0.15
0.15
TOMATO PLANT 30-DAY-OLD
0.37
0.10
0.10
CANDLE
0.25
0.32
MAIL PIGEON
0.36
0.21
0.15
CARRIER
0.195
0.24
0.24
VERTICAL FARMING NUTRIENT SOLUTION PUMP
0.09
0.30
0.25
FAX MACHINE
0.22
0.31
0.02
LAPTOP
0.38
0.31
0.34
CCTV CAMERA
0.17
0.39
0.24
AMAZON PARCEL
0.30
0.33
0.33
FILE STORAGE BOX
0.18
0.30
0.39
ELECTRICAL TYPEWRITER
0.3
0.30
INTERNET SUBMARINE CABLE
0.33
1.0
0.72
MEDICAL DRONE
0.07
0.49
0.22
SEED TRAY
0.40
0.40
0.20
SONAR

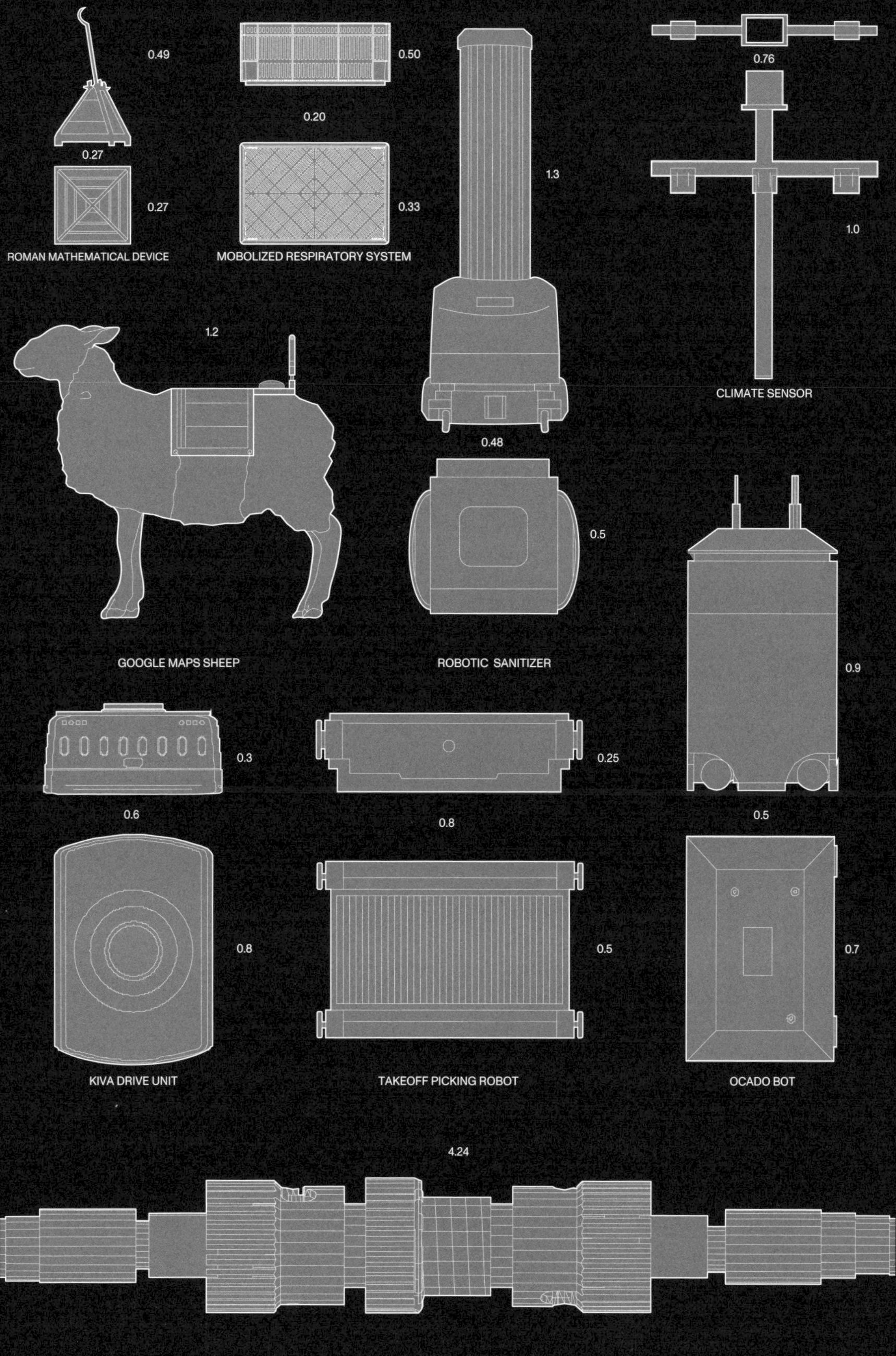
0.49
0.27
0.27
ROMAN MATHEMATICAL DEVICE
0.50
0.20
0.33
MOBOLIZED RESPIRATORY SYSTEM
1.3
0.48
0.5
ROBOTIC SANITIZER
0.76
1.0
CLIMATE SENSOR
1.2
GOOGLE MAPS SHEEP
0.9
0.5
0.7
OCADO BOT
0.3
0.6
0.8
KIVA DRIVE UNIT
0.25
0.8
0.5
TAKEOFF PICKING ROBOT
4.24
SUBMARINE AMPLIFIER

1 : 200

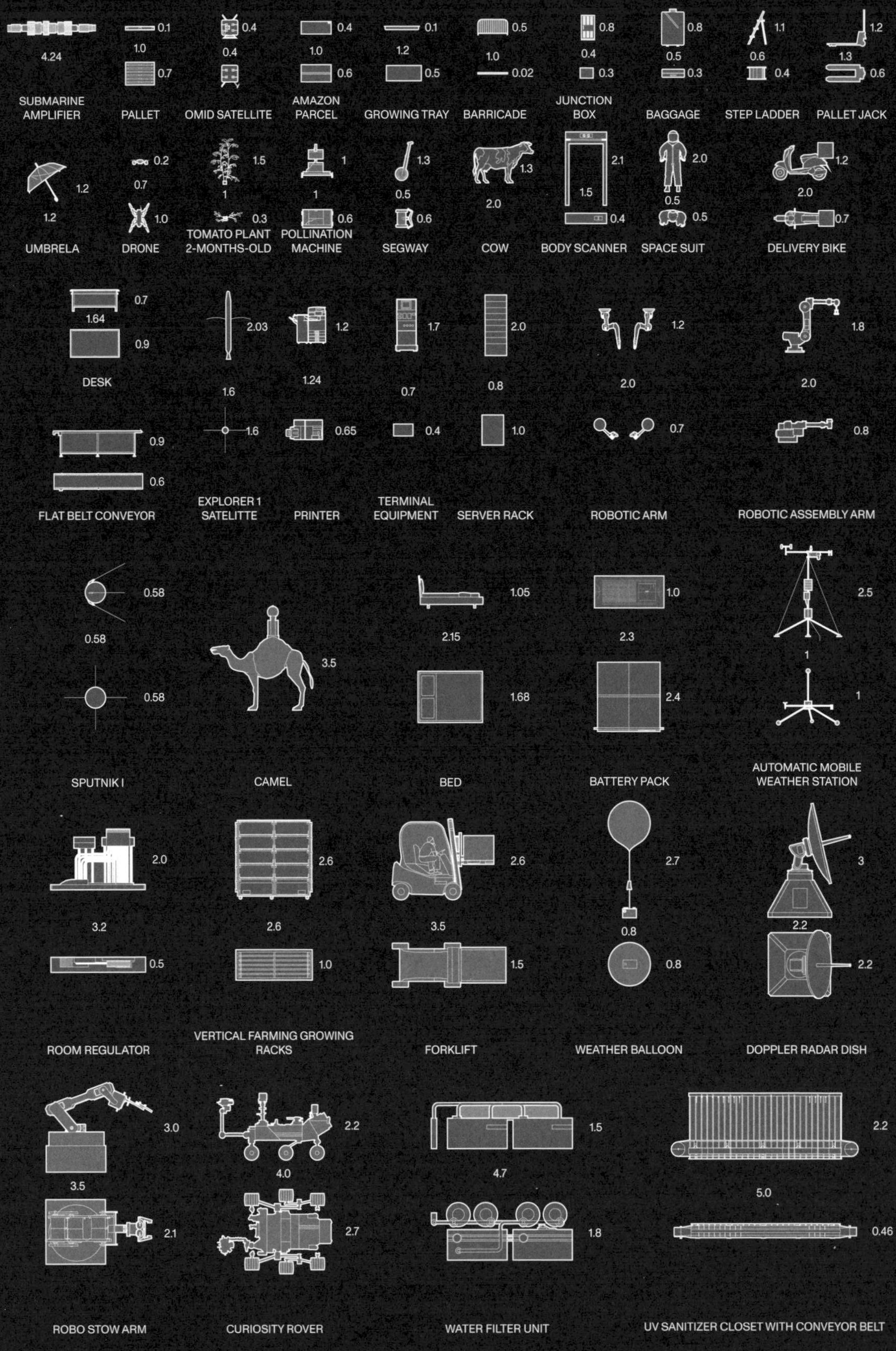

0.1 1.0 0.7
PALLET
4.24
SUBMARINE AMPLIFIER
0.4 0.4
OMID SATELLITE
0.4 1.0 0.6
AMAZON PARCEL
0.1 1.2 0.5
GROWING TRAY
0.5 1.0 0.02
BARRICADE
0.8 0.4 0.3
JUNCTION BOX
0.8 0.5 0.3
BAGGAGE
1.1 0.6 0.4
STEP LADDER
1.2 1.3 0.6
PALLET JACK
1.2 1.2
UMBRELA
0.2 0.7 1.0
DRONE
1.5 1 0.3
TOMATO PLANT 2-MONTHS-OLD
1 1 0.6
POLLINATION MACHINE
1.3 0.5 0.6
SEGWAY
1.3 2.0
COW
2.1 1.5 0.4
BODY SCANNER
2.0 0.5 0.5
SPACE SUIT
1.2 2.0 0.7
DELIVERY BIKE
0.7 1.64 0.9
DESK
0.9 0.6
FLAT BELT CONVEYOR
2.03 1.6 1.6
EXPLORER 1 SATELITTE
1.2 1.24 0.65
PRINTER
1.7 0.7 0.4
TERMINAL EQUIPMENT
2.0 0.8 1.0
SERVER RACK
1.2 2.0 0.7
ROBOTIC ARM
1.8 2.0 0.8
ROBOTIC ASSEMBLY ARM
0.58 0.58 0.58
SPUTNIK I
3.5
CAMEL
1.05 2.15 1.68
BED
1.0 2.3 2.4
BATTERY PACK
2.5 1 1
AUTOMATIC MOBILE WEATHER STATION
2.0 3.2 0.5
ROOM REGULATOR
2.6 2.6 1.0
VERTICAL FARMING GROWING RACKS
2.6 3.5 1.5
FORKLIFT
2.7 0.8 0.8
WEATHER BALLOON
3 2.2 2.2
DOPPLER RADAR DISH
3.0 3.5 2.1
ROBO STOW ARM
2.2 4.0 2.7
CURIOSITY ROVER
1.5 4.7 1.8
WATER FILTER UNIT
2.2 5.0 0.46
UV SANITIZER CLOSET WITH CONVEYOR BELT

1:200

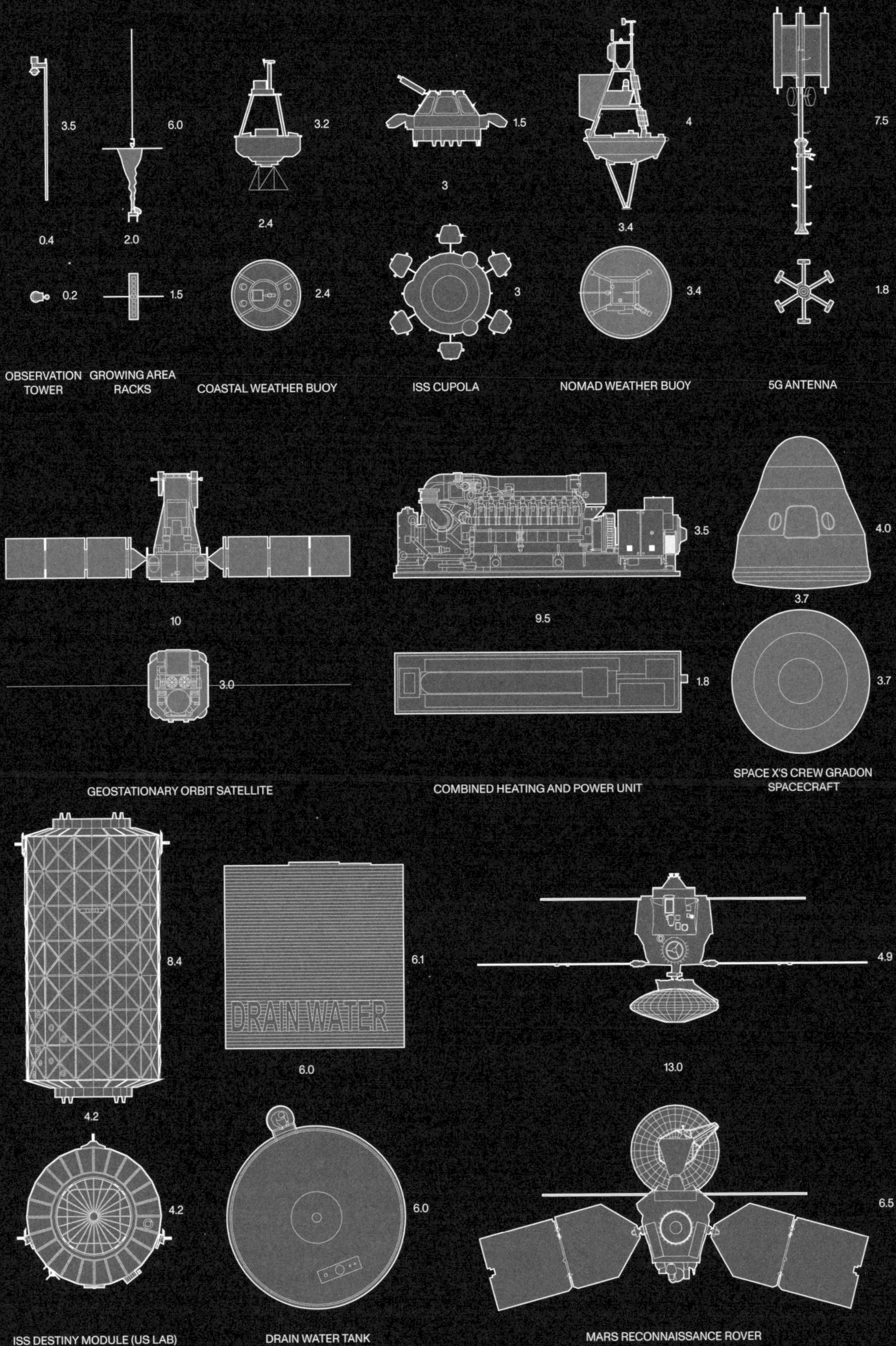
3.5
0.4
0.2
OBSERVATION TOWER
6.0
2.0
1.5
GROWING AREA RACKS
3.2
2.4
2.4
COASTAL WEATHER BUOY
1.5
3
3
ISS CUPOLA
4
3.4
3.4
NOMAD WEATHER BUOY
7.5
1.8
5G ANTENNA
10
3.0
GEOSTATIONARY ORBIT SATELLITE
3.5
9.5
1.8
COMBINED HEATING AND POWER UNIT
4.0
3.7
3.7
SPACE X'S CREW GRADON SPACECRAFT
8.4
4.2
4.2
ISS DESTINY MODULE (US LAB)
DRAIN WATER
6.1
6.0
6.0
DRAIN WATER TANK
4.9
13.0
6.5
MARS RECONNAISSANCE ROVER

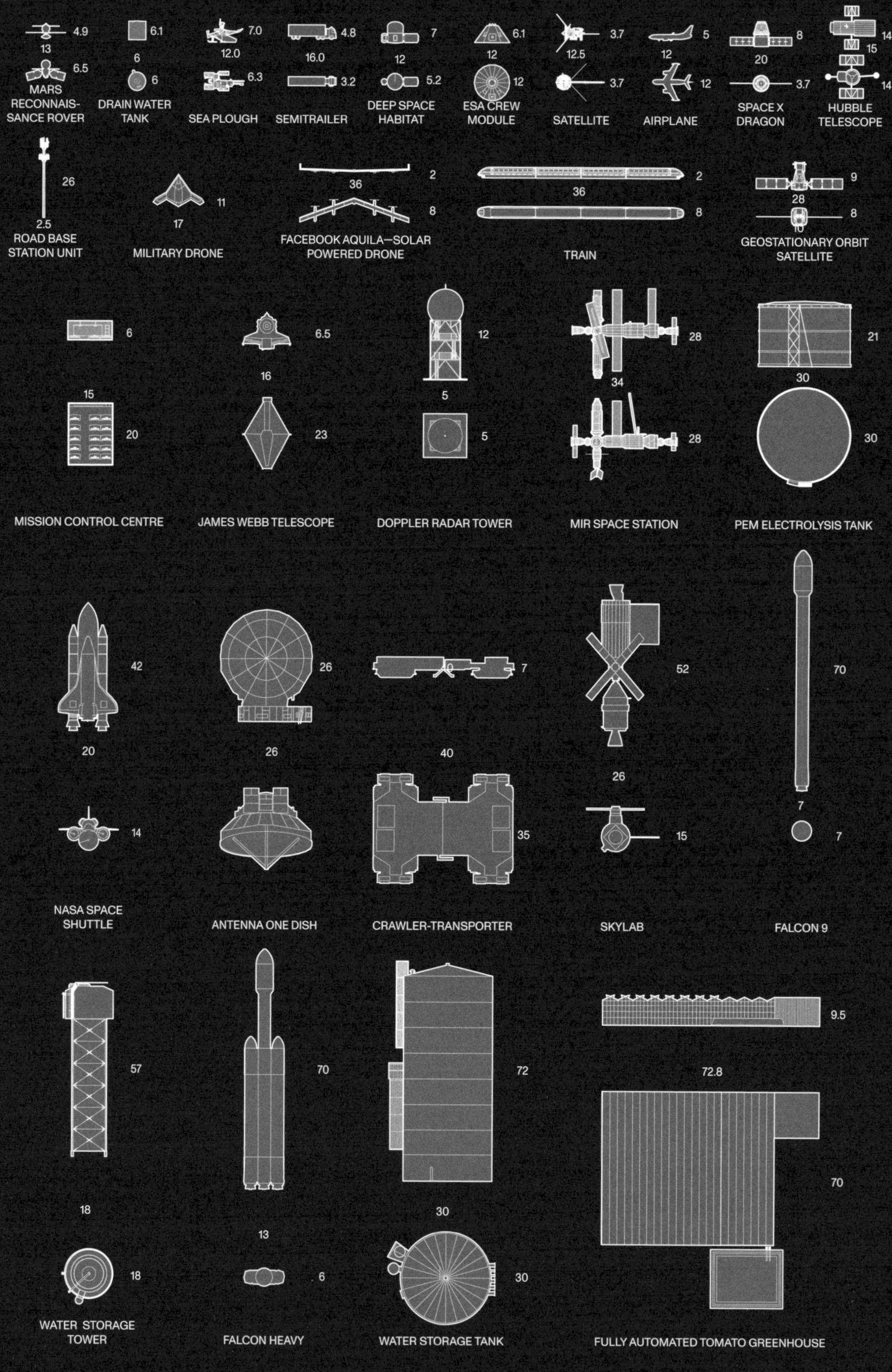

1 : 2000
MARS RECONNAISSANCE ROVER
DRAIN WATER TANK
SEA PLOUGH
SEMITRAILER
DEEP SPACE HABITAT
ESA CREW MODULE
SATELLITE
AIRPLANE
SPACE X DRAGON
HUBBLE TELESCOPE
ROAD BASE STATION UNIT
MILITARY DRONE
FACEBOOK AQUILA—SOLAR POWERED DRONE
TRAIN
GEOSTATIONARY ORBIT SATELLITE
MISSION CONTROL CENTRE
JAMES WEBB TELESCOPE
DOPPLER RADAR TOWER
MIR SPACE STATION
PEM ELECTROLYSIS TANK
NASA SPACE SHUTTLE
ANTENNA ONE DISH
CRAWLER-TRANSPORTER
SKYLAB
FALCON 9
WATER STORAGE TOWER
FALCON HEAVY
WATER STORAGE TANK
FULLY AUTOMATED TOMATO GREENHOUSE

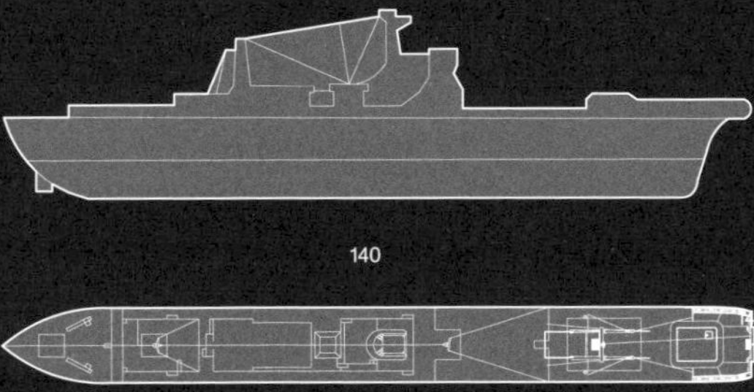


CABLE LAYING SHIP

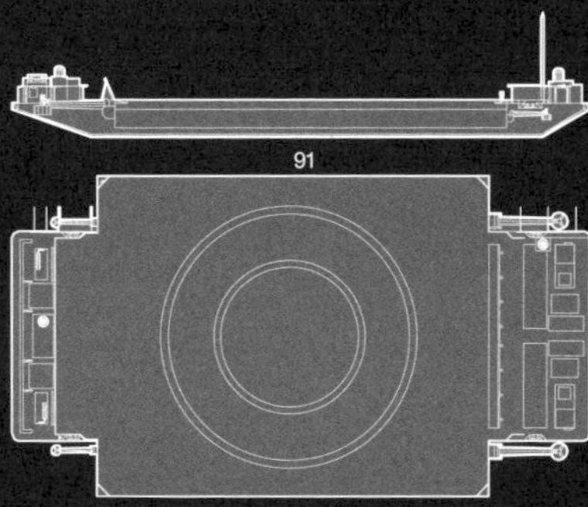


AUTONOMOUS SPACEPORT DRONE SHIP

SATURN V

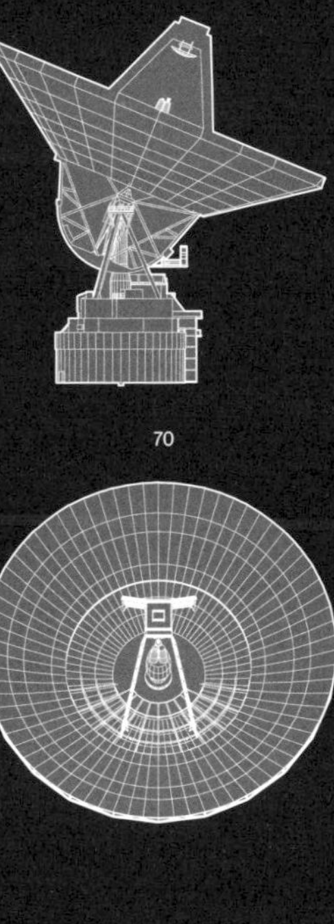


DEEP SPACE NETWORK ANTENNA

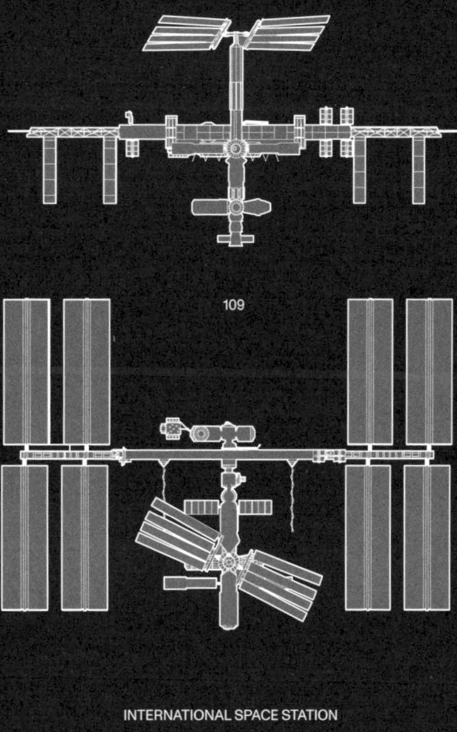


INTERNATIONAL SPACE STATION

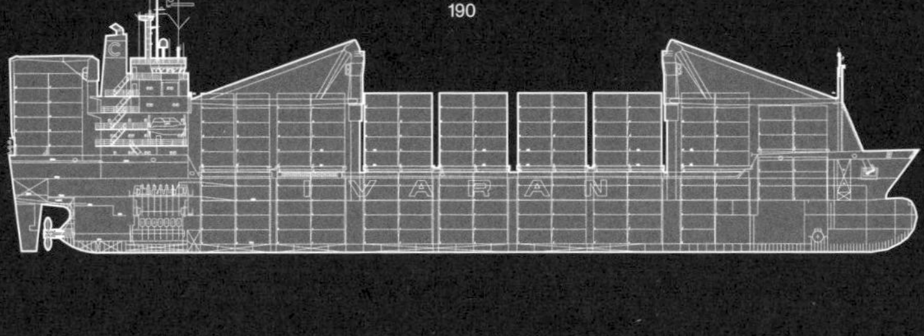


CARGO SHIP

# ON GIZMOS

ON GIZMOS, GADGETS, OBJECTS AND DATA

Kees Kaan

The curators of this wonderful publication asked me to reflect on the gizmo catalogue, a fantastic section of the book. It features a range of objects, from astronaut suits and spacecrafts to pens and printers, beautifully drawn in scales 1:200, 1:200 or 1:20, and 1:2, with top and side views. Any reflection on data, AI, new technology versus proven technology, or the transition from analogue to digital generally raises more questions than answers. My contribution to this book does the same.

*Datapolis* investigates the data revolution's impact on humans and our physical environment. What is the effect of the ostensibly ephemeral cloud that hosts the virtual world? How does it influence our habits, behaviour and the things we surround ourselves with?

We have built an immense physical infrastructure, apparently invisible, hiding in the countryside, oceans and desolate places, to perpetuate the myth of people and things being wirelessly connected. *Datapolis* unmasks the cloud and exposes its physical impact. City and countryside are changing due to the disruptive effect of the 'delivery society' we are creating, whether it concerns bringing food, clothing, energy or ideas to our homes. Every wish is fulfilled at the touch of a button. But at what cost? How does this translate into a design brief or building requirements? What will be radically different in future buildings, and what will stay the same?

Simply put, what is the physical impact of data? This question is critical to architecture, a slow profession by nature. Buildings facilitate societal change and reflect them once they have already been crystallized. Architecture is a result rather than a driver of change, consolidating instead of initiating it. This means that we are now gazing at the immense complexes that the new technology sector generates. As the semiconductor is reduced to the nanoscale, getting smaller and smaller, data centres, production facilities, offices and warehouses produce giant structures devoid of people.

*Datapolis*'s catalogue of gizmos is an overview of instruments, tools, means of transport or communication. These objects are time-bound and testify to human ambitions and expectations. They tell the story of the birth and rise of the new tech era, witnessing the innocence of the first computer and spacecraft to the swell of the current wave of change. They are also designed objects.

Museums worldwide exhibit artefacts related to food, work, war and domesticity that tell us something about the

# GADGETS, G

culture that produced and used them. We are interested in stuff that helps us understand our culture and, ultimately ourselves. Coins, jugs, tools, weapons and works of art are silent witnesses of humanity.

Humans surround themselves with objects. We find them in the layers of soil, but also orbiting the globe. The gizmos are a vestige of our development, and that wave swells exponentially. The objects we make are designed not just around our bodies, but for abstract activities, including required furniture, equipment and tools. Humans stay more or less the same in their basic needs and requirements, physical proportions, strength and weaknesses; the change is evolutionary and slow. But as the performance level of buildings towards the body rises, demanding higher comfort, safety and wellbeing, the gizmos change much quicker. Our buildings and cities transform while our bodies stay the same.

Not only do people surround themselves with stuff, they also drag it into their buildings. A functional office is a building where you can set up desks and chairs or distribute conference tables properly. A functional building is not just made according to the human body, but according to the logic of the tools it uses and all its surrounding processes and requirements.

The new connected world causes a considerable shift in those work and life processes and the gizmos that come with it. Today's body has acquired a set of personal tools. Processes and tools are increasingly pointing towards the self and the I. Steve Jobs saw it clearly—gizmos must become personalized tools merged with our identity. Maybe this is a first step towards integrating the body with the gizmo. The I-tool is a step towards the cyborg.

To optimally facilitate the workings of those small personal gadgets, the tech sector is building ample facilities in rural and industrial areas and hiding infrastructure in oceans and space. The freedom of the influencer in the city is facilitated by giant complexes in the hinterland, -sea and -space. There is an unstoppable simultaneous process of shrinking to nanotools and growing to megacomplexes. Both are subject to permanent innovation and either scaling up or downsizing radically.

Architects keep their eyes on the scale of 20 to 200, and the in-between where we try to maintain our precious cities and rural areas because that is where we live. It is our habitat that these new giants should not invade. The middle scale (normal buildings) is still developing in a noncommittal formalism. Exciting, iconic buildings, which are actually immensely boring non-exciting inventions, must disguise the fact that mid-scale development lags behind the two extremes. At this level, the radical urgency that characterizes new tech infrastructures has so far been lacking, or people are obsessively holding on to what they have because it's their core habitat, a place of comfort and joy—it is a gigantic dichotomy that is becoming a real issue.

Meanwhile, the city as we are used to it is gradually becoming untenable. The comfort and joy are not sustainable, the energy is inefficient, the waste is piling up and the food production is out of balance. There is a crisis because the city and the hinterland, mutually dependent, are out of balance.

Here we are, working and living in high-density areas, in buildings overcrowded with bodies and mobile devices served from empty buildings elsewhere buzzing with installations. We were able to manage the pandemic thanks to these virtual connections facilitated from those large empty buildings. We could largely continue our societal necessities, despite the personal physical isolation that the lockdown forced on us.

Is this a step towards a metaverse that will generally give some relief to the physical world? *Datapolis* does not yet indicate a development in that direction; the people's physical world and the footprint created are still expanding. But our built environment is our daily habitat. It is shaped by people over time and continuously changes and updates through complex processes driven by societal, economic, political and technical forces. In this setup, the role of design is crucial. Keeping in mind the best interest of society and human life, one could assume that if more accurate, complete and unbiased information is available in this ongoing process, our ability to shape our habitat will be better too. If this assumption is correct, then open data—impartial and fair—is essential.

This was my answer when Georg Vrachliotis invited me to The New Open conference on 27 and 28 October 2022. Tech is causing unmanageable, potentially ruinous disruptions and scale wrinkles in the built environment, but now we must also assume that it will help us fix those problems? Is the very industrial complex that is magnifying the problems also the key to solving them?

That is why I have added: 'If this assumption is correct.' The next step could be that the built environment gradually becomes more interactive, anticipating and performative. This performative environment, with its inhabitants, humans or other forms of life in all their diversities, might constitute a new cyber intelligence. Sounds like science fiction.

When it comes to design, we are guided by our values. They are in a permanent state of change while we, as a society, try to figure out what works and what needs to be updated. This means that design is evolutionary, but if so—do we also keep making the same mistakes?

Is knowledge built upon big data a guarantee for better solutions for real-world problems, or could it also generate huge collective mistakes?

While I'm thinking about the gizmos in this book, an e-invitation to participate in the event of Communities of Tacit Knowledge drops in my mailbox. The community investigates Architecture and Its Ways of Knowing and is organizing an exhibition and debate on the Object. They're asking me to choose an object that, for me, exemplifies the way tacit knowledge operates in architecture. Tacit is such an abstract concept.

Bringing it to the surface and making it visible and tangible through objects could be a method to understand architecture better and find out where specific knowledge is hiding.

I wonder if tacit knowledge might be the opposite of artificial intelligence. Or are they just different forms of knowledge? Tacit knowledge is unspoken knowledge that allows us to do or know things without understanding why and how we got that ability. Kind of like riding a bicycle.

Artificial intelligence is the man-made ability of machines to learn and apply knowledge. Would a machine be able to deploy tacit knowledge? Can the machine do or know something without first knowing the why and how?

This is a question for one of our AI scientists, so I decided to talk to Seyran Khademi, the director of the AI lab in our department. It seemed only fitting to do it through a gizmo in WhatsApp.

KK Do you think a machine can develop tacit knowledge?
SK I think the challenge with tacit knowledge is transferring it to other humans. So even if a machine could generate it, I would not know how to visualize/represent it, so basically, how to show it.
KK Well, I understood that it is knowledge or ability that we cannot trace back to its origin; we cannot explain precisely how it works, like riding a bike.
SK That is exactly how AI learns from data, but I doubt it can transfer tacit knowledge to explicit knowledge. AI learns from data because we cannot explicitly program computers to do complex tasks like vision (or riding a bike) :)

She says that computer scientists don't know how to write a program to make a computer visually recognize a cat just by seeing it. They can repeatedly show a million or more images of a cat to the machine. The machine will, from now on, recognize a cat by comparing the data. Next, the machine cannot recognize a dog and needs to see a million images of dogs. That doesn't sound very 'intelligent', does it?

It brings me back to the question. If we knew all there is to know, if all existing human knowledge built up over thousands of years would be broken down into data and become accessible for AI, would it prevent us from making the same eternal mistakes?

KK then you are saying that AI is a form of tacit knowledge.
SK I would say tacit information. The knowledge still needs to be extracted.

Then I bring it forward—the big question:

KK Do you think AI could also lead to big collective mistakes because it uses data that was implicitly extracted from human errors, tacit mistakes?
SK that is a very hot research topic.
SK for sure, it can magnify our mistakes for the worse.
KK so many scientists are on this topic
SK yes.

A couple of days later I ask Seyran permission to use the conversation for this contribution to *Datapolis*. So, I send her the complete written text above.

SK Also, there is a question the interviewer asks I suppose, and I quote: 'She says that computer scientists don't know how to write a program to make a computer visually recognize a cat just by seeing it. They can repeatedly show a million or more images of a cat to the machine. The machine will, from now on, recognize a cat by comparing the data. Next, the machine cannot recognize a dog and needs to see a million images of dogs. That doesn't sound very 'intelligent', does it?"

The answer is: What do you consider intelligence? If we assume that humans are intelligent then this is how our brain learns. If a kid did not see a dog before, she would not recognize it. We as intelligent species learn to perceive our world using our biological sensors, modern AI borrows the same process, however, AI 'perceives' the digital world abstracted in the data.

Call me if you have any other questions. So, I did call her by mail.

KK Great Seyran, I will work it in the text. Of course, humans must also learn everything piece by piece, bit by bit, but do children learn quicker (with fewer examples) than machines or do machines also more quickly learn quicker the dog after having learned the cat?
KK sorry for bothering you with these questions...
SK My favourite questions :)

Yes that is true and our machines are still an infant in that sense, the human brain is able to do multitask learning, e.g., knowing German makes it easier to learn Dutch. Also, we are better at generalizing unseen examples, seeing a dog for the first time, a kid can categorize it as an animal or even a pet.

Many researchers are also working on multitask learning in AI, inspired by how our brain works again by learning from data (data-driven) rather than explicit algorithms (knowledge-driven). However, machines are scalable in terms of processing power so analysing lots of data, unlike the human brain, which has not changed much in thousand years.

Epilogue

This conversation opens so many doors to new questions. The new AI program that was initiated by Delft University of Technology is very exciting. I feel privileged to be able to work with all these brains.

Today we need to cope with a lot of information while designing, and if we want it to be state of the art, pure humanistic design talent is no longer enough. Of course experience is important, but we need to access new layers of information/knowledge simultaneously. We might come closer to the actual concept of a 'knowledge of everything' by connecting many brains to generate a new collective knowing. Connecting brains requires ultra-clear lines of communication and exchange of information. I have a hunch that new technologies, big data and AI are based on universal templates. Using those could prove beneficial to this improved communication and collaboration of brains. Design research collaboration between practitioners and academics could establish new models of collecting information and designing.

If humans have a future on this planet, we need to change how we deal with our built environment; we cannot continue to rely on the interventions of individual geniuses. We need a collective effort and all the knowledge in the world to achieve higher standards and better results.

We must change our way of working as architects, not our way of being architects.

Glossary

| | |
|---|---|
| Gizmo | A gadget, especially one whose real name and use has been forgotten. It's a weird kind of gadget. |
| Gadget | A small technological or mechanical object, innovative and appealing, a wanna-have. |
| Object | A material thing that can be seen and touched. A person to whom a certain action is specified. |
| Design object | An expressed solution to a real-world problem. |
| Ready-made | A mass-produced object displayed as a work of art. |
| Objective | A concept of truth independent from individual subjectivity. Without need for interpretation. |
| Subject | A thing or person being discussed or described |
| Subjective | Based on or influenced by. |
| Tacit | Understood or implied without being stated. |
| Implicit | Suggested but not directly expressed. |
| Explicit | Stated clearly and in detail, leaving no room for doubt. |
| Symbolic | Implicitly referring to another thing. |
| Specific | Something peculiarly adapted. |
| Generic | Related to a characteristic of a whole group. |
| Intelligence | The ability to learn and apply knowledge. |
| Artificial | Made by humans. |
| Stuff | Miscellaneous objects with possessive personal effects. |

# ATLAS OF D

ATLAS OF DATA LANDSCAPES
Paul Swagerman

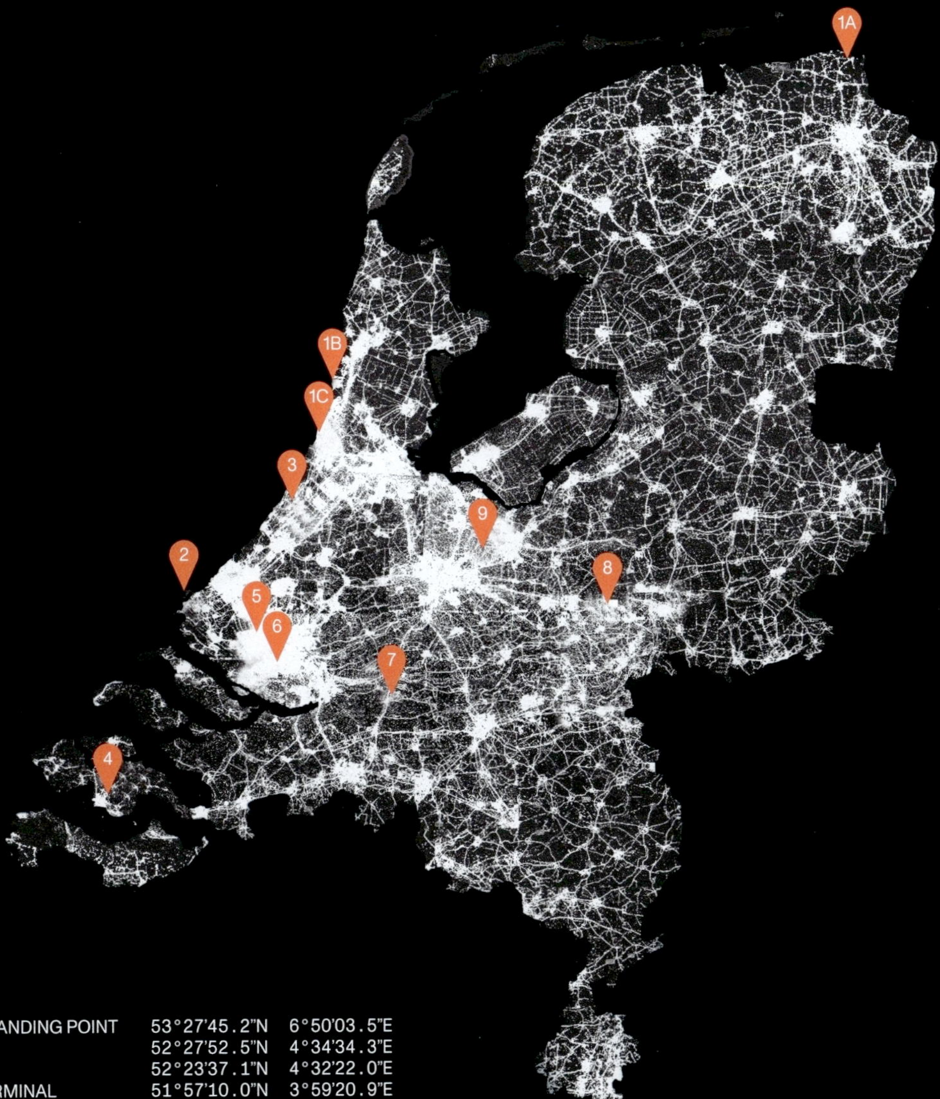


| | | | |
|---|---|---|---|
| 1A | CABLE LANDING POINT | 53°27'45.2"N | 6°50'03.5"E |
| 1B | | 52°27'52.5"N | 4°34'34.3"E |
| 1C | | 52°23'37.1"N | 4°32'22.0"E |
| 2 | RWG TERMINAL | 51°57'10.0"N | 3°59'20.9"E |
| 3 | ESTEC | 52°13'13.0"N | 4°25'16.0"E |
| 4 | DATABUNKER | 51°30'07.9"N | 3°54'25.9"E |
| 5 | FLOATING FARM | 51°54'46.4"N | 4°24'50.5"E |
| 6 | N+P RECYCLING | 51°53'00.3"N | 4°25'48.4"E |
| 7 | BOL.COM | 51°42'12.7"N | 5°04'04.2"E |
| 8 | BACKUPNED | 51°59'18.2"N | 5°50'15.6"E |
| 9 | WHITE SPOTS | 52°09'11.6"N | 5°17'41.6"E |

WIRELESS NETWORK DENSITY, MAP OF THE NETHERLANDS
Date Range: 2001–2023
Source: wigle.net
Drawing by Léa Alapini

# CABLE LAN

# DING POINT

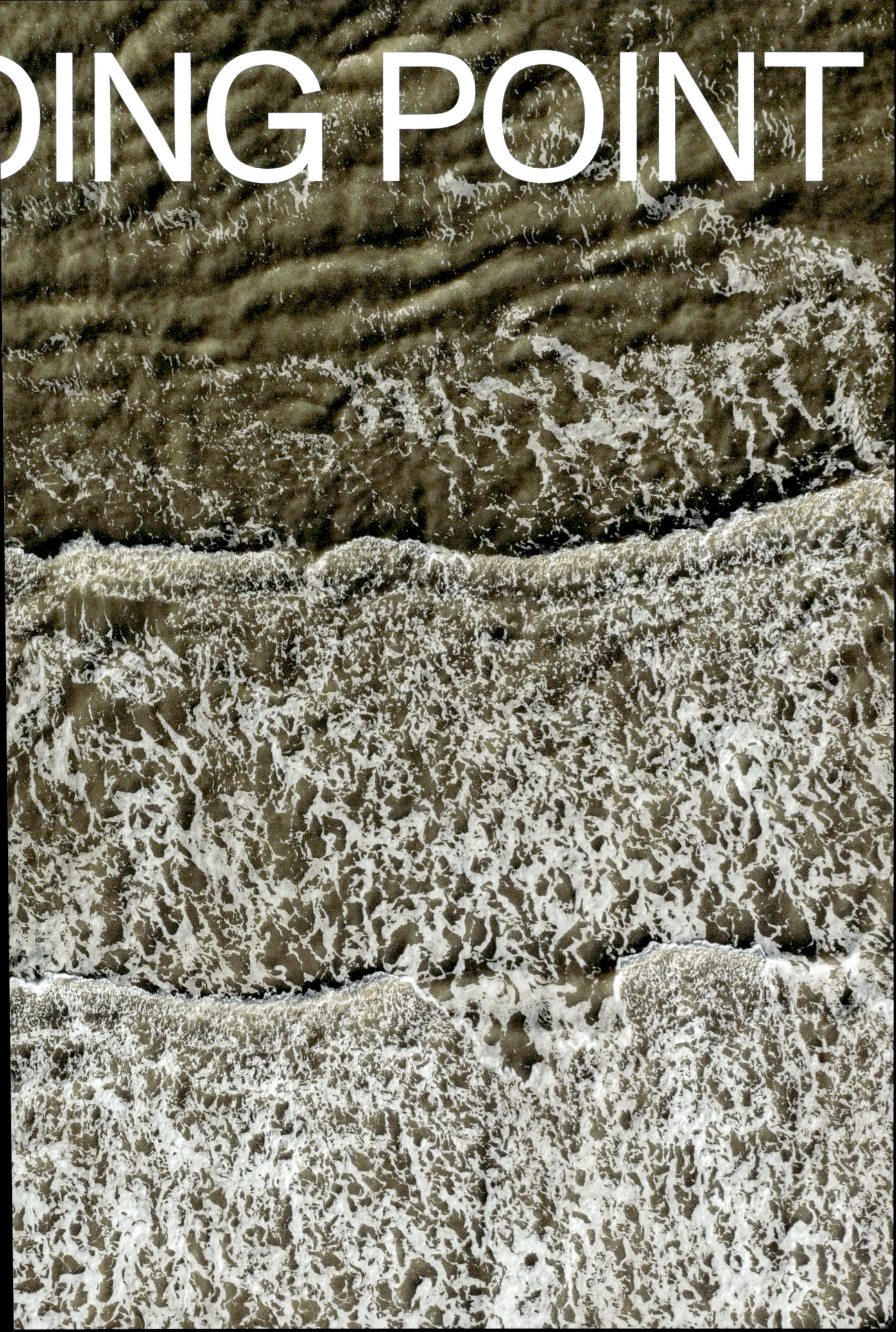

The shore break of Zandvoort, NL. Circe North, Zeus and Concerto submarine communications cable landing point.

↑ Circe North, 1999, length: 203 km. The landing point at the other end of the cable: Lowestoft, UK. Zeus, 2022, length: n.a., other landing point: Lowestoft, UK.

← Eemshaven, NL. COBRAcable, 2019. Length: 304 km. The landing point at the other end of the cable: Endrup, DK ↑ IJmuiden, NL. Scylla cable, 2021, length: 204 km, the landing point at the other end of the cable: Lowestoft, UK.

# RWG TERMINAL

Datapolis

RWG container terminal, the largest fully automated container transhipment in the world. Trucks without cargo move in, are identified by a scanner, and just-in-time pick up their brightly coloured container, the exact one out of 50,000. This is the only place where there's a human presence.

Beyond the trucks, a truly fascinating choreography of autonomous vehicles (ATV's) takes place, in a strictly prohibited area enclosed by fences. In the pouring rain they quietly follow an invisible grid in an almost confident way, picking up or delivering containers from the deep-sea vessels. They intelligently sort out where to store each one for short or long stay. When out of power, they pause for a minute in their garages, their batteries are swapped behind a closed roller door and then they continue their duty, 24:7, year round. In theory, people never have to intervene.

↑ AGV's (automatic guided vehicles) drive into the battery exchange station for a battery swap. In ten minutes, they are ready again for six to nine hours of driving.

RWG
RWG
RWG
528
RWG
RWG
RWG

↑ Thirteen deep-sea cranes, three barge/feeder cranes and two rail cranes are used to move containers from barges, feeders and deep-sea vessels onto the container storage area.

TEREX | GOTTWALD
RWG
AS 201
AS 202
tex
VL 514

↑ Fifty Automatic stacking cranes move 2.35 million TEU (Twenty Foot Equivalent Unit, the dimension of a container) annually, which is approximately 90.5 million $m^3$.

EVERGREEN
Hapag-Lloyd
HMM
tex
VAN DER WINDT
SCANIA
22

↑ Around 80% of the containers, stored on RWG, are distributed by truck and 20% by train.

# ESTEC

## ESTEC

In stark contrast to the endless surrounding residential areas, the ESTEC (European Space Research and Technology Centre) is wedged between the city of Noordwijk and the dunes. From the entrance gate and reception desk, it's already clear that this is a distinct area, marked by an outdoor exhibition of dysfunctional satellites and other unnameable objects. This is where, prior to launch, instruments get tested in the conditions of space, improper and unnatural to Earth. The guide speaks quickly with her French accent, and continuously calls her colleagues about the possible and impossible spots to visit and capture. Every door is opened with a pass, and behind each door is another door, sometimes marked with a red lamp: test active. We walk through a maze of boxes, interconnected by winding walkways and sizable parking lots. There's a massive deep-black tube where the sun's reflection is simulated, abstract rooms that isolate all noise distortion, an immense shaking platform, and a large microwave-sized apparatus that simulates years of exposure to the sun. At the end is a decor of a Martian landscape for robots, which are sometimes let out in the nearby dunes.

← Nitrogen tanks cool down the space environment simulation testing facilities below -173.15°C, corresponding to the shade's temperature in space. ↑ Flight model of IXV (Intermediate eXperimental Vehicle), for testing critical systems and technologies for return missions from low Earth orbit.

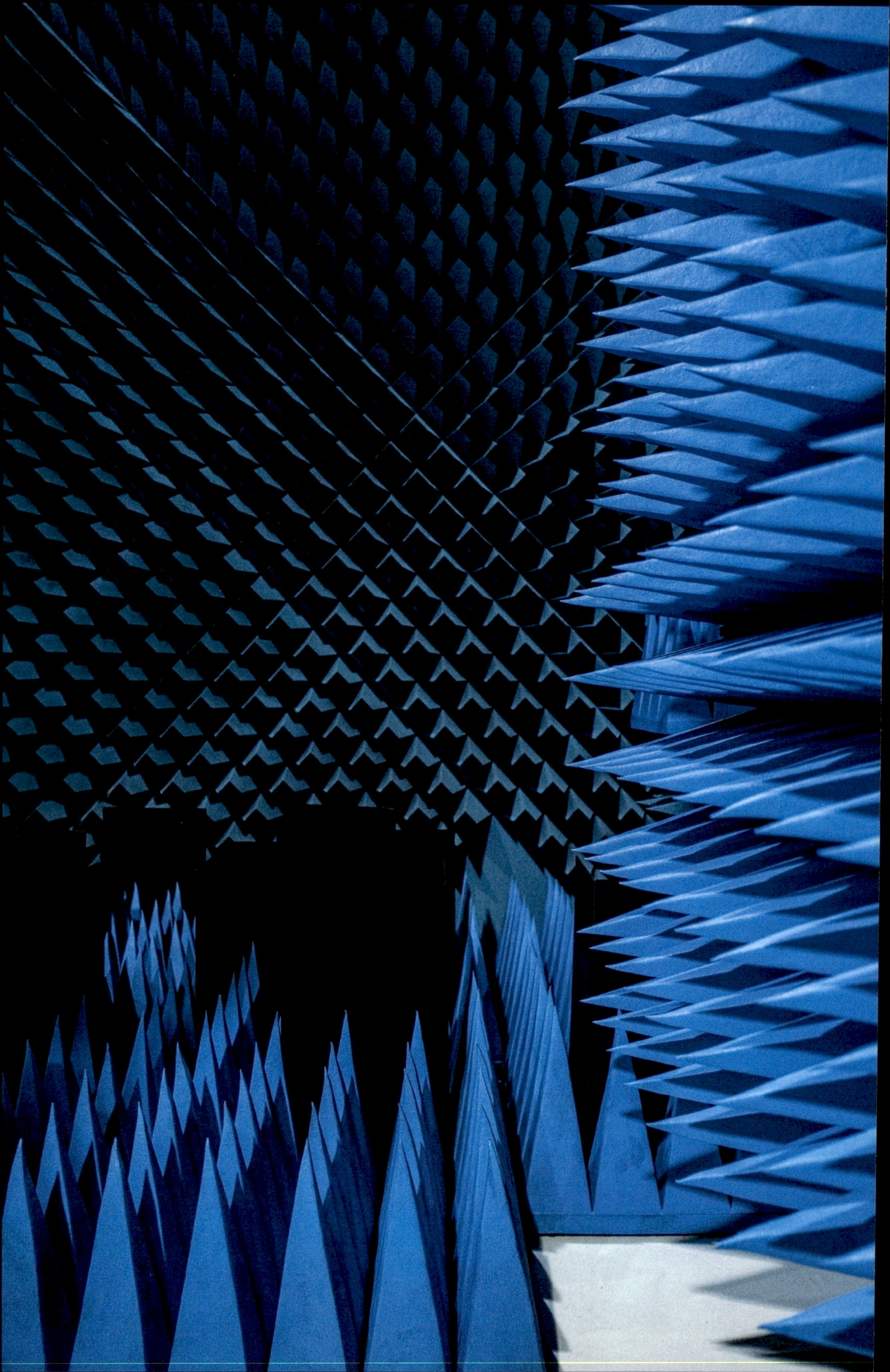

← EMC anechoic chamber. The chamber's design blocks out outside sound and absorbs acoustic echoes to simulate a space environment. ↑ A half-scale ExoMars rover is tested on terrain representing Mars's surface, which is made of sand and volcanic rocks, taken from different sites in Europe.

← The solar beam entrance exposes the satellite to horizontal solar beam of 6-m diameter with excellent uniformity and very long- and short-term stabilities. ↑ Inside Europe's largest thermal vacuum chamber, wherein a mock-up of ESA's exoplanet detecting spacecraft named PLATO is being tested.

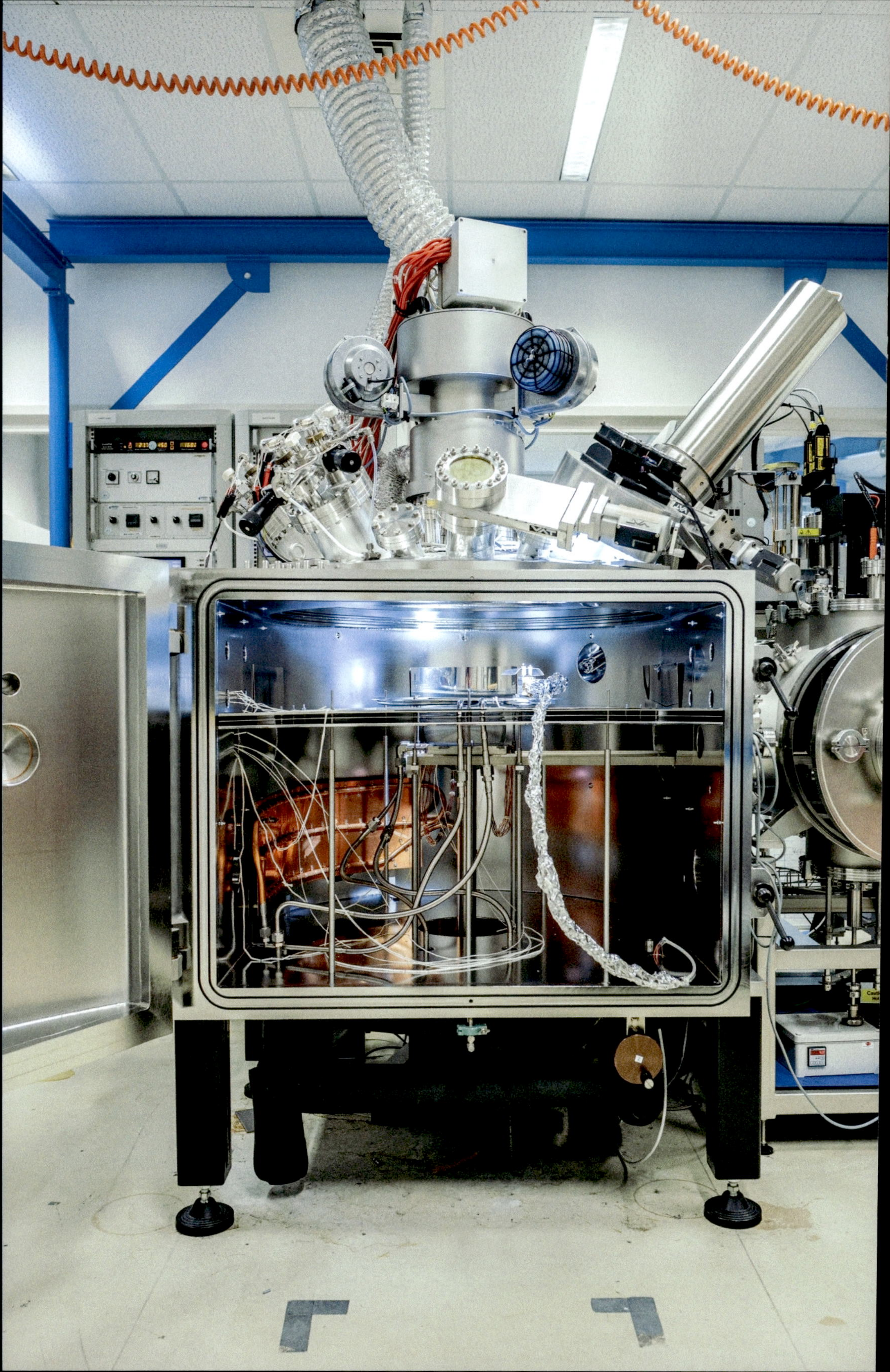

← STAR II testing facility where temperature-controlled samples are exposed to radiation sources after which the thermos-optical properties are measured in-situ without breaking the vacuum. ↑ ESA's Large Space Simulator, the largest thermal vacuum chamber in Europe.

REXROTH

← HYDRA, the largest satellite shaker in Europe is a multi-axis vibration test facility for satellites. It can generate vibrations equivalent to an earthquake of 7.5 on the Richter scale. ↑ HYDRA, the 5.5 m octagon-shaped test table of 21 tonnes is excited by eight powerful hydraulic actuators (four vertical and four horizontal).

# DATABUNKE

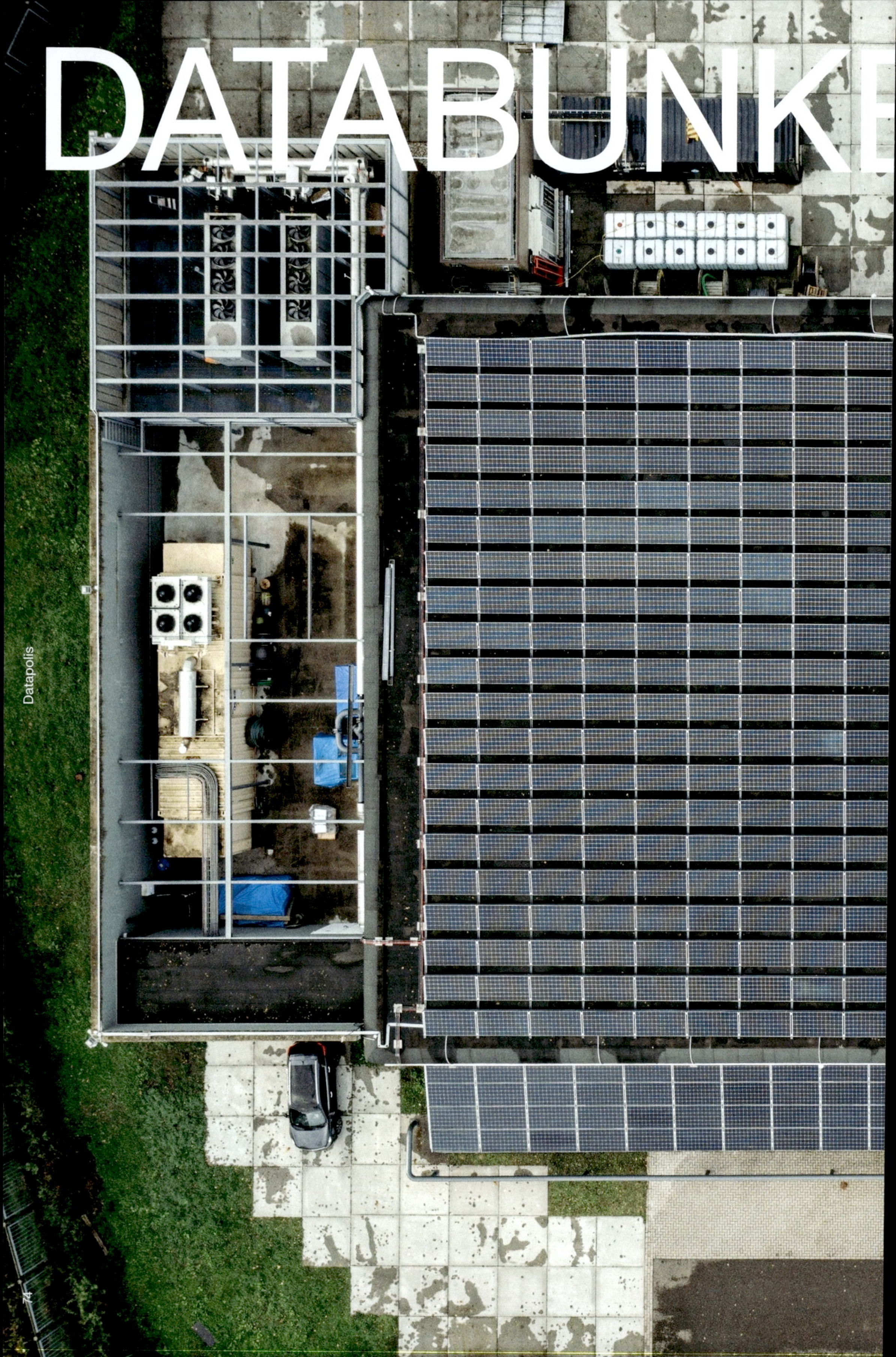

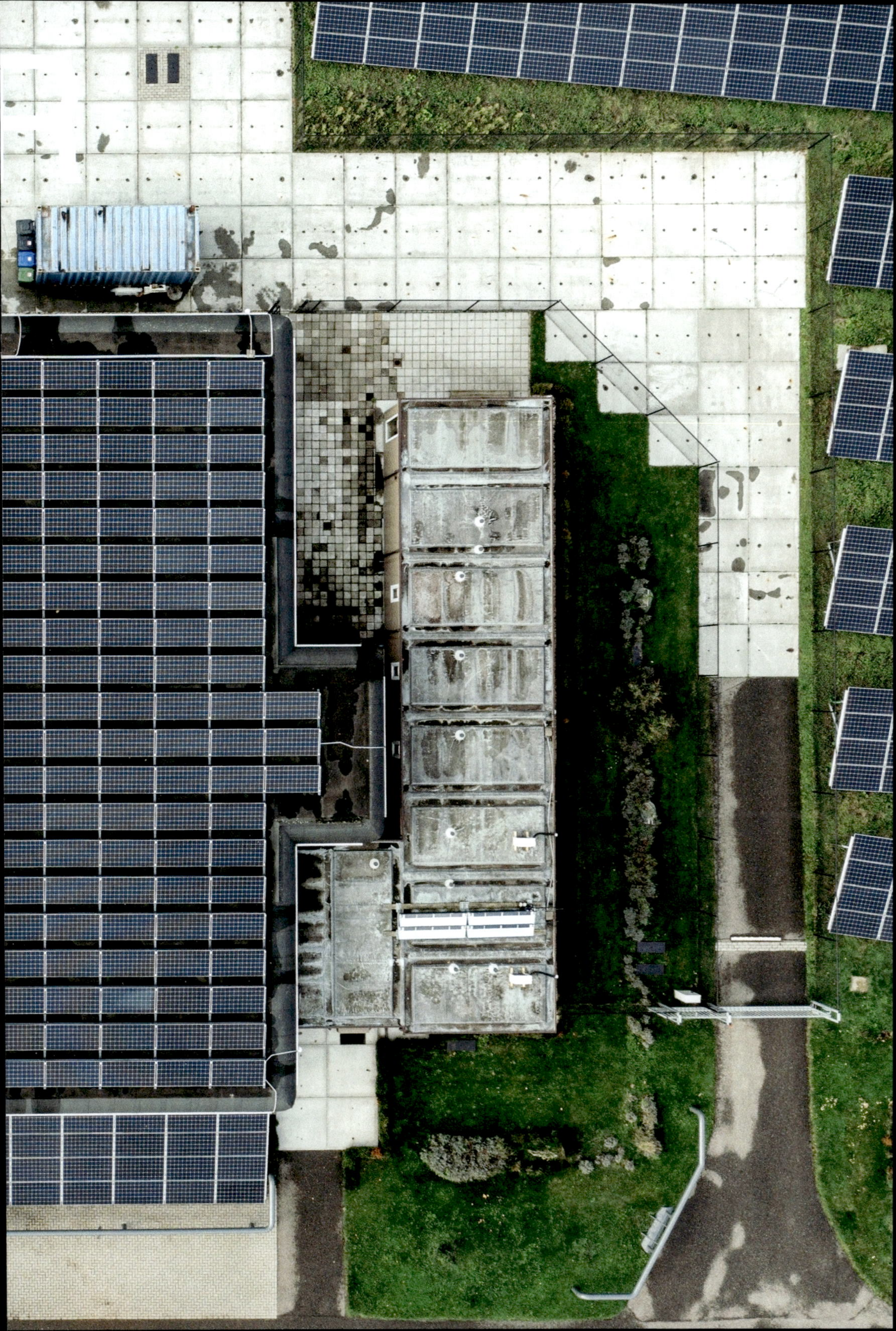

303 PV panels on the roof, 64 in front and 448 in the field produce over 150 MWh yearly. 815 PV solar panels produce over 150 MWh electricity yearly, exceeding the consumption of the data centre.

served as a governmental communication headquarters, are clearly visible on the ceiling, walls and floors. The massive concrete box seems to have undergone a surgical operation with tubes for cooling water, data- and power cables passing through the walls, roof and between the spaces, PV-panel converters, security cameras, H-steel frames and small antennas, all clicked onto the concrete surface like addons. Nothing indicates that this is a place where data is stored for a specific type of client that requires the highest level of data security, and I wonder if data is more protected when encapsuled in a 5-m-thick layer of concrete. It is fascinating to witness the layers of time and the fleeting technological systems that have, for now, nestled in a former top-secret call centre.

← Entrance with solar panel array and ram raid barriers. ↑ A 'hot corridor' in one of the co-location rooms, containing server racks under a black steel ceiling.

↖ Transformers transform high voltage 10kV to 400V. ← Ducts and air conditioners on the exterior wall, for cooling the servers and to air-condition the former command and control room. ↑ The painted red and blue tubes connect to the super chillers and are part of the cooling system of the Databunker.

# FLOATING F

ARM

In the Floating Farm, in the old harbour of Rotterdam, the cows are housed on the semi-transparent upper level and the floating solar panels supply electricity.

Datapolis

↑ The produced milk, yoghurt, butter and cheese at the farm are being transported to a van that distributes them in and around the city of Rotterdam, to restaurants and directly to consumers.

URIN
SOURCE OF
WATER AND NUTRIENS

← Through the funnel, the cow's manure and urine fall from the upper-level stable into the lower-level big bags and the containers. ↑ Cows are fed three times a day through the feed-belt with residual products from the city like bread, beer broth and vegetables.

# N+P RECYC

# ING

Screening drum, wind-shifter and magnetic separator are interconnected by conveyor belts.

## N+P RECYCLING

Entering the waste separation machine is like entering a wondrous game environment. The massive hall is spanned by a beltline; like a 'bridge' between the torn bags that enter through cut-out holes in the steel plates, it divides the temple of recycling into different sections. In the space interconnected by grids, stairways and steel H-profiles, one is allowed to follow the green marked railings and catch a glimpse of a scene that resembles one of Piranesi's prison drawings, in reality. Overwhelmed by the quantity and speed of the 12-step plastic recycling process, I almost wonder why this is not accessible as a theme park attraction.

All of the German Orange machines are connected to each other by homogenous tubes. Generating a constant vibration and exceptional soundscape through the structure, they scan, suck, blow or magnetically attract any object that is not made of plastic. While his private kitchen is orange, the director thinks orange is an ugly colour; he would have ordered all of the machines in blue if there had been a choice.

'Why are you as architects visiting this place? When I think about architecture, I think about buildings,' the director of the automated plastic separation plant asked. For him it is such a self-evident fact that the 400 × 30 × 30-m volume situated in the southern part of the port of Rotterdam is not architecture, but merely a system—an infrastructural machine, something that could not have any other appearance than what it is constructed of. It is so instrumental and logical that it could not be architecture.

← The N+P recycling centre in Waalhaven, Rotterdam receives 500 tons of waste daily. ↑ The main hall. The optical separator separates beverage cartons, the magnetic separator and aluminium separator remove metals from unsifted waste.

3F43
3F41

← The wind-shifter separates lightweight foil from plastic. ↑↑ Incoming waste. Sifting starts in the supply hall where waste from the acceptance area is spread with cranes and bulldozers. ↑ The semi-finished product. Approximately 1,000 bales are stored daily in the storage area before turning into raw materials or alternative fuels.

# BOL.COM

The distribution centre measures a ground area of 100,000 m$^2$ and a floor area over 240,000 m$^2$ in three floors.

BOL.COM

A very wide duotone-coloured box with a blue and grey stripe across its entire width looms up on the outskirts of Waalwijk, one of the many industrial zones in the Netherlands. It is located, as if with a wink, at the end of the street called Pakketweg—which means 'package road' in Dutch. A large open, former agricultural field on the right side of the mega-box is bordered by a dike, or something resembling a mountain of waste. A battery of windmills and a field of PV panels cover the hill.

Every now and then a truck drives back and forth, but nothing indicates that the most successful logistics operation in the Netherlands is taking place here. The peak of inbound and outbound traffic is at night, when the flow is least hindered by traffic jams.

It is difficult to see the size of the 20-hectare complex in its entirety at a glance, with its endless number of loading and unloading areas for trucks, it is almost a landscape in the landscape. A box full of boxes wherein only one principle is facilitated: the flow of goods to people. It is not the only term that refers to the feeling of an airport, the whole complex is reminiscent of an airport luggage pickup hall with its endless conveyor belts that swing through the entire complex like a roller coaster. The entrance where I got in contains the most strict security checks I have ever experienced; the ultimate check is that the IMEI code of my phone is registered so that I can be tracked in any location within the complex. The two colours of the exterior are complemented inside by a bright coding system of red, yellow and blue. It suddenly becomes clear that this is the physical reality behind the click on the 'buy now' button. The initiation of a linear process of scanning, sorting, folding, wrapping, packaging, labelling, shipping and the mechanical logic of bodies who move between the machines and rolling belts.

↑ Twenty-five Packing machines pack the single-item orders as customized as possible. The dimension of each item is scanned, then the cardboard is cut and folded into the exact size of the object.

<< NO ORDER >>

↖ Totes containing small to medium items are mechanically stored and retrieved via conveyor belts when an item needs to be picked for an order. ← Belt with integrated scanner to manually sort a customer's order with products stored in the warehouse. ↑ The outgoing package section contains 86 chutes and two conveyor belts atop each other to increase sorting capacity on a smaller footprint.

← The new generation of vertically oriented packaging system. ↑ Double carton infeed for the carton wrap package cutting machine. Each feed has a different width to reduce the carton waste.

# BACKUPNE

The central scanning point is located between two tape rooms, where the full registration of the in- and outgoing tapes takes place.

BACKUPNED

Tall grass waves in front of a generic white bungalow. The association of a private house is confusing; the oversized entry is just to the left of it. The heated driveway with its steep angle of inclination is used daily by small vans, following a strict schedule. They pick up and deliver hundreds of similar, sealed suitcases of which is unclear what the exact contents are.

Nothing indicates that possibly the most classified information is stored here. Because of the Covid situation, Backupned got rid of its pre-war office villa that merely served representation purposes, and operates solely from its cold-war atomic bunker situated in a pleasantly green, hilly and quiet neighbourhood in the Arnhem region. I enter a long hallway with a thick steel door, a blue LED light and screen with projections of security cameras resemble a constructed film scene of a science fiction movie. 'Everything should be organized; so structured, clean, aligned in a consistent way, otherwise clients will not gain enough trust,' the director explains his reasoning behind the organization of this peculiar backup facility. Rows of tapes, colour-coded by client and labelled with codes, make it feel like an anachronistic VHS video-rental store. 'But tape is far from an outdated data storage medium. Innovation is still going strong and demand is becoming higher and higher.' This is the opposite of on-demand; here, data is stored that might one day be summoned, and maybe with extreme urgency—when needed, there's a chopper available to deliver throughout Europe within a timeframe of hours.

← The main presentation room. Tapes are stored in TAP-racks, 50 racks can store 600 tapes.
↑ The backup facility is a bunker, disguised under a bungalow: a legacy of the Cold War.

One of the main halls containing approximately 35.000 tapes. Empty shelves are reserved for customers to anticipate on their data growth.

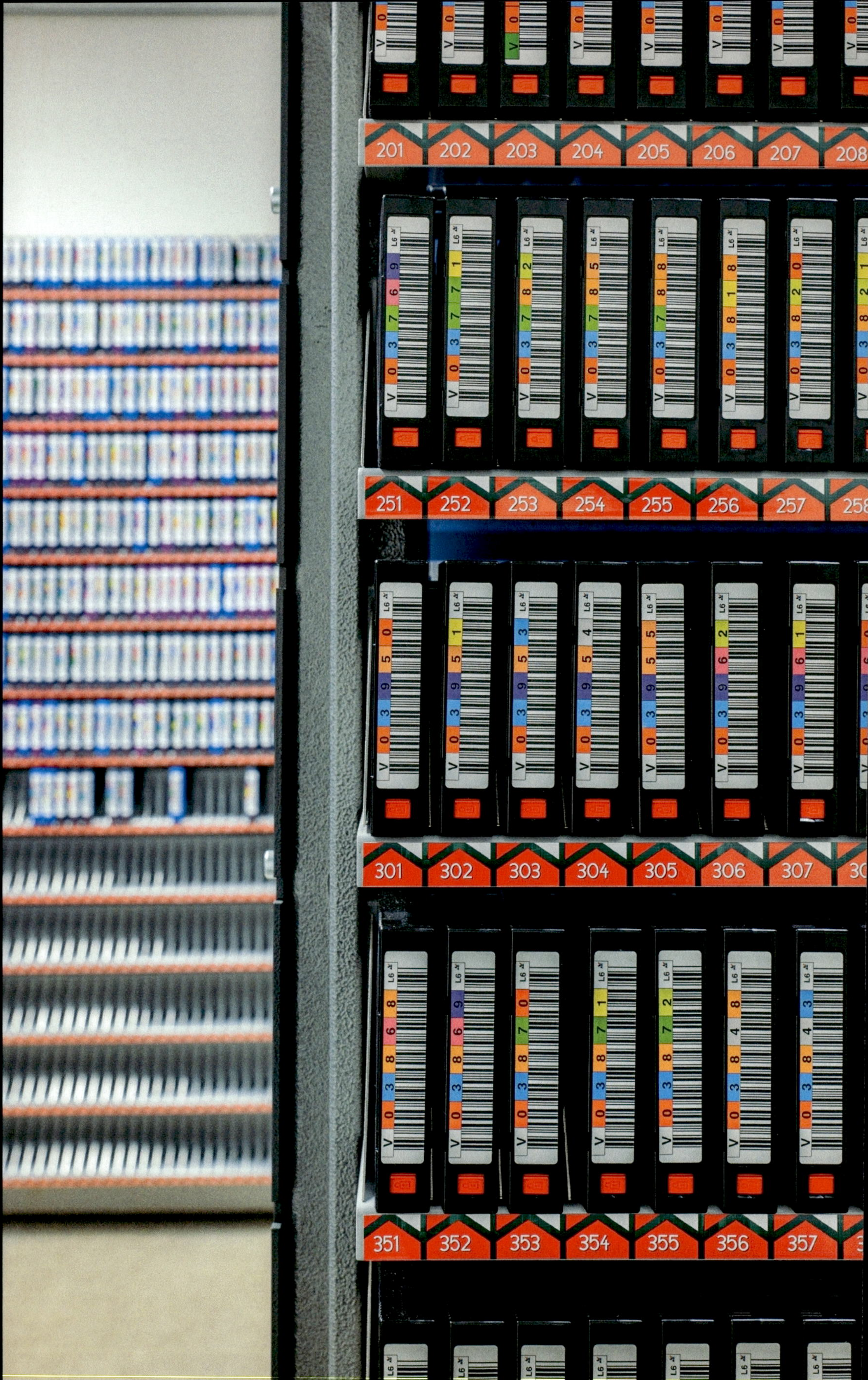
201 202 203 204 205 206 207 208
251 252 253 254 255 256 257
301 302 303 304 305 306 307
351 352 353 354 355 356 357

← A LTO tape is a magnetic tape data storage technology with the capacity of 45 TB. Originally it was developed in the late 1990s as an open standard alternative to the proprietary magnetic tape data

# WHITE SPO

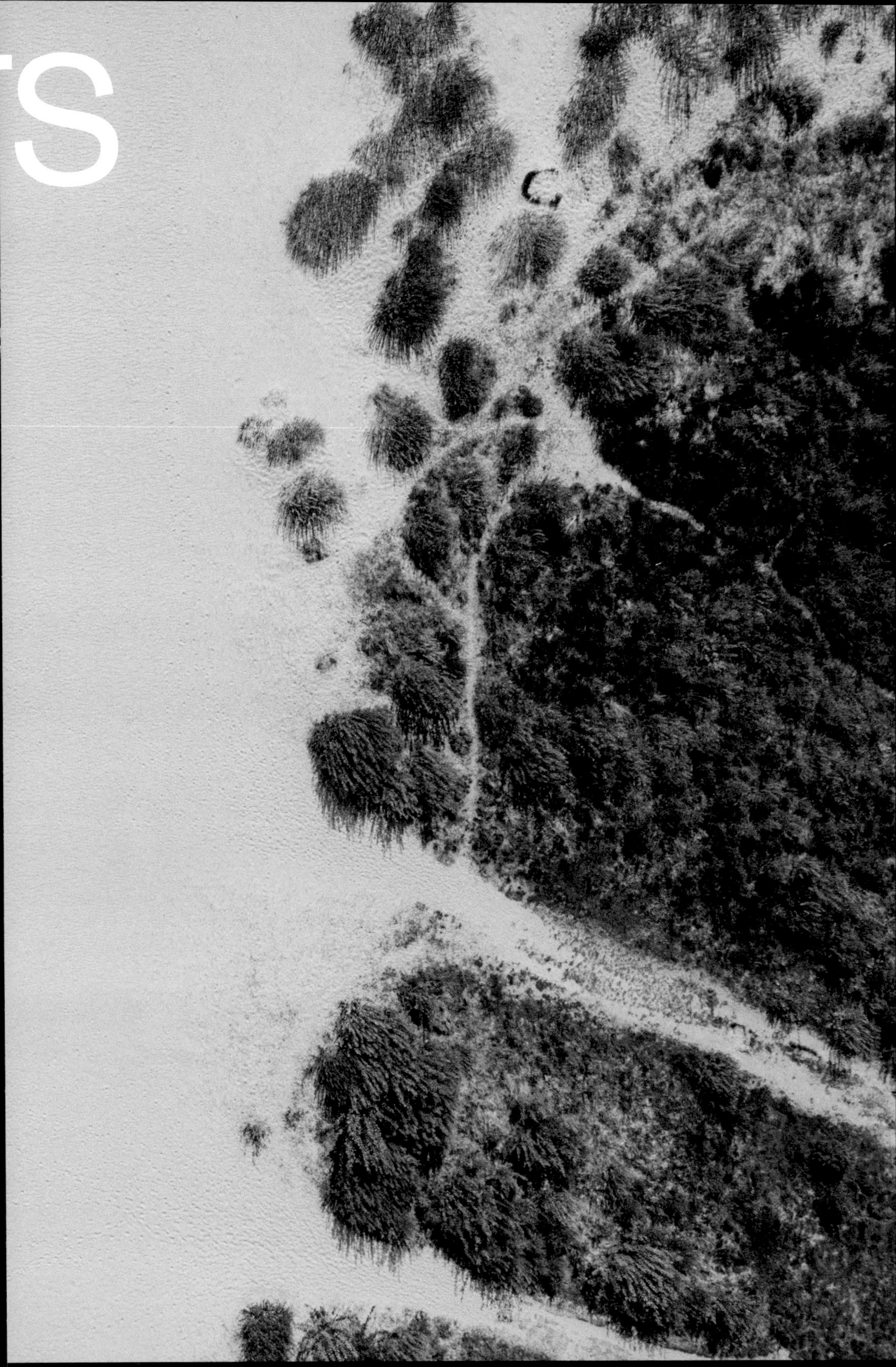

Soesterduinen, NL. Nearest phone cell tower: T-Mobile, 1,957 m.

WHITE SPOTS

In one of the sparse original sand drifts of the Netherlands, I meet a small group of middle-aged women who are there for an afternoon gathering. The woeste grond (an unproductive landscape that has not been cultivated), is a vast dune landscape in almost precisely the centre of the country, burning the eyes in the sun at its highest point. The group of women, who suffer from Electromagnetic Hypersensitivity (EHS), gather physically; their meeting is arranged by email conversations over a landline instead of via Wi-Fi or cellular communication. To make sure I don't interfere with their attempt to hide from the cloud of data waves, I activate flight mode on my iPhone. Beforehand, I check for nearby radiation with the White Spots app. A virtual forest of red spikes indicates that the closest transmission towers are around 550 m away.

I'm curious about all the practicalities that one encounters in a world of invisible and intangible signals that are impossible to avoid. And before I get to observe their walk we fall into an extensive conversation. They explain how they found each other in the shared ambition of creating awareness in the outside world of their illness. Among those who should take their sensitivity into account, are architects. This is already the case in Germany, where triple-paned glass and a strong separation of Wi-Fi signals in different rooms in houses or hotels lead to their radiation-measuring instrument showing much lower results. Martine says the most basic feeling is that of not being protected, being under constant stress, and one of the women talks about how the only place she can escape to is the Egyptian desert. Suddenly I understand why they chose this location as their weekly escape from data.

← Soesterduinen, NL, Nearest phone cell tower: Vodafone Libertel, 1,257 m. ↑ Soesterduinen, NL. Nearest phone cell tower: T-Mobile, 1,085 m.

# ESSAYS: ON

## ESSAYS: ONTOLOGY AND AGENCY OF DATA [INFRASTRUCTURE]

### A Trilogy: Earth, Grafts, Race



### New Reflections: Infrastructure, Space, Representation

# EARTH

EARTH First published in: Kate Crawford, *Atlas of AI* (Yale University Press., 2021).
Kate Crawford

The Boeing 757 banks right over San Jose on its final approach to San Francisco International Airport. The left wing drops as the plane lines up with the runway, revealing an aerial view of the tech sector's most iconic location. Below are the great empires of Silicon Valley. The gigantic black circle of Apple's headquarters is laid out like an uncapped camera lens, glistening in the sun. Then there's Google's head office, nestled close to NASA's Moffett Federal Airfield. This was once a key site for the US Navy during the Second World War and the Korean War, but now Google has a 60-year lease on it, and senior executives park their private planes here. Arrayed near Google are the large manufacturing sheds of Lockheed Martin, where the aerospace and weapons manufacturing company builds hundreds of orbital satellites destined to look down on the activities of Earth. Next, by the Dumbarton Bridge, appears a collection of squat buildings that are home to Facebook, ringed with massive parking lots close to the sulfuric salt ponds of the Ravenswood Slough. From this vantage point, the nondescript suburban cul-de-sacs and industrial midrise skyline of Palo Alto betray little of its true wealth, power and influence. There are only a few hints of its centrality in the global economy and in the computational infrastructure of the planet.

I'm here to learn about artificial intelligence and what it is made from. To see that, I will need to leave Silicon Valley altogether.

From the airport, I jump into a van and drive east. I cross the San Mateo-Hayward Bridge and pass by the Lawrence Livermore National Laboratory, where Edward Teller directed his research into thermonuclear weapons in the years after the Second World War. Soon the Sierra Nevada foothills rise beyond the Central Valley towns of Stockton and Manteca. Here the roads start winding up through the tall granite cliffs of the Sonora Pass and down the eastern side of the mountains towards grassy valleys dotted with golden poppies. Pine forests give way to the alkaline waters of Mono Lake and the parched desert landforms of the Basin and Range. To refuel, I pull into Hawthorne, Nevada, site of the world's biggest ammunition depot, where the US Army stores armaments in dozens of dirt-covered ziggurats that populate the valley in neat rows. Driving along Nevada State Route 265 I see a lone vOrtAC in the distance, a large bowling pin-shaped radio tower that was designed for the era before GPS. It has a single function: it broadcasts 'I am here' to all passing aircraft, a fixed point of reference in a lonely terrain.

My destination is the unincorporated community of Silver Peak in Nevada's Clayton Valley, where about 125 people live, depending on how you count. The mining town, one of the oldest in Nevada, was almost abandoned in 1917

Fig. 1 Silver Peak Lithium Mine. Photograph by Kate Crawford

after the ground was stripped bare of silver and gold. A few gold rush buildings still stand, eroding under the desert sun. The town may be small, with more junked cars than people, but it harbours something exceedingly rare. Silver Peak is perched on the edge of a massive underground lake of lithium. The valuable lithium brine under the surface is pumped out of the ground and left in open, iridescent green ponds to evaporate. From miles away, the ponds can be seen when they catch the light and shimmer. Up close, it's a different view. Alien-looking black pipes erupt from the ground and snake along the salt- encrusted earth, moving in and out of shallow trenches, ferrying the salty cocktail to its drying pans.

Here, in a remote pocket of Nevada, is a place where the stuff of AI is made.

## Mining for AI

Clayton Valley is connected to Silicon Valley in much the way that the nineteenth-century goldfields were to early San Francisco. The history of mining, like the devastation it leaves in its wake, is commonly overlooked in the strategic amnesia that accompanies stories of technological progress. As historical geographer Gray Brechin points out, San Francisco was built from the gains of pulling gold and silver out of the lands of California and Nevada in the 1800s.[1] The city is made from mining. Those same lands had been taken from Mexico under the Treaty of Guadalupe Hidalgo in 1848 at the end of the Mexican-American War, when it was already clear to the settlers that these would be highly valuable goldfields. It was a textbook example, Brechin observes, of the old adage that 'commerce follows the flag, but the flag follows the pick'.[2] Thousands of people were forced from their homes during this substantial territorial expansion of the United States. After America's imperial invasion, the miners moved in. The land was stripped until the waterways were contaminated and the surrounding forests destroyed.

Since antiquity, the business of mining has only been profitable because it does not have to account for its true costs: including environmental damage, the illness and death of miners and the loss to the communities it displaces. In 1555, Georgius Agricola, known as the father of mineralogy, observed that 'it is clear to all that there is greater detriment from mining than the value of the metals which the mining produces'.[3] In other words, those who profit from mining do so only because the costs must be sustained by others, those living and those not yet born. It is easy to put a price on precious metals, but what is the exact value of a wilderness, a clean stream, breathable air, the health of local communities? It was never estimated, and thus an easy calculus emerged: extract everything as rapidly as possible. It was the 'move fast and break things' of a different time. The result was that the Central Valley was decimated, and as one tourist observed in 1869: 'Tornado, flood, earthquake and volcano combined could hardly make greater havoc, spread wider ruin and wreck than [the] gold-washing operations ... There are no rights which mining respects in California. It is the one supreme interest.'[4]

As San Francisco drew enormous wealth from the mines, it was easy for its populace to forget where it all came from. The mines were located far from the city they enriched, and this remoteness allowed city dwellers to remain ignorant of what was happening to the mountains, rivers and labourers that fed their fortunes. But small reminders of the mines are all around. The city's new buildings used the same technology that came from deep within the Central Valley for transport and life support. The pulley systems that carried miners down into the mine shafts were adapted and turned upside down to transport people in elevators to the top of the city's high-rises.[5] Brechin suggests that we should think of the skyscrapers of San Francisco as inverted minescapes. The ores extracted from holes in the ground were sold to create the stories in the air; the deeper the extractions went, the higher the great towers of office work stretched into the sky.

San Francisco is enriched once more. Once it was gold ore that underwrote fortunes; now it is the extraction of substances like white lithium crystal. It's known in mineral markets as 'grey gold'.[6] The technology industry has become a new supreme interest, and the five biggest companies in the world by market capitalization have offices in this city: Apple, Microsoft, Amazon, Facebook and Google. Walking past the start-up warehouses in the SoMa district where miners in tents once lived, you can see luxury cars, venture capital-backed coffee chains and sumptuous buses with tinted windows running along private routes, carrying workers to their offices in Mountain View or Menlo Park.[7] But only a short walk away is Division Street, a multilane thoroughfare between SoMa and the Mission district, where rows of tents have returned to shelter people who have nowhere to go. In the wake of the tech boom, San Francisco now has one of the highest rates of street homelessness in the United States.[8] The United Nations special rapporteur on adequate housing called it an 'unacceptable' human rights violation, due to the thousands of homeless residents denied basic necessities of water, sanitation and health services in contrast to the record number of billionaires who live nearby.[9] The greatest benefits of extraction have been captured by the few.

In this chapter we'll traverse across Nevada, San Jose and San Francisco, as well as Indonesia, Malaysia, China and Mongolia: from deserts to oceans. We'll also walk the spans of historical time, from conflict in the Congo and artificial black lakes in the present day to the Victorian passion for white latex. The scale will shift, telescoping from rocks to cities, trees to megacorporations, transoceanic shipping lanes to the atomic bomb. But across this planetary supersystem we will see the logics of extraction, a constant drawdown of minerals, water and fossil fuels, undergirded by the violence of wars, pollution, extinction and depletion. The effects of large-scale computation can be found in the atmosphere, the oceans, the Earth's crust, the deep time of the planet and the brutal impacts on disadvantaged populations around the world. To understand it all, we need a panoramic view of the planetary scale of computational extraction.

## Landscapes of Computation

I'm driving through the desert valley on a summer afternoon to see the workings of this latest mining boom. I ask my phone to direct me to the perimeter of the lithium ponds, and it replies from its awkward perch on the dashboard, tethered by a white USB cable. Silver Peak's large, dry lake bed was formed millions of years ago during the late Tertiary Period. It's surrounded by crusted stratifications pushing up into ridgelines containing dark limestones, green quartzites and grey and red slate.[10] Lithium was discovered here after the area was scoped for strategic minerals like potash during the Second World War. This soft, silvery metal was mined in only modest quantities for the next 50 years, until it became highly valuable material for the technology sector.

In 2014, Rockwood Holdings, Inc., a lithium mining operation, was acquired by the chemical manufacturing company Albemarle Corporation for $6.2 billion. It is the only operating lithium mine in the United States. This makes Silver Peak a site of intense interest to Elon Musk and the many other tech tycoons for one reason: rechargeable batteries. Lithium is a crucial element for their production. Smartphone batteries, for example, usually contain about three-tenths of an ounce of it. Each Tesla Model S electric car needs about one hundred thirty-eight pounds of lithium for its battery pack.[11] These kinds of batteries were never intended to supply a machine as power hungry as a car, but lithium batteries are currently the only mass-market option available.[12] All of these batteries have a limited lifespan; once degraded, they are discarded as waste.

About 200 miles north of Silver Peak is the Tesla Gigafactory. This is the world's largest lithium battery plant. Tesla is the number-one lithium-ion battery consumer in the world, purchasing them in high volumes from Panasonic and Samsung and repackaging them in its cars and home chargers. Tesla is estimated to use more than 28,000 tons of lithium hydroxide annually—half of the planet's total consumption.[13] In fact, Tesla could more accurately be described as a battery business than a car company.[14] The imminent shortage of such critical minerals as nickel, copper and lithium poses a risk for the company, making the lithium lake at Silver Peak highly desirable.[15] Securing control of the mine would mean controlling the US domestic supply.

As many have shown, the electric car is far from a perfect solution to carbon dioxide emissions.[16] The mining, smelting, export, assemblage and transport of the battery supply chain has a significant negative impact on the environment and, in turn, on the communities affected by its degradation. A small number of home solar systems produce their own energy. But for the majority of cases, charging an electric car necessitates taking power from the grid, where currently less than a fifth of all electricity in the United States comes from renewable energy sources.[17] So far none of this has dampened the determination of auto manufacturers to compete with Tesla, putting increasing pressure on the battery market and accelerating the removal of diminishing stores of the necessary minerals.

Global computation and commerce rely on batteries. The term 'artificial intelligence' may invoke ideas of algorithms, data and cloud architectures, but none of that can function without the minerals and resources that build computing's core components. Rechargeable lithium-ion batteries are essential for mobile devices and laptops, in-home digital assistants and data centre backup power. They undergird the Internet and every commerce platform that runs on it, from banking to retail to stock market trades. Many aspects of modern life have been moved to 'the cloud' with little consideration of these material costs. Our work and personal lives, our medical histories, our leisure time, our entertainment, our political interests—all of this takes place in the world of networked computing architectures that we tap into from devices we hold in one hand, with lithium at their core.

The mining that makes AI is both literal and metaphorical. The new extractivism of data mining also encompasses and propels the old extractivism of traditional mining. The stack required to power artificial intelligence systems goes well beyond the multi-layered technical stack of data modelling, hardware, servers and networks. The full-stack supply chain of AI reaches into capital, labour and Earth's resources—and from each, it demands an enormous amount.[18] The cloud is the backbone of the artificial intelligence industry, and it's made of rocks and lithium brine and crude oil.

In his book *A Geology of Media,* theorist Jussi Parikka suggests we think of media not from Marshall McLuhan's point of view—in which media are extensions of the human senses—but rather as extensions of Earth.[19] Computational media now participate in geological (and climatological) processes, from the transformation of the Earth's materials into infrastructures and devices to the powering of these new systems with oil and gas reserves. Reflecting on media and technology as geological processes enables us to consider the radical depletion of non-renewable resources required to drive the technologies of the present moment. Each object in the extended network of an AI system, from network routers to batteries to data centres, is built using elements that required billions of years to form inside the Earth.

From the perspective of deep time, we are extracting Earth's geological history to serve a split second of contemporary technological time, building devices like the Amazon Echo and the iPhone that are often designed to last for only a few years. The Consumer Technology Association notes that the average smartphone life span is a mere 4.7 years.[20] This obsolescence cycle fuels the purchase of more devices, drives up profits and increases incentives for the use of unsustainable extraction practices. After a slow process of development, these minerals, elements and materials then go through an extraordinarily rapid period of excavation, processing, mixing, smelting and logistical transport—crossing thousands of miles in their transformation. What begins as ore removed from the ground, after the spoil and the tailings are discarded, is then made into devices that are used and discarded. They ultimately end up buried in e-waste dumping grounds in places like Ghana and Pakistan. The lifecycle of an AI system from birth to death has many fractal supply chains: forms of exploitation of human labour and natural resources and massive concentrations of corporate and geopolitical power. And all along the chain, a continual, large-scale consumption of energy keeps the cycle going.

The extractivism on which San Francisco was built is echoed in the practices of the tech sector based there today.[21] The massive ecosystem of AI relies on many kinds of extraction: from harvesting the data made from our daily activities and expressions, to depleting natural resources and to exploiting labour around the globe so that this vast planetary network can be built and maintained. And AI extracts far more from us and the planet than is widely known. The Bay Area is a central node in the mythos of AI, but we'll need to traverse far beyond the United States to see the many-layered legacies of human and environmental damage that have powered the tech industry.

## The Mineralogical Layer

The lithium mines in Nevada are just one of the places where the materials are extracted from the Earth's crust to make AI. There are many such sites, including the Salar in southwest Bolivia—the richest site of lithium in the world and thus a site of ongoing political tension—as well as places in central Congo, Mongolia, Indonesia and the Western Australia deserts. These are the other birthplaces of AI in the greater geography of industrial extraction. Without the minerals from these locations, contemporary computation simply does not work. But these materials are in increasingly short supply.

In 2020, scientists at the US Geological Survey published a short list of 23 minerals that are a high 'supply risk' to manufacturers, meaning that if they became unavailable, entire industries—including the tech sector—would grind to a halt.[22] The critical minerals include the rare earth elements dysprosium and neodymium, which are used inside iPhone speakers and electric vehicle motors; germanium, which is used in infrared military devices for soldiers and in drones; and cobalt, which improves performance for lithium-ion batteries.

There are 17 rare earth elements: lanthanum, cerium, praseodymium, neodymium, promethium, samarium, europium, gadolinium, terbium, dysprosium, holmium, erbium, thulium, ytterbium, lutetium, scandium and yttrium. They are processed and embedded in laptops and smartphones, making those devices smaller and lighter. The elements can be found in colour displays, speakers, camera lenses, rechargeable batteries, hard drives and many other components. They are key elements in communication systems, from fibreoptic cables and signal amplification in mobile communication towers to satellites and GPS technology. But extracting these minerals from the ground often comes with local and geopolitical violence. Mining is and always has been a brutal undertaking. As Lewis Mumford writes:

> Mining was the key industry that furnished the sinews of war and increased the metallic contents of the original capital hoard, the war chest: on the other hand, it furthered the industrialization of arms, and enriched the financier by both processes.[23]

To understand the business of AI, we must reckon with the war, famine and death that mining brings with it.

Recent US legislation that regulates some of those 17 rare earth elements only hints at the devastation associated with their extraction. The 2010 Dodd-Frank Act focused on reforming the financial sector in the wake of the 2008 financial crisis. It included a specific provision about so-called *conflict minerals,* or natural resources extracted in a conflict zone and then sold to fund the conflict. Companies using gold, tin, tungsten and tantalum from the region around the Democratic Republic of the Congo now had a reporting requirement to track where those minerals came from and whether the sale was funding armed militia in the region.[24] Like 'conflict diamonds', the term 'conflict minerals' masks the profound suffering and prolific killing in the mining sector. Mining profits have financed military operations in the decades-long Congo- area conflict, fuelling the deaths of thousands and the displacement of millions.[25] Furthermore, working conditions inside the mines have often amounted to modern slavery.[26]

It took Intel more than four years of sustained effort to develop basic insight into its own supply chain.[27] Intel's supply chain is complex, with more than 16,000 suppliers in over 100 countries providing direct materials for the company's production processes, tools and machines for their factories, as well as their logistics and packaging services.[28] In addition, Intel and Apple have been criticized for auditing only smelters—not the actual mines—to determine the conflict-free status of minerals. The tech giants were assessing smelting plants outside of Congo, and the audits were often performed by locals. So even the conflict-free certifications of the tech industry are now under question.[29]

Dutch-based technology company Philips has also claimed that it was working to make its supply chain 'conflict-free'. Like Intel, Philips has tens of thousands of suppliers, each of which provides component parts for the company's manufacturing processes.[30] Those suppliers are themselves linked downstream to thousands of component manufacturers acquiring treated materials from dozens of smelters. The smelters in turn buy their materials from an unknown number of traders who deal directly with both legal and illegal mining operations to source the various minerals that end up in computer components.[31]

According to the computer manufacturer Dell, the complexities of the metals and mineral supply chains pose almost insurmountable challenges to the production of conflict-free electronics components. The elements are laundered through such a vast number of entities along the chain that sourcing their provenance proves impossible—or so the end-product manufacturers claim, allowing them a measure of plausible deniability for any exploitative practices that drive their profits.[32]

Just like the mines that served San Francisco in the nineteenth century, extraction for the technology sector is done by keeping the real costs out of sight. Ignorance of the supply chain is baked into capitalism, from the way businesses protect themselves through third-party contractors and suppliers to the way goods are marketed and advertised to consumers. More than plausible deniability, it has become a well-practiced form of bad faith: the left hand cannot know what the right hand is doing, which requires increasingly lavish, baroque and complex forms of distancing.

While mining to finance war is one of the most extreme cases of harmful extraction, most minerals are not sourced from direct war zones. This doesn't mean, however, that they are free from human suffering and environmental destruction. The focus on conflict minerals, though important, has also been used to avert focus from the harms of mining writ large. If we visit the primary sites of mineral extraction for computational systems, we find the repressed stories of acid-bleached rivers and deracinated landscapes and the extinction of plant and animal species that were once vital to the local ecology.

## Black Lakes and White Latex

In Baotou, the largest city in Inner Mongolia, there is an artificial lake filled with toxic black mud. It reeks of sulphur and stretches as far as the eye can see, covering more than five and a half miles in diameter. The black lake contains more than 180 million tons of waste powder from ore processing.[33] It was created by the waste runoff from the nearby Bayan Obo mines, which is estimated to contain almost 70 per cent of the world's reserves of rare earth minerals. It is the largest deposit of rare earth elements on the planet.[34]

China supplies 95 per cent of the world's rare earth minerals. China's market domination, as the writer Tim Maughan observes, owes far less to geology than to the country's willingness to take on the environmental damage of extraction.[35] Although rare earth minerals like neodymium and cerium are relatively common, making them usable requires the hazardous process of dissolving them in large volumes of sulfuric and nitric acid. These acid baths yield reservoirs of poisonous waste that fill the dead lake in Baotou. This is just one of the places that are brimming with what environmental studies scholar Myra Hird calls 'the waste we want to forget'.[36]

To date, the unique electronic, optical and magnetic uses of rare earth elements cannot be matched by any other metals, but the ratio of usable minerals to waste toxins is extreme. Natural resource strategist David Abraham describes the mining in Jiangxi, China, of dysprosium and terbium, which are used in a variety of high-tech devices. He writes: 'Only 0.2 percent of the mined clay contains the valuable rare earth elements. This means that 99.8 percent of earth removed in rare earth mining is discarded as waste, called "tailings", that are dumped back into the hills and streams', creating new pollutants like ammonium.[37] In order to refine one ton of these rare earth elements, 'the Chinese Society of Rare Earths estimates that the process produces 75,000 liters of acidic water and one ton of radioactive residue'.[38]

About 3,000 miles south of Baotou are the small Indonesian islands of Bangka and Belitung, off the coast of Sumatra. Bangka and Belitung produce 90 per cent of Indonesia's tin, used in semiconductors. Indonesia is the world's second-largest producer of the metal, behind China.

Fig. 2. *Palaquium gutta*

Indonesia's national tin corporation, PT Timah, supplies companies such as Samsung directly, as well as solder makers Chernan and Shenmao, which in turn supply Sony, LG and Foxconn—all suppliers for Apple, Tesla and Amazon.[39]

On these small islands, grey-market miners who are not officially employed sit on makeshift pontoons, using bamboo poles to scrape the seabed before diving underwater to suck tin from the surface by drawing their breath through giant, vacuumlike tubes. The miners sell the tin they find to middlemen, who also collect ore from miners working in authorized mines, and they mix it together to sell to companies like Timah.[40] Completely unregulated, the process unfolds beyond any formal worker or environmental protections. As investigative journalist Kate Hodal reports:

> Tin mining is a lucrative but destructive trade that has scarred the island's landscape, bulldozed its farms and forests, killed off its fish stocks and coral reefs, and dented tourism to its pretty palm-lined beaches. The damage is best seen from the air, as pockets of lush forest huddle amid huge swaths of barren orange earth. Where not dominated by mines, this is pockmarked with graves, many holding the bodies of miners who have died over the centuries digging for tin.[41]

The mines are everywhere: in backyards, in the forest, by the side of the road, on the beaches. It is a landscape of ruin.

It is a common practice of life to focus on the world immediately before us, the one we see and smell and touch every day. It grounds us where we are, with our communities and our known corners and concerns. But to see the full supply chains of AI requires looking for patterns in a global sweep, a sensitivity to the ways in which the histories and specific harms are different from place to place and yet are deeply interconnected by the multiple forces of extraction.

We can see these patterns across space, but we can also find them across time. Transatlantic telegraph cables are the essential infrastructure that ferries data between the continents, an emblem of global communication and capital. They are also a material product of colonialism, with its patterns of extraction, conflict and environmental destruction. At the end of the nineteenth century, a particular Southeast Asian tree called *Palaquium gutta* became the centre of a cable boom. These trees, found mainly in Malaysia, produce a milky white natural latex called gutta-percha. After English scientist Michael Faraday published a study in the *Philosophical Magazine* in 1848 about the use of this material as an electrical insulator, gutta-percha rapidly became the darling of the engineering world. Engineers saw gutta-percha as the solution to the problem of insulating telegraphic cables to withstand harsh and varying conditions on the ocean floor. The twisted strands of copper wire needed four layers of the soft, organic tree sap to protect them from water incursion and carry their electrical currents.

As the global submarine telegraphy business grew, so did demand for *Palaquium gutta* tree trunks. The historian John Tully describes how local Malay, Chinese and Dayak workers were paid little for the dangerous work of felling the trees and slowly collecting the latex.[42] The latex was processed and then sold through Singapore's trade markets to the British market, where it was transformed into, among other things, lengths upon lengths of submarine cable sheaths that wrapped around the globe. As media scholar Nicole Starosielski writes: 'Military strategists saw cables as the most efficient and secure mode of communication with the colonies—and, by implication, of control over them.'[43] The routes of submarine cables today still mark out the early colonial networks between the centres and the peripheries of empire.[44]

A mature *Palaquium gutta* could yield around 11 ounces of latex. But in 1857, the first transatlantic cable was around 1,800 miles long and weighed 2,000 tons—requiring about 250 tons of gutta-percha. To produce just one ton of this material required around 900,000 tree trunks. The jungles of Malaysia and Singapore were stripped; by the early 1880s, the *Palaquium gutta* had vanished. In a last-ditch effort to save their supply chain, the British passed a ban in 1883 to halt harvesting the latex, but the tree was all but extinct.[45]

The Victorian environmental disaster of gutta-percha, at the dawn of the global information society, shows how the relations between technology and its materials, environments and labour practices are interwoven.[46] Just as Victorians precipitated ecological disaster for their early cables, so do contemporary mining and global supply chains further imperil the delicate ecological balance of our era.

There are dark ironies in the prehistories of planetary computation. Currently large-scale AI systems are driving forms of environmental, data and human extraction, but from the Victorian era onward, algorithmic computation emerged out of desires to manage and control war, population and climate change. The historian Theodora Dryer describes how the founding figure of mathematical statistics, English scientist Karl Pearson, sought to resolve uncertainties of planning and management by developing new data architectures including standard deviations and techniques of correlation and regression. His methods were, in turn, deeply imbricated with race science, as Pearson—along with his mentor, the statistician and founder of eugenics Sir Francis Galton—believed that statistics could be 'the first step in an enquiry into the possible effect of a selective process upon any character of a race'.[47]

As Dryer writes:

> By the end of the 1930s, these data architectures—regression techniques, standard deviation, and correlations—would become dominant tools used in interpreting social and state information on the world stage. Tracking the nodes and routes of global trade, the interwar 'mathematical statistics movement' became a vast enterprise.[48]

This enterprise kept expanding after the Second World War, as new computational systems were used in domains such as weather forecasting during periods of drought to eke out more productivity from large-scale industrial farming.[49] From this perspective, algorithmic computing, computational statistics and artificial intelligence were developed in the twentieth century to address social and environmental challenges, but would later be used to intensify industrial extraction and exploitation and further deplete environmental resources.

## The Myth of Clean Tech

Minerals are the backbone of AI, but its lifeblood is still electrical energy. Advanced computation is rarely considered in terms of carbon footprints, fossil fuels and pollution; metaphors like 'the cloud' imply something floating and delicate within a natural, green industry.[50] Servers are hidden in nondescript data centres, and their polluting qualities are far less visible than the billowing smokestacks of coal-fired power stations. The tech sector heavily publicizes its environmental policies, sustainability initiatives and plans to address climate-related problems using AI as a problem-solving tool. It is all part of a highly produced public image of a sustainable tech industry with no carbon emissions. In reality, it takes a gargantuan amount of energy to run the computational infrastructures of Amazon Web Services or Microsoft's Azure, and the carbon footprint of the AI systems that run on those platforms is growing.[51]

As Tung-Hui Hu writes in *A Prehistory of the Cloud*: 'The cloud is a resource-intensive, extractive technology that converts water and electricity into computational power, leaving a sizable amount of environmental damage that it then displaces from sight.'[52] Addressing this energy-intensive infrastructure has become a major concern. Certainly, the industry has made significant efforts to make data centres more energy efficient and to increase their use of renewable energy. But already, the carbon footprint of the world's computational infrastructure has matched that of the aviation industry at its height, and it is increasing at a faster rate.[53] Estimates vary, with researchers like Lotfi Belkhir and Ahmed Elmeligi estimating that the tech sector will contribute 14 per cent of global greenhouse emissions by 2040, while a team in Sweden predicts that the electricity demands of data centres alone will increase about fifteenfold by 2030.[54]

By looking closely at the computational capacity needed to build AI models, we can see how the desire for exponential increases in speed and accuracy is coming at a high cost to the planet. The processing demands of training AI models, and thus their energy consumption, is still an emerging area of investigation. One of the early papers in this field came from AI researcher Emma Strubell and her team at the University of Massachusetts Amherst in 2019. With a focus on trying to understand the carbon footprint of natural language processing (NLP) models, they began to sketch out potential estimates by running AI models over hundreds of thousands of computational hours.[55] The initial numbers were striking. Strubell's team found that running only a single NLP model produced more than 660,000 pounds of carbon dioxide emissions, the equivalent of five gas-powered cars over their total lifetime (including their manufacturing) or 125 round-trip flights from New York to Beijing.[56]

Worse, the researchers noted that this modelling is, at minimum, a baseline optimistic estimate. It does not reflect the true commercial scale at which companies like Apple and Amazon operate, scraping Internet-wide datasets and feeding their own NLP models to make AI systems like Siri and Alexa sound more human. But the exact amount of energy consumption produced by the tech sector's AI models is unknown; that information is kept as highly guarded corporate secrets. Here, too, the data economy is premised on maintaining environmental ignorance.

In the AI field, it is standard practice to maximize computational cycles to improve performance, in accordance with a belief that bigger is better. As Rich Sutton of DeepMind describes it: 'Methods that leverage computation are ultimately the most effective, and by a large margin.'[57] The computational technique of brute-force testing in AI training runs, or systematically gathering more data and using more computational cycles until a better result is achieved, has driven a steep increase in energy consumption. OpenAI estimated that since 2012, the amount of compute used to train a single AI model has increased by a factor of ten every year. That's due to developers 'repeatedly finding ways to use more chips in parallel, and being willing to pay the economic cost of doing so'.[58] Thinking only in terms of economic cost narrows the view on the wider local and environmental price of burning computation cycles as a way to create incremental efficiencies. The tendency towards 'compute maximalism' has profound ecological impacts.

Data centres are among the world's largest consumers of electricity.[59] Powering this multilevel machine requires grid electricity in the form of coal, gas, nuclear or renewable energy. Some corporations are responding to growing alarm about the energy consumption of large-scale computation, with Apple and Google claiming to be carbon neutral (which means they offset their carbon emissions by purchasing credits) and Microsoft promising to become carbon negative by 2030. But workers within the companies have pushed for reductions in emissions across the board, rather than what they see as buying indulgences out of environmental guilt.[60] Moreover, Microsoft, Google and Amazon all license their AI platforms, engineering workforces and infrastructures to fossil fuel companies to help them locate and extract fuel from the ground, which further drives the industry most responsible for anthropogenic climate change.

Beyond the United States, more clouds of carbon dioxide are rising. China's data centre industry draws 73 per cent of its power from coal, emitting about 99 million tons of CO2 in 2018.[61] And electricity consumption from China's data centre infrastructure is expected to increase by two-thirds by 2023.[62] Greenpeace has raised the alarm about the colossal energy demands of China's biggest technology companies, arguing that 'China's leading tech companies, including Alibaba, Tencent, and GDS, must dramatically scale up clean energy procurement and disclose energy use data'.[63] But the lasting impacts of coal-fired power are everywhere, exceeding any national boundaries. The planetary nature of resource extraction and its consequences goes well beyond what the nation-state was designed to address.

Water tells another story of computation's true cost. The history of water use in the United States is full of battles and secret deals, and as with computation, the deals made over water are kept close. One of the biggest US data centres belongs to the National Security Agency (NSA) in Bluffdale, Utah. Open since late 2013, the Intelligence Community Comprehensive National Cybersecurity Initiative Data Center is impossible to visit directly. But by driving up through the adjacent suburbs, I found a cul-de-sac on a hill thick with sagebrush, and from there I was afforded a closer view of the sprawling 1.2-million-square-foot facility. The site has a kind of symbolic power of the next era of government data capture, having been featured in films like *Citizenfour* and pictured in thousands of news stories about the NSA. In person, though, it looks non-descript and prosaic, a giant storage container combined with a government office block.

The struggle over water began even before the data centre was officially open, given its location in drought-parched Utah.[64] Local journalists wanted to confirm whether the estimated consumption of 1.7 million gallons of water per day was accurate, but the NSA initially refused to share usage data, redacted all details from public records, and claimed that its water use was a matter of national security. Anti-surveillance activists created handbooks encouraging the end of material support of water and energy to surveillance, and they strategized that legal controls over water usage could help shut down the facility.[65] But the city of Bluffdale had already made a multiyear deal with the NSA, in which the city would sell water at rates well below the average in return for the promise of economic growth the facility might bring to the region.[66] The geopolitics of water are now deeply combined with the mechanisms and politics of data centres, computation and power—in every sense. From the dry hillside that overlooks the NSA's data repository, all the contestation and obfuscation about water makes sense: this is a landscape with a limit, and water that is used to cool servers is being taken away from communities and habitats that rely on it to live.

Just as the dirty work of the mining sector was far removed from the companies and city dwellers who profited most, so the majority of data centres are far removed from major population hubs, whether in the desert or in semi-industrial exurbs. This contributes to our sense of the cloud being out of sight and abstracted away, when in fact it is material, affecting the environment and climate in ways that are far from being fully recognized and accounted for. The cloud is of the Earth, and to keep it growing requires expanding resources and layers of logistics and transport that are in constant motion.

## The Logistical Layer

So far, we have considered the material stuff of AI, from rare earth elements to energy. By grounding our analysis in the specific materialities of AI—the things, places and people—we can better see how the parts are operating within broader systems of power. Take, for example, the global logistical machines that move minerals, fuel, hardware, workers and consumer AI devices around the planet.[67] The dizzying spectacle of logistics and production displayed by companies like Amazon would not be possible without the development and widespread acceptance of a standardized metal object: the cargo container. Like submarine cables, cargo containers bind the industries of global communication, transport and capital, a material exercise of what mathematicians call 'optimal transport'—in this case, as an optimization of space and resources across the trade routes of the world.

Standardized cargo containers (themselves built from the basic earth elements of carbon and iron forged as steel) enabled the explosion of the modern shipping industry, which in turn made it possible to envision and model the planet as a single massive factory. The cargo container is the single unit of value—like a piece of Lego—that can travel thousands of miles before meeting its final destination as a modular part of a greater system of delivery. In 2017, the capacity of container ships in seaborne trade reached nearly 250 million deadweight tons of cargo, dominated by giant shipping companies including Maersk of Denmark, the Mediterranean Shipping Company of Switzerland and France's CMA CGM Group, each owning hundreds of container vessels.[68] For these commercial ventures, cargo shipping is a relatively cheap way to navigate the vascular system of the global factory, yet it disguises far larger external costs. Just as they tend to neglect the physical realities and costs of AI infrastructure, popular culture and media rarely cover the shipping industry. The author Rose George calls this condition 'sea blindness'.[69]

In recent years, shipping vessels produced 3.1 per cent of yearly global carbon dioxide emissions, more than the total produced by Germany.[70] In order to minimize their internal costs, most container shipping companies use low-grade fuel in enormous quantities, which leads to increased amounts of airborne sulphur and other toxic substances. One container ship is estimated to emit as much pollution as 50 million cars, and 60,000 deaths every year are attributed indirectly to cargo-ship-industry pollution.[71]

Even industry-friendly sources like the World Shipping Council admit that thousands of containers are lost each year, sinking to the ocean floor or drifting loose.[72] Some carry toxic substances that leak into the oceans; others release thousands of yellow rubber ducks that wash ashore around the world over decades.[73] Typically, workers spend almost six months at sea, often with long working shifts and without access to external communications.

Here, too, the most severe costs of global logistics are borne by the Earth's atmosphere, the oceanic ecosystem and low-paid workers. The corporate imaginaries of AI fail to depict the lasting costs and long histories of the materials needed to build computational infrastructures or the energy required to power them. The rapid growth of cloud-based computation, portrayed as environmentally friendly, has paradoxically driven an expansion of the frontiers of resource extraction. It is only by factoring in these hidden costs, these wider collections of actors and systems, that we can understand what the shift towards increasing automation will mean. This requires working against the grain of how the technological imaginary usually works, which is completely untethered from earthly matters. Like running an image search of 'AI', which returns dozens of pictures of glowing brains and blue-tinted binary code floating in space, there is a powerful resistance to engaging with the materialities of these technologies. Instead, we begin with the earth, with extraction, and with

Fig. 3. The ruins at Blair. Photograph by Kate Crawford

the histories of industrial power and then consider how these patterns are repeated in systems of labour and data.

## AI as Megamachine

In the late 1960s, the historian and philosopher of technology Lewis Mumford developed the concept of the *megamachine* to illustrate how all systems, no matter how immense, consist of the work of many individual human actors.[74] For Mumford, the Manhattan Project was the defining modern megamachine whose intricacies were kept not only from the public but even from the thousands of people who worked on it at discrete, secured sites across the United States. A total of 130,000 workers operated in complete secrecy under the direction of the military, developing a weapon that would kill (by conservative estimates) 237,000 people when it hit Hiroshima and Nagasaki in 1945. The atomic bomb depended on a complex, secret chain of supply, logistics and human labour.

Artificial intelligence is another kind of megamachine, a set of technological approaches that depend on industrial infrastructures, supply chains and human labour that stretch around the globe but are kept opaque. We have seen how AI is much more than databases and algorithms, machine learning models and linear algebra. It is metamorphic: relying on manufacturing, transportation and physical work; data centres and the undersea cables that trace lines between the continents; personal devices and their raw components; transmission signals passing through the air; datasets produced by scraping the Internet; and continual computational cycles. These all come at a cost.

We have looked at the relations between cities and mines, companies and supply chains, and the topographies of extraction that connect them. The fundamentally intertwined nature of production, manufacturing and logistics reminds us that the mines that drive AI are everywhere: not only sited in discrete locations but diffuse and scattered across the geography of the Earth, in what Mazen Labban has called the 'planetary mine'.[75] This is not to deny the many specific locations where technologically driven mining is taking place. Rather, Labban observes that the planetary mine expands and reconstitutes extraction into novel arrangements, extending the practices of mines into new spaces and interactions around the world.

Finding fresh methods for understanding the deep material and human roots of AI systems is vital at this moment in history, when the impacts of anthropogenic climate change are already well under way. But that's easier said than done. In part, that's because many industries that make up the AI system chain conceal the ongoing costs of what they do. Furthermore, the scale required to build artificial intelligence systems is too complex, too obscured by intellectual property law, and too mired in logistical and technical complexity for us to see into it all. But the aim here is not to try and make these complex assemblages transparent: rather than trying to see *inside* them, we will be connecting *across* multiple systems to understand how they work in relation to each other.[76] Thus, our path will follow the stories about the environmental and labour costs of AI and place them in context with the practices of extraction and classification braided throughout everyday life. It is by thinking about these issues together that we can work towards greater justice.

I make one more trip to Silver Peak. Before I reach the town, I pull the van over to the side of the road to read a weather-beaten sign. It's Nevada Historical Marker 174, dedicated to the creation and destruction of a small town called Blair. In 1906, the Pittsburgh Silver Peak Gold Mining Company bought up the mines in the area. Anticipating a boom, land speculators purchased all of the available plots near Silver Peak along with its water rights, driving prices to record artificial highs. So the mining company surveyed a couple of miles north and declared it the site for a new town: Blair. They built a hundred-stamp cyanide mill for leachmining, the biggest in the state, and laid the Silver Peak railroad that ran from Blair Junction to the Tonopah and Goldfield main line. Briefly, the town thrived. Many hundreds of people came from all over for the jobs, despite the harsh working conditions. But with so much mining activity, the cyanide began to poison the ground, and the gold and silver seams began to falter and dry up. By 1918, Blair was all but deserted. It was all over within 12 years. The ruins are marked on a local map—just a 45-minute walk away.

It's a blazing hot day in the desert. The only sounds are the metallic reverberations of cicadas and the rumble of an occasional passenger jet. I decide to start up the hill. By the time I reach the collection of stone buildings at the top of the long dirt road, I'm exhausted from the heat. I take shelter inside the collapsed remains of what was once a gold miner's house. Not much is left: some broken crockery, shards of glass bottles, a few rusted tins. Back in Blair's lively years, multiple saloons thrived nearby and a two-storey hotel welcomed visitors. Now it's a cluster of broken foundations.

Through the space where a window used to be, the view stretches all the way down the valley. I'm struck by the realization that Silver Peak will also be a ghost town soon. The current draw on the lithium mine is aggressive in response to the high demand, and no one knows how long it will last. The most optimistic estimate is 40 years, but the end may come much sooner. Then the lithium pools under the Clayton Valley will be exsanguinated—extracted for batteries that are destined for landfill. And Silver Peak will return to its previous life as an empty and quiet place, on the edge of an ancient salt lake, now drained.

Notes

1 Gray Brechin, *Imperial San Francisco: Urban Power, Earthly Ruin* (Berkeley: University of California Press, 2007).
2 Ibid., 29.
3 Agricola quoted ibid., 25.
4 Quoted in ibid., 50.
5 Ibid., 69.
6 See, for example: Kate Davies and Liam Young, *Tales from the Dark Side of the City: The Breastmilk of the Volcano, Bolivia and the Atacama Desert Expedition* (London: Unknown Fields, 2016) and 'A Grey Goldmine: Recent Developments in Lithium Extraction in Bolivia and Alternative Energy Projects', Council on Hemispheric Affairs, 17 November 2009.
7 For more on the street-level changes in San Francisco, see: Anne B. Bloomfield, 'A History of the California Historical Society's New Mission Street Neighborhood', *California History* 74/4 (1995–1996), 372–393.
8 'Street Homelessness', San Francisco Department of Homelessness and Supportive Housing, http://hsh.sfgov.org/street-homelessness/. See also: 'Counterpoints: An Atlas of Displacement and Resistance', Anti-Eviction.
9 Alastair Gee, 'San Francisco or Mumbai? UN Envoy Encounters Homeless Life in California', *The Guardian*, 22 January 2018.
10 H.W. Turner published a detailed geological survey of the Silver Peak area in July 1909. In beautiful prose, Turner extolled the geological variety within what he described as 'slopes of cream and pink tuffs, and little hillocks of a bright brick red', H.W. Turner, 'Contribution to the Geology of the Silver Peak Quadrangle, Nevada', *Bulletin of the Geological Society of America* 20/1 (1909), 228.
11 Fred Lambert, 'Breakdown of Raw Materials in Tesla's Batteries and Possible Breaknecks', *electrek*, 1 November 2016.
12 Kevin Bullis, 'Lithium-Ion Battery', *MIT Technology Review*, 19 June 2012.
13 Bloomberg, 'Chinese Lithium Giant Agrees to Three-Year Pact to Supply Tesla', *Industry Week*, 21 September 2018.
14 Ellen Wald, 'Tesla Is a Battery Business, Not a Car Business', Forbes, 15 April 2017.
15 Ernest Scheyder, 'Tesla Expects Global Shortage of Electric Vehicle Battery Minerals', *Reuters*, 2 May 2019.
16 Lizzie Wade, 'Tesla's Electric Cars Aren't as Green as You Might Think', *Wired*, 31 March 2016.
17 '2019 Sustainable Energy in America Factbook', Business Council for Sustainable Energy, 11 February 2019, https://www.bcse.org/wp -content/uploads/2019-Sustainable-Energy-in-America-Factbook.pdf; and 'What Is U.S. Electricity Generation by Energy Source?', U.S. Energy Information Administration, https://www.eia.gov/tools/faqs/faq.php?id=427&t=21.
18 Meredith Whittaker et al., *AI Now Report 2018* (New York: AI Now Institute, 2018).
19 Jussi Parikka, *A Geology of Media* (Minneapolis: University of Minnesota Press, 2015), vii-viii; Marshall McLuhan, *Understanding Media* (Cambridge, MA: MIT Press, 1994).
20 Chris Ely, 'The Life Expectancy of Electronics', *Consumer Technology Association*, 16 September 2014.
21 Sandro Mezzadra and Brett Neilson use the term 'extractivism' to name the relation between different forms of extractive operations in contemporary capitalism, which we see repeated in the context of the AI industry. Sandro Mezzadra and Brett Neilson, 'On the Multiple Frontiers of Extraction: Excavating Contemporary Capitalism', *Cultural Studies* 31/2–3 (2017), 185–204.
22 Nedal Nassar et al., 'Evaluating the Mineral Commodity Supply Risk of the US Manufacturing Sector', *Science Advances* 6/8 (2020), eaa8647.
23 Lewis Mumford, *Technics and Civilization* (Chicago: University of Chicago Press, 2010), 74.
24 See, for example: Melvin Ayogu and Zenia Lewis, 'Conflict Minerals: An Assessment of the Dodd-Frank Act', Brookings Institution, 3 October 2011, https://www.brookings.edu/opinions/conflict-minerals-an-assessment-of-the-dodd-frank-act/.
25 Jason Burke, 'Congo Violence Fuels Fears of Return to 90s Bloodbath', *The Guardian*, 30 June 2017.
26 'Congo's Bloody Coltan', Pulitzer Center on Crisis Reporting, 6 January 2011, https://pulitzercenter.org/reporting/congos-bloody-coltan.
27 Ibid.
28 'Transforming Intel's Supply Chain with Real-Time Analytics', Intel whitepaper, September 2017, https://www.intel.com/content/dam/www/public/us/en/documents/white-papers/transforming-supply-chain-with-real-time-analytics-whitepaper.pdf.
29 See, for example, an open letter from 70 signatories that criticizes the limitations of the so-called conflict-free certification process: 'An Open Letter', https://ethuin.files.wordpress.com/2014/09/09092014-open-letter-final-and-list.pdf.
30 'Responsible Minerals Policy and Due Diligence', Philips, https://www.philips.com/a-w/about/company/suppliers/supplier-sustainability/our-programs/responsible-sourcing-of-minerals.html.
31 In *The Elements of Power*, David S. Abraham describes the invisible networks of rare metals traders in global electronics supply chains: 'The network to get rare metals from the mine to your laptop travels through a murky network of traders, processors, and component manufacturers. Traders are the middlemen who do more than buy and sell rare metals: they help to regulate information and are the hidden link that helps in navigating the network between metals plants and the components in our laptops.' David S. Abraham, *The Elements of Power: Gadgets, Guns, and the Struggle for a Sustainable Future in the Rare Metal Age* (New Haven: Yale University Press, 2017), 89.
32 'Responsible Minerals Sourcing', Dell, https://www.dell.com/learn/us/en/uscorp1/conflict-minerals?s=corp.
33 Zhiyi Liu, 'Chinese Mining Dump Could Hold Trillion-Dollar Rare Earth Deposit', *China Dialogue*, 14 December 2012.
34 34 'Bayan Obo Deposit, ... Inner Mongolia, China', Mindat.org, https://www.mindat.org/loc-720.html.
35 Tim Maughan, 'The Dystopian Lake Filled by the World's Tech Lust', *BBC Future*, 2 April 2015.
36 M.J. Hird, 'Waste, Landfills, and an Environmental Ethics of Vulnerability', *Ethics and the Environment* 18/1 (2013), 105–124.
37 Abraham, *The Elements of Power*, op. cit. (note 31), 175.
38 Ibid., 176.
39 Cam Simpson, 'The Deadly Tin inside Your Smartphone', *Bloomberg*, 24 August 2012.
40 Kate Hodal, 'Death Metal: Tin Mining in Indonesia', *The Guardian*, 23 November 2012.
41 Ibid.
42 John Tully, 'A Victorian Ecological Disaster: Imperialism, the Telegraph, and Gutta-Percha', *Journal of World History* 20/4 (2009), 559–579.
43 Nicole Starosielski, *The Undersea Network* (Durham, NC: Duke University Press, 2015).
44 See: Nick Couldry and Ulises A. Mejías, *The Costs of Connection: How Data Is Colonizing Human Life and Appropriating It for Capitalism* (Stanford, CA: Stanford University Press, 2019), 46.
45 Ibid., 574.
46 For a superb account of the history of undersea cables, see: Starosielski, *The Undersea Network*, op. cit. (note 43).
47 Theodora Dryer, *Designing Certainty: The Rise of Algorithmic Computing in an Age of Anxiety 1920–1970* (PhD. dissertation, University of California, San Diego, 2019), 45.
48 Ibid., 46.
49 Ibid., 266–268.
50 More people are now drawing attention to this problem, including researchers at AI Now. See: Roel Dobbe and Meredith Whittaker, 'AI and Climate Change: How They're Connected, and What We Can Do about It', *Medium* (blog), 17 October 2019, https://medium.com/@AINowInstitute/ai-and-climate-change-how-theyre-connected-and-what-we-can-do-about-it-6aa8d0f5b32c.
51 See, as an example of early scholarship in this area: Nathan Ensmenger, 'Computation, Materiality, and the Global Environment', *IEEE Annals of the History of Computing* 35/3 (2013), 80.
52 Tung-Hui Hu, *A Prehistory of the Cloud* (Cambridge, MA: MIT Press, 2015), 146.
53 Nicola Jones, 'How to Stop Data Centres from Gobbling Up the World's Electricity', *Nature*, 12 September 2018. Some progress has been made towards mitigating these concerns through greater energy efficiency practices, but significant long-term challenges remain. Eric Masanet et al., 'Recalibrating Global Data Center Energy-Use Estimates', *Science* 367/6481 (2020), 984–986.
54 L. Belkhir and A. Elmeligi, 'Assessing ICT Global Emissions Footprint: Trends to 2040 and Recommendations', *Journal of Cleaner Production* 177 (2018), 448–463; Anders A.E. Andrae and Tomas Edler, 'On Global Electricity Usage of Communication Technology: Trends to 2030', *Challenges* 6/1 (2015), 117–157.
55 Emma Strubell, Anaya Ganesh and Andrew McCallum, 'Energy and Policy Considerations for Deep Learning in NLP', in: *Proceedings of the 57th Annual Meeting of the Association for Computational Linguistics* (Florence, Italy: Association for Computational Linguistics, 2019), 3645–3650.
56 Ibid.
57 Rich Sutton, 'The Bitter Lesson', blog, 13 March 2019, http://www.incompleteideas.net/IncIdeas/BitterLesson.html.

58 'AI and Compute', Open AI, blog, 16 May 2018, https://openai.com/blog/ai-and-compute/.
59 Gary Cook et al., *Clicking Clean: Who Is Winning the Race to Build a Green Internet?* (Washington, DC: Greenpeace, 2017).
60 Shirin Ghaffary, 'More Than 1,000 Google Employees Signed a Letter Demanding the Company Reduce Its Carbon Emissions', *Recode*, 4 November 2019. See also: Apple, 'Apple Commits to Be 100 Percent Carbon Neutral for Its Supply Chain and Products by 2030', 21 July 2020; Roger Harrabin, 'Google Says Its Carbon Footprint Is Now Zero,' *BBC News*, 14 September 2020; Brad Smith, 'Microsoft Will Be Carbon Negative by 2030,' *Official Microsoft Blog*, 20 January 2020, https://blogs.microsoft.com/blog/2020/01/16/microsoft-will-be-carbon-negative-by-2030/.
61 'Powering the Cloud: How China's Internet Industry Can Shift to Renewable Energy', Greenpeace, 9 September 2019.
62 Ibid.
63 Ibid.
64 Mel Hogan, 'Data Flows and Water Woes: The Utah Data Center', *Big Data and Society* (December 2015).
65 'Off Now: How Your State Can Help Support the Fourth Amendment', Off-Now.org, https://s3.amazonaws.com/TAChandbooks/OffNow-Handbook.pdf.
66 Nate Carlisle, 'Shutting Off NSA's Water Gains Support in Utah Legislature', *Salt Lake Tribune*, 20 November 2014.
67 Materiality is a complex concept, and there is a lengthy literature that contends with it in such fields as STS, anthropology and media studies. In one sense, materiality refers to what Leah Lievrouw describes as 'the physical character and existence of objects and artifacts that makes them useful and usable for certain purposes under particular conditions'. Lievrouw quoted in: Tarleton Gillespie, Pablo J. Boczkowski and Kristen A. Foot (eds.), *Media Technologies: Essays on Communication, Materiality, and Society* (Cambridge, MA: MIT Press, 2014), 25. But as Diana Coole and Samantha Frost write: 'Materiality is always something more than 'mere' matter: an excess, force, vitality, relationality, or difference that renders matter active, self-creative, productive, unproductive.' Diana Coole and Samantha Frost (eds.), *New Materialisms: Ontology, Agency, and Politics* (Durham, NC: Duke University Press, 2012), 9.
68 United Nations Conference on Trade and Development, *Review of Maritime Transport*, 2017, https://unctad.org/en/PublicationsLibrary/rmt2017_en .pdf.
69 Rose George, *Ninety Percent of Everything: Inside Shipping, the Invisible Industry That Puts Clothes on Your Back, Gas in Your Car, and Food on Your Plate* (New York: Metropolitan Books, 2013).
70 Zoë Schlanger, 'If Shipping Were a Country, It Would Be the Sixth-Biggest Greenhouse Gas Emitter', *Quartz*, 17 April 2018.
71 John Vidal, 'Health Risks of Shipping Pollution Have Been "Underestimated"', *The Guardian*, 9 April 2009.
72 'Containers Lost at Sea: 2017 Update', World Shipping Council, 10 July 2017, http://www.worldshipping.org/industry-issues/safety/Containers _Lost_at_Sea_-_2017_Update_FINAL_July_10.pdf.
73 Guy Adams, 'Lost at Sea: On the Trail of Moby-Duck', *Independent*, 27 February 2011.
74 Lewis Mumford, *The Myth of the Machine, Vol. 1: Technics and Human Development* (New York: Harcourt Brace Jovanovich, 1967).
75 Mazen Labban, 'Deterritorializing Extraction: Bioaccumulation and the Planetary Mine', *Annals of the Association of American Geographers* 104/3 (2014), 560–576. For an expansion on this idea, see: Martin Arboleda, *Planetary Mine: Territories of Extraction under Late Capitalism* (London: Verso, 2020).
76 Mike Ananny and Kate Crawford, 'Seeing without Knowing: Limitations of the Transparency Ideal and Its Application to Algorithmic Accountability', *New Media and Society* 20/3 (2018), 973–989.

# A CITY IS N

A CITY IS NOT A COMPUTER First published in: Shannon Mattern, *A City Is Not a Computer: Other Urban Intelligences* (Princeton, New Jersey: Princeton University Press, 2021).

Shannon Mattern

The spring and summer of 2020 seemed to upend much of what we thought we knew about cities. A global pandemic, shelter-in-place rules, economic collapse, broken supply chains, skyrocketing unemployment, widely circulating videos of police violence (which we'd seen before, of course) and amplified reminders of entrenched and systematic racism precipitated a host of global adaptations and reckonings. Industries and corporations that had long justified their metropolitan locations by celebrating the proximity of complementary services and suppliers, and the concentration of intellectual and creative minds, transitioned overnight to telecommuting. Planned density, mass transit, packed restaurants and performing arts venues—all central to the appeal, vibrancy and purported sustainability of urbanism—were hastily framed as liabilities. Libraries, museums, performing arts venues and restaurants were shuttered, and some analysts predicted that a good proportion of those institutions and businesses would never again reopen. Parks and sidewalks and other spaces that afforded space for circulation and distancing proved their critical importance—and their paucity. Maintenance and service workers revealed their essential nature, and their vulnerability too.

Some critics, planners and city administrators were asking ontological questions (based on flawed presuppositions) about the urban: If density is a risk, if embodied sociality is expendable, and if culture is digitizable, what makes a city *a city*? Equally resonant, amid all the infection graphs and protest videos, were epistemological inquiries: What kinds of knowledge does a city foster?[1] How do its infrastructures facilitate the creation, sharing and storage of information? And what does a city allow us to know about ourselves?

Consider some concrete examples: the closure of educational and cultural institutions raised questions about what roles they play and what values they champion in their cities. What happens, for instance, when school moves online and a large portion of the student body doesn't have reliable access to the Internet at home and the libraries are closed? Folks who had never before thought much about police abolition or systemic racism started caring about their cities' police budgets and reading Black authors to make sense of the spatialization and institutionalization of injustice. Communities began looking anew at the statues populating their plazas, wondering if these embodiments of cultural heritage honestly and inclusively reflect their fraught histories. Stir-crazy quarantined apartment dwellers pried themselves away from Netflix and Animal Crossing and took their bodies—and voices and placards—to the streets, calling for a reckoning with racial injustices and demanding redress. Urbanists became newly invested

T A COMPU

in epidemiology, wondering about the ethical and methodological affordances and limitations of data extracted from contact tracing apps produced by Big Tech companies, who were themselves under public and congressional scrutiny for their broad reach, influence and consolidation. City public health officials acknowledged that digital health surveillance alone wasn't sufficient: we also needed ethically attuned *qualitative* methods and intensive, pavement-pounding fieldwork to track the virus. The fall and winter then cast new light on long festering epistemological problems that encompassed and extended beyond the urban scale: from crises of confidence in election data and the processes by which they're derived, to the metastasization of conspiracy theories and the potential roles of our underfunded public media and public schools in fending them off.

Metropolitan newspapers, podcasts, Twitter threads and a veritable wave of Zoom symposia tackled these questions—all of which, in one way or another, engaged with what a city *knows*, and through what spatial and technological means it creates that knowing. Could this moment of entangled crises, our quarantined interlocutors wondered, reflect a dramatic shift in the ways we think about, design, administer and steward our cities? Policy and planning (and academic book publishing) typically move slowly—so the answers to these questions, as they're made manifest on the ground, will reveal themselves over the next several years, and most likely into the next few decades. But one decisive urban action occurred amid the pandemic that does prove emblematic, and perhaps symptomatic, of larger changes: Sidewalk Labs walked away from Toronto. (Granted, they were, to some degree, pushed out the door.)

In 2015, before it metamorphosed into Alphabet, Google launched Sidewalk Labs, an 'urban innovation' division dedicated to solving urban problems with 'forward-thinking design and cutting-edge technology', like networked sensors and automated systems. The following year I wrote about Sidewalk, in relation to New York City's 'smart' Hudson Yards development, where the company established its headquarters.[2] In 2017 the company, under the leadership of Dan Doctoroff, former deputy mayor of New York City and CEO of Bloomberg LP, was commissioned to develop the Quayside area on Toronto's waterfront. As he articulated in a 2016 *Sidewalk Talk* post, Doctoroff had been wondering: 'What would a city look like if you started from scratch in the internet era—if you built a city "from the internet up?"'[3] In many public presentations since Sidewalk's launch, he offered a bit of revisionist history that positioned his company as a catalyst for the next revolution in urban infrastructure:

> Looking at history, one can make the argument that the greatest periods of economic growth and productivity have occurred when we have integrated innovation into the physical environment, especially in cities. The steam engine, electricity grid, and automobile all fundamentally transformed urban life, but we haven't really seen much change in our cities

since before World War II. If you compare pictures of cities from 1870 to 1940, it's like night and day. If you make the same comparison from 1940 to today, hardly anything has changed. Thus it's not surprising that, despite the rise of computers and the internet, growth has slowed and productivity increases are so low. ... So our mission is to accelerate the process of urban innovation.[4]

The team offered a plan that incorporated many of the greatest hits of twenty-first-century urbanism—pedestrian- and bike-friendly streets, affordable housing, sustainable construction—with high-tech interventions like generative design tools, enhanced fibreoptics, and a 'comprehensive digital map of the public realm', much like a dashboard.[5] Despite their attempts to accelerate urban innovation, circulation and connectivity within the Quayside development itself, Doctoroff's team encountered, through the design process, the (often productive) slowness and friction of government bureaucracy and democratic deliberation.[6] They moved at the speed of argument and paperwork rather than that of computation, whose efficiency many technologists assume to be an inherent good. After a long, messy process plagued by controversy over financing, governance, data privacy and a host of other concerns, Doctoroff took again to *Sidewalk Talk* in May 2020 to announce that the Quayside project was pulling the plug. The purported cause: the 'unprecedented economic uncertainty' wrought by Covid-19.[7]

Those economic projections could be computed, leading Doctoroff's team to deem the project a bad bet. What was less amenable to computational modelling, however, and more likely to have shaped the project's outcome, was public engagement and reception.



While the failure is certainly due in part to a changed world," local Sidewalk critic and public technology expert Bianca Wiley writes, "this explanation brushes under the rug years of sustained public involvement in the project, from supporters and critics alike. From its inception, the project failed to appreciate the extent to which cities remain strongholds of democracy.[8]

The project's demise seemed to mark the end of an era. Sidewalk Toronto was only the most visible of several smart city projects that had been abandoned, derailed, mothballed or run out of town in the preceding years. High-profile developments like Songdo, in South Korea, or Masdar, in the United Arab Emirates (UAE), had fallen from the headlines. New Yorkers thwarted the construction of Amazon's HQ2, the company's East Coast headquarters, in Queens. Alphabet walked away from its Google Fiber project in Louisville, Kentucky, and cut back its services in other municipalities.[9]

Designers, planners, engineers, investors, technologists, developers and entrepreneurial city leaders had for years espoused visions of an urban future in which embedded sensors, ubiquitous cameras and beacons, networked smartphones, dashboards and omniscient operating systems would produce unprecedented efficiency, seamless connectivity, convenience and, for the especially well connected, the realization of what Aaron Bastani calls 'fully automated luxury communism' (or capitalism: the smart city could be made to support a range of ideologies). Yet the appeal of such an urban vision, and the belief that it's even possible, seems to have waned.[10]

Fig. 1 Generative design platform on display at 307, Sidewalk Toronto's showroom/'experimental work space'

The technical means through which that fantasy was to be realized, however, readily lent themselves to appropriation. A ruthless Trump presidency, Covid-19 and the Movement for Black Lives uprisings provided a fertile new context for these tools and platforms to reposition themselves, to rebrand for a post-pandemic, post-white-supremacist future (if such a thing ever arrives). Gadgets that were once sold as necessary equipment for autonomous vehicles and drone deliveries could be justified as (and have, in fact, long served as) critical infrastructure for public health and security surveillance.[11] Smart technologies often furnish convenient stopgap solutions; they provide a quick, and often lucrative, targeted fix that absolves leaders of the responsibility to investigate and resolve the root causes for health and racial injustices and systemic breakdowns.

Silicon Valley moves fast and breaks things, but cities, if responsibly designed and administered, can't afford such negligence—even if multiple converging crises seem to necessitate the rapid prototyping of urban solutions. Over the next several years we'll see what becomes of our smart city dreams, how those cameras and cables will be (re)deployed, what governments will cement or sever relationships with Big Tech contractors, and if and how all the lessons of Covid-19 and #BlackLivesMatter and the 2020 election crisis will stick with us. What will we have learned from the bodies gathered together in protest, the move to emergency online learning and telehealth, the spread of medical and election misinformation, the distribution of aid through care networks, and the new visibility of supply chains and service workers? Will we still aspire to build cities from the Internet up, to design schools and offices and streets to accommodate the administrative logics of algorithms, to apply computational tools in tracking sick, noncompliant or otherwise marginalized urban subjects? Will we sustain Doctoroff's vision of the city-as-computer? Or will other models, metaphors or modes of operation arise—and what will their epistemological and political implications be?

## Urban Metaphors

For well over half a century, major tech companies have been transforming the cities where they've been headquartered—San Francisco, Seattle, Shenzhen—skewing demographics,

Fig. 2 Poster illustrating the various sensors planned for deployment in Quayside, on display on 307

concentrating intellectual capital, exacerbating congestion and driving up real estate prices.[12] It wasn't until the turn of the millennium, however, that these corporations enjoyed the luxury of building new cities on a blank slate. Digital urban developments arose across the Americas, Africa, Asia and the Middle East, where companies like Cisco, Siemens and IBM partnered with real estate developers and governments to build smart cities tabula rasa. The trend continues: in late spring 2020, even after the world followed Toronto's cautionary tale, Tencent, China's largest Internet company, commissioned the architectural firm NBBJ to design Net City, a master-planned district along Shenzhen's Dachanwan Port.[13]

With programmers and planners working together, design teams imagined transforming the idealized topology of the web or the 'Internet of Things' into urban form. Programmer and tech writer Paul McFedries explains this thinking:

> The city is a computer, the streetscape is the interface, you are the cursor, and your smartphone is the input device. This is the user-based, bottom-up version of the city-as-computer idea, but there's also a top-down version, which is systems-based. It looks at urban systems such as transit, garbage, and water and wonders whether the city could be more efficient and better organized if these systems were 'smart'.[14]

The city-as-computer model has spun off more nuanced subcategories, including the city-as-platform and the city-as-operating system.[15] Each of these metaphors suggests different sites and agencies for urban planning, administration and citizenship; they focus our attention on either figure or ground, structure or infrastructure, interface or code. These metaphors also suggest different locations or degrees for planners', administrators' and citizens' potential interventions into city making, the ways information flows through an urban system, and what kinds of civic data are operable or 'executable', or even *valued*.

While projects like Sidewalk Labs were conceived in an age of cloud computing and artificial intelligence, they're rooted in earlier reveries. Ever since the Internet was little more than a few linked nodes, urbanists, technologists and sci-fi writers and filmmakers have envisioned cybercities and e-topias embodying computational or networked form.[16] Modernist designers and futurists discerned morphological parallels between urban forms and circuit boards. Just as new modes of telecommunication have reshaped physical terrains and political economies, new computational methods have informed urban planning, modelling and administration.[17]

Modernity is good at renewing metaphors, from the city-as-machine, to the city-as-organism or -ecology, to the city-as-cyborgian-merger of the technological and the organic.[18] Those metaphors, which have evolved to reflect shifts in the *zeitgeist*, go a long way in determining how we conceive of urban planning, urban form, governance, maintenance, citizenship and so forth. The city-as-machine, for instance, is the sum of its divisible parts. It focuses on capitalist growth, strives for efficiency and can be repaired piecemeal. Some argue that the machinic city has a history much deeper than the machine age itself, as evidenced by use of grid layouts, linear patterns and regular geometric forms (like the tree) since ancient times, and by the use of standardized patterns for colonial urban development.[19]

Then we have the city-as-organism, which reflects the influence of the natural sciences in the eighteenth and nineteenth centuries. This model frames the city as a self-contained, autonomous entity composed of differentiated parts, or organs, serving complementary functions. As architect Kevin Lynch explains, the city-as-organism, like the city-as-machine, does have 'differentiated parts', but they 'work together and influence each other in subtle ways. Form and function are indissolubly linked, and the function of the whole is complex, not to be understood simply by knowing the nature of the parts.'[20]

A more specific conception of the city-as-organism is that of the city-as-biophysical-body, with its own circulatory, respiratory and nervous systems and waste streams.[21] In the mid-nineteenth century, as industrialization transformed cities and spurred their growth, physicians were developing new theories of infectious disease (for example, miasma, filth), complete with scientific models and maps that depicted cities as unhealthy. City planners and health officials joined forces to advocate for sanitation reform, zoning, new infrastructures, street improvements and urban parks.[22] Healthy buildings and cities were associated with certain phenotypical expressions, although designers didn't always agree on the ideal form. Frederick Law Olmsted's parks, Daniel Burnham's City Beautiful movement, Ebenezer Howard's Garden Cities, 1920s zoning ordinances, modernist social housing projects and sanatoriums: all promised reform yet produced distinct morphologies.[23]

Historian Jennifer Light describes how, in the early twentieth century, the new urban professions 'sought to deliver the prestige and predictability of the natural sciences and scientific management of nature to the development of urban research and policy'.[24] Some urbanists were originally drawn to the laboratory model of chemistry and physics, before acknowledging that its experimental methods didn't lend themselves to ethical application in the urban realm.[25] Ultimately, the field research methods and conceptual models of plant and animal ecologists proved more fitting, and framed cities as ecosystems in need of scientific management and conservation. Yet urbanist and GIS methodologist Leah Meisterlin addresses the continued appeal, and persistent limitations, of the lab metaphor in an age of algorithmic planning:

Not only does the city fail to produce the necessary conditions for controlled inquiry, it also produces the opposite in abundance ... That the city is not a lab places obvious and important limitations on our findings when research is understood through laboratory-science frameworks.[26]

## Dreams of an Informatic Urbanism

There's no shortage of recent 'think pieces' calling for '*systems* thinking' in urban design.[27] Yet much of this discourse can trace its genealogy back to post–Second World War cybernetics, which was itself informed by computational and militaristic models.[28] Urban scientists envisioned the city as an informational or logistical system and city planning as a science of communication, information and control. Cybernetic approaches to urbanism are the clear precursors to contemporary calls for cities built from the Internet up.

The city-as-computer model is appealing because it frames the messiness of urban life as programmable and subject to rational order. Computing, Ranjodh Dhaliwal writes, has been variously associated with the digital, the binary (1s and 0s), the algorithmic, the automated and the addressable—each implying a form of order and control.[29] Anthropologist Hannah Knox explains: 'As technical solutions to social problems, information and communications technologies encapsulate the promise of order over disarray ... as a path to an emancipatory politics of modernity.'[30] And there are echoes of the premodern too. The computational city draws power from an urban imaginary that goes back millennia, to the city as an apparatus for record keeping and information management.

We've long conceived of our cities as knowledge repositories and data processors, and they've always functioned as such. Lewis Mumford observed that when the wandering rulers of the European Middle Ages settled in capital cities, they installed a 'regiment of clerks and permanent officials' and established all manner of paperwork and policies (deeds, tax records, passports, fines, regulations), which necessitated a new urban apparatus, the office building, to house its bureaus and bureaucracy.[31] The classic example is the Uffizi offices in Florence, designed by Giorgio Vasari in the mid-sixteenth century, which provided an architectural template copied in cities around the world. 'The repetitions and regimentations of the bureaucratic system'—the work of data processing, formatting and storage—left a 'deep mark', as Mumford put it, on the early modern city.[32]

Yet the city's informational role began even earlier than that. Writing and urbanization developed concurrently in the ancient world, and those early scripts—on clay tablets, mud-brick walls and landforms of various types—were used to record transactions, mark territory, celebrate ritual and embed contextual information in the landscape.[33] Mumford described the city as a fundamentally communicative space, rich in information:

Through its concentration of physical and cultural power, the city heightened the tempo of human intercourse and translated its products into forms that could be stored and reproduced. Through its monuments, written records, and orderly habits of association, the city enlarged the scope of all human activities extending them backwards and forwards in time. By means of its storage facilities (buildings, vaults, archives, monuments, tablets, books), the city became capable of transmitting a complex culture from generation to generation, for it marshaled together not only the physical means but the human agents needed to pass on and enlarge this heritage. That remains the greatest of the city's gifts. As compared with the complex human order of the city, our present ingenious electronic mechanisms for storing and transmitting information are crude and limited.[34]

Mumford's city is an assemblage of media forms (vaults, archives, monuments, physical and electronic records, oral histories, lived cultural heritage), agents (architectures, institutions, media technologies, people) and functions (storage, processing, transmission, reproduction, contextualization, operationalization).[35] It is a large, complex and varied epistemological and bureaucratic apparatus. It's an information processor, to be sure, but it is also more than that.

Were he alive today, Mumford would reject the creeping notion that the city is simply the Internet writ large. He would remind us that the processes of city making are more complicated than writing parameters for rapid spatial optimization. He would inject history and happenstance. *The city is not a computer.* This seems an obvious truth, but it has been challenged (again) by technologists and political actors who speak as if they could reduce urban planning and administration—and the logistics of public health and policing—to algorithms and addressable agents.[36] Why should we care about debunking obviously false metaphors? It matters because the metaphors give rise to technical models, which inform design processes, which in turn shape knowledges and politics, not to mention material cities. The sites and systems where we locate the city's informational functions—the places where we see information processing, storage, and transmission 'happening' in the urban landscape—shape larger understandings of urban intelligence.

## Urban Epistemological Ecologies

The idea of the city as an information-processing machine has in recent years manifested as a cultural obsession with urban sites of data storage and transmission. Scholars, artists and designers write books, conduct walking tours and make maps of Internet infrastructures. We take pleasure in pointing at nondescript buildings that hold thousands of whirring servers, at surveillance cameras, camouflaged antennae and hovering drones, and localizing the city's computation in these particular sites and systems.[37] Yet such work runs the risk of reifying and essentializing information, even depoliticizing it. When we treat data as a 'given' (which is, in fact, the etymology of the word), we see it in the abstract, as an urban fixture like traffic or crowds. We need to shift our gaze and look at data in context, at the life cycle of urban information, distributed within a varied ecology of urban sites and subjects that interact with it in multiple ways. We need to see data's human, institutional and technological creators, its curators, cleaners and preservers, owners and brokers, 'users', hackers and critics. As Mumford understood, there is more than information *processing* going on here. Urban information is *made*, commodified, accessed, secreted, politicized, operationalized, preserved and erased.

But where? Can we point to the chips and drives, cables and warehouses—the specific urban architectures and infrastructures—where this expanded ecology of information management resides and operates? For years I've written and taught classes about the challenges of reducing complicated technical and intellectual structures to their material, geographic manifestations, that is, mapping 'where the data live'.[38] Yet such exercises can be useful in identifying points of entry to the larger system. It's not only the infrastructural object that matters but also the personnel and paperwork and protocols, the machines and management practices, the conduits and cultural variables that shape terrain within the larger ecology of urban information.

So the next time you're staring up at a surveillance camera, or the assemblage of contraptions perched atop a telephone pole, ask how it got there, how it generates or filters data—not only how the equipment operates technically but also what information it claims to be harvesting, and through what methodology—and whose interests it serves.[39] And don't let the totalizing idea of the city-as-computer blind you to the countless other forms of data and sites of intelligence-generation in the city: municipal agencies and departments, universities, hospitals, laboratories and corporations. Each of these sites has a distinctive orientation towards urban intelligence. Let's consider a few of the more public ones.

First, the municipal archive. Most cities today have archives that contain records of administrative activity, finances, land ownership, property taxes, legislation and labour. The archives of ancient Mesopotamian and Egyptian cities held similar material, although historians debate whether ancient record-keeping practices served similar documentary functions.[40] Archives have served to ensure financial accountability and symbolically legitimize governing bodies and colonial rulers, while also erasing the heritage of previous regimes and conquered populations. They monumentalize a culture's historical consciousness and intellectual riches. In the modern age, they also support journalism, genealogy and scholarship.[41] The 'information' inherent in archives resides not solely in the content of their records, but also in their very existence, provenance and organization (there's much to be learned about the ideals of a culture by examining its archival forms), and even in their omissions and erasures.[42]

Of course, not all archives are ideologically equal. Community archives validate the personal histories and intellectual contributions of diverse publics, particularly those whose lives have traditionally been omitted from the historical record. Meanwhile, law enforcement agencies and customs and immigration offices are networked with geographically distributed National Security Agency and other federal repositories, which tend to *overrepresent* those same marginalized subjects. These archives are not of the same species, nor do they 'process' 'data' in the same fashion.

Practices and politics of curation and access have historically distinguished archives from another key site of urban information: libraries. Whereas archives collect unpublished materials and attend primarily to their preservation and security, libraries collect published materials and prioritize making those materials intelligible and accessible to patrons. In practice, such distinctions are fuzzy and contested, especially today, as many archives seek to be more public-facing. Nevertheless, these two institutions embody different knowledge regimes and ideologies.

Fig. 3 The removal of a statue of Thomas 'Stonewall' Jackson a Confederate general, from Richmond, Virginia, in summer 2020

Modern libraries and librarians have sought to empower patrons to access information across platforms and formats, and to critically assess bias, privacy and other issues under the rubric of 'information literacy'.[43] They build a critical framework around their resources, often in partnership with schools, universities and other cultural and civic organizations. Further, libraries perform vital symbolic functions, embodying the city's commitment to its intellectual heritage (which may include heritage commandeered through imperial, settler colonial, racist and sexist activities).

Similarly, the city's museums reflect its commitment to knowledge in embodied form, to its artifacts and material culture. Again, such institutions are open to ideological critique. Acquisition policies, display practices and access protocols are immediate and tangible, and they reflect particular cultural and intellectual politics. We see these issues reflected again in public art, particularly statuary. The defacement and removal of colonialist and Confederate statues in the wake of the 2020 Movement for Black Lives protests sparked debate over the historiographic, pedagogical and moral value of these public installations. The members of the Society of Architectural Historians, who are typically champions for the preservation of historic resources, instead 'supported and encouraged the removal of Confederate monuments' because, unlike other aspects of the built environment that might serve as painful but instructive reminders of past mistakes, these monuments 'do not serve as catalysts for a cleansing public conversation, but rather express white supremacy and dominance, causing discomfort and distress to African-American citizens who utilize the public spaces these monuments occupy'.[44] The removal, erasure or defacement of particular records—as well as the refusal to collect and store particular data or knowledge in the first place—also demonstrate how our material cities shape the politics of knowledge and memory.

Just as important as the data stored and accessed on city servers, in archival boxes, on library shelves and on museum walls are the forms of urban intelligence that cannot be easily contained, framed, catalogued or placed on a pedestal. What place-based 'information' doesn't fit on a shelf or in a database? What are the non-textual, unrecordable forms of cultural memory? These questions are especially relevant for marginalized populations, Indigenous cultures and postcolonial regions. Urbanist Paul Goodwin calls upon designers, planners and scholars to study Black communities; to acknowledge how their distinctive 'spatial knowledges, discourses and strategies' represent productively 'dissident' spatial interventions; and to use these insights to imagine and make cities differently.[45] Performance studies scholar Diana Taylor urges us to pay particular attention to ephemeral, performative forms of knowledge, such as dance, ritual, food, sport and oral culture.[46] These forms cannot be reduced to 'information', nor can they be 'processed', stored or transmitted via fibre-optic cable. Yet they are vital urban intelligences that live within bodies, minds and communities.

Finally, consider data of the environmental, ambient, 'immanent' kind. Malcolm McCullough has shown that our cities are full of fixed architectures, persistent terrains and reliable environmental patterns that anchor all the unstructured data and image streams that float on top.[47] What can we learn from the 'nonsemantic information' inherent in shadows, wind, rust, in the signs of wear on a well-trodden staircase, the creaks of a battered bridge—all the indexical messages of our material environments? I'd argue that the intellectual value of this ambient, immanent information exceeds its function as stable ground, or substrate, for the city's digital flux. Environmental data are just as much figure as they are ground; they are local, too, as Yanni Loukissas reminds us. They remind us of necessary truths: that urban intelligence comes in multiple forms, that it is produced within environmental as well as cultural contexts, that it is reshaped over the *longue durée* by elemental exposure and urban development, and that it can be lost or forgotten.[48] These data remind us to think on a climatic scale, a geologic scale, as opposed to the scale of financial markets, transit patterns, and data processing.

## The Case against 'Information Processing'

Here's some geologic insight from T. S. Eliot's 1934 poem *The Rock*:

Where is the Life we have lost in living?
Where is the wisdom we have lost in knowledge?
Where is the knowledge we have lost in the information?[49]

Management theorist Russell Ackoff took Eliot's idea one step further, proposing the now famous (and widely debated) hierarchy: Data < Information < Knowledge < Wisdom.[50] Each level of processing implies an extraction of utility from the level before. Thus, contextualized or patterned data can be called information. Or, to quote philosopher and computer scientist Frederick Thompson, information is 'a product that results from applying the processes of organization to the raw material of experience, much like steel is obtained from iron ore'. Swapping the industrial metaphor for an artistic one, he writes: 'Data are to the scientist like the colors on the palette of the painter. It is by the artistry of his theories that we are informed. It is the organization that is the information.'[51] Thompson's mixed metaphors suggest that there are multiple ways of turning data into information and knowledge into wisdom.

Yet the term 'information processing', whether employed within computer science, cognitive psychology or urban design, typically refers to *computational* methods. Cyberneticians Warren McCulloch and Walter Pitts proposed a computational theory of mind in 1943.[52] Mathematician John von Neumann's *The Computer and the Brain*, published posthumously in 1958, extended the theory, as have several philosophers, neuroscientists, psychologists, and technologists since then.[53] But 'by viewing the brain as a computer that passively responds to inputs and processes data,' psychologist/geneticist Matthew Cobb writes, 'we forget that it is an active organ, part of a body that is intervening in the world and which has an evolutionary past that has shaped its structure and function.'[54] Neuroscientist Yuri Danilov agrees that the brain 'is *not a computer.* The brain is not doing *any* programming.'[55] Technological development, he argues, is 'morphing how our understanding of the brain works'. Neuroscientist Karl Lashley acknowledged this tendency—to draw cognitive metaphors from prevailing technology—as early as 1951:

> Descartes was impressed by the hydraulic figures in the royal gardens and developed a hydraulic theory of the action of the brain. We have since had telephone theories, electrical field theories, and now, theories based on the computing machines and automatic rudders ... The similarities in such comparisons are the product of an oversimplification of the problems of behavior.[56]

The brain-as-computer is just the latest link in a long chain of metaphors that powerfully shape scientific endeavour and medical practice in their own images.

The city-as-computer model likewise conditions urban design, planning, policy and administration—even residents' everyday experience—in ways that hinder the development of healthy, just and resilient cities. We have seen that urban ecologies 'process' data by means that are not strictly digital, addressable or algorithmic, and that not all urban intelligences can be called 'information'. One can't 'process' the local cultural effects of long-term weather patterns or derive insights from the generational evolution of a neighbourhood without a degree of sensitivity that exceeds mere computation; urban intelligence of this kind involves site-based experience, participant observation and sensory engagement. Indigenous knowledge holders and scientists and Western scientists possess non-translatable but potentially complementary ways of thinking about climate change and urban resilience, for instance.[57] I have written elsewhere about the vast material archives of climate science, and the embodied knowledges involved in creating, analysing and preserving that data.[58]

We need new models for thinking about cities that *do not compute*, and we need new terminology. In contemporary urban discourses, where 'data' rhetoric is often frothy and fetishistic, we seem to have lost critical perspective on how urban data become meaningful spatial information or translate into place-based knowledge. We need to expand our *repertoire* (to borrow a term from Diana Taylor) of urban intelligences, to draw upon the wisdom of information scientists and theorists, archivists, librarians, intellectual historians, cognitive scientists, philosophers, folklorists and others who think about the management of information and the production of knowledge.[59] They can help us better understand the breadth of intelligences that are integrated within our cities, which would be greatly impoverished if they were to be rebuilt, or built anew, with computational logic as their prevailing epistemology.

We could also be better attuned to the lifecycles of urban information resources—to their creation, curation, provision, preservation and destruction—and to the assemblages of urban sites and subjects that make up our cities' intellectual ecologies. 'If we think of the city as a long-term construct, with more complex behaviors and processes of formation, feedback, and processing,' architect Tom Verebes proposes, then we can imagine it as an organization, or even an organism, that can learn.[60] Urbanists and designers are already drawing on concepts and methods from artificial intelligence research: neural nets, cellular processes, evolutionary algorithms, mutation and evolution.[61] Perhaps quantum entanglement and other computer science breakthroughs could reshape the way we think about urban information too. Yet we must be cautious to avoid translating this interdisciplinary intelligence into a new urban formalism.

Instead of more gratuitous parametric modelling, we need to think about urban epistemologies that embrace memory and history; that recognize spatial intelligence as sensory, experiential and cross-generational; that consider other species' ways of knowing; that appreciate the wisdom of local crowds and communities; that acknowledge the information embedded in the city's facades, flora, statuary and stairways; that aim to integrate forms of distributed cognition paralleling our brains' own distributed cognitive processes and the logics of more-than-human ecologies.

'If cities are computers for humans,' artist and teacher Taeyoon Choi writes, 'they run buggy software and often fail to compile.'[62] We must also recognize the shortcomings in models that presume the objectivity of urban data and conveniently delegate critical, often ethical decisions to the machine. We humans *make* urban information by various means: through sensory experience, through long-term exposure to a place and, yes, by systematically filtering data. It's essential to make space in our cities for those diverse methods of knowledge production and stewardship. And we have to grapple with the political and ethical implications of our methods and models, embedded in all acts of planning and design. *City making* is always, simultaneously, an enactment of *city knowing*—which cannot be reduced to computation.

Notes

1 Neil Brenner and Christian Schmid, 'Toward a New Epistemology of the Urban', *City* 9/2–3 (2015).
2 Shannon Mattern, 'Instrumental City', *Places Journal* (April 2016).
3 Daniel L. Doctoroff, 'Reimagining Cities from the Internet Up', *Sidewalk Talk*, 30 November 2016. See also: Shannon Mattern, 'Post-It Note City', *Places Journal* (February 2020).
4 Doctoroff, 'Reimagining Cities', op. cit. (note 3).
5 *Sidewalk Toronto, Master Innovation and Development Plan*, Chapter 2, 'Public Realm' (2019), 187, https://www.sidewalktoronto.ca/documents/.
6 Michiel de Lange, email to author, 22 June 2020; Nanna Verhoeff, email to author, 22 June 2020; Michiel de Lange, 'Report Workshop Critical Making of Frictional Urban Interfaces' (2018), https://urbaninterfaces.sites.uu.nl/report-workshop-critical-making-of-frictional-urban-interfaces/; Eric Gordon and Gabriel Mugar, *Meaningful Inefficiencies: Civic Design in an Age of Digital Expediency* (New York: Oxford University Press, 2020).
7 Dan Doctoroff, 'Why We're No Longer Pursuing the Quayside Project—and What's Next for Sidewalk Labs', *Sidewalk Talk*, 7 May 2020; Ellen P. Goodman, 'Sidewalk Toronto Goes Sideways: Five Lessons for Digital Governance,' *Protego Press*, 20 June 2020; Josh O'Kane, 'Sidewalk's End: How the Downfall of a Toronto "Smart City" Plan Began Long Before COVID-19', *Globe and Mail*, 24 May 2020.
8 Bianca Wiley, 'In Toronto, Google's Attempt to Privative Government Fails—For Now', *Boston Review*, 13 May 2020.
9 Kaleigh Rogers, 'Kansas City Was First to Embrace Google Fiber, Now Its Broadband Future is "TBD"', *Motherboard*, 30 August 2017; Chris Welch, 'Google Fiber Is Leaving Louisville in Humiliating Setback', *Verge*, 7 February 2019; Kyle Wiggers, 'Google Fiber Eliminates 100Mbps Plan for New Customers', *VentureBeat*, 4 December 2019.
10 Sommer Mathis and Alexandra Kanik, 'Why You'll Be Hearing a Lot Less About "Smart Cities"', *City Monitor*, 18 February 2021.
11 Against a 'harrowing backdrop of mass death', technosolutionism was being repackaged and 'sold to us on the dubious promise that these technologies are the only possible way to pandemic-proof our lives'—and, I'd add, to maintain order in the face of racial unrest and police misconduct. Naomi Klein, 'Screen New Deal', *Intercept*, 8 May 2020. James Kilgore, an activist and expert on mass incarceration, predicted that, 'as local authorities look for ways to defund policing, new technologies may be one of the carceral "solutions" they employ to dilute the demands for transformation.' James Kilgore, 'Big Tech Is Using the Pandemic to Push Dangerous New Forms of Surveillance', *Truthout*, 22 June 2020. Embracing the city as a computationally managed environment could be our salvation from the entangled challenges of 2020. New York governor Andrew Cuomo seemed to think so; he enlisted Microsoft founder Bill Gates; Eric Schmidt, former CEO of Google and current liaison between Silicon Valley and the Defense Department; and the famously technocratic Michael Bloomberg to lead task forces charged with 'reimagining' a post-Covid New York. Techno-solutions proposed to address the pandemic could easily be extended well beyond the period of crisis. Yet these are not future visions: robust networks of surveillance technologies have been operational in many cities for decades. Emily Birnbaum and Issie Lapowsky, 'Microsoft, Amazon, and IBM Express "Solidarity": Should They End Police Contracts?', *Protocol*, 3 June 2020; Kevin Rogan, 'Digital Contact Tracing Is the New "Smart" Frontier of Urban Surveillance', *Failed Architecture*, 24 June 2020.
12 See: Susie Cagle, 'Why One Silicon Valley City Said "No" to Google', *Next City*, 11 May 2015; Sean Hollister, 'Welcome to Googletown', *Verge*, 26 February 2014; Robert H. Kargon and Arthur P. Molella, *Invented Edens: Techno-Cities of the Twentieth Century* (Cambridge, MA: MIT Press, 2008); Jennifer Light, *From Warfare to Welfare: Defense Intellectuals and Urban Problems in Cold War America* (Baltimore: Johns Hopkins University Press, 2003); Chris Morris-Lent, 'How Amazon Swallowed Seattle', *Gawker*, 18 August 2015; Margaret O'Mara, *Cities of Knowledge: Cold War Science and the Search for the Next Silicon Valley* (Princeton, NJ: Princeton University Press, 2005).
13 Konrad Putzier, 'China's Tencent Plays Master Builder as Tech Firm Plans "Net City"', *Wall Street Journal*, 9 June 2020.
14 Paul McFedries, 'The City as System [Technically Speaking],' *IEEE Spectrum* 51/4 (April 2014), 36.
15 Sarah Barns, *Platform Urbanism: Negotiating Platform Ecosystems in Connected Cities* (London: Palgrave Macmillan, 2020); David Bollier, *The City as Platform: How Digital Networks Are Changing Urban Life and Governance* (Washington, DC: Aspen Institute, 2016); Stephen Goldsmith and Neil Kleiman, *A New City O/S: The Power of Open, Collaborative, and Distributed Governance* (Washington, DC: Brookings Institution Press, 2017); Andrés Luque-Ayala and Simon Marvin, *Urban Operating Systems: Producing the Computational City* (Cambridge, MA: MIT Press, 2020). Luque-Ayala and Marvin propose a model of the smart city as an assemblage of inputs and outputs, hardware and software. 'A computer scientist we interviewed highlighted the similarities between this illustration of a city's intelligent operations center ... and a computer processing unit (CPU). Pointing to the hardware-centric nature of the terminology used, he identified analogies between this way of representing urban operations and ways of representing computers themselves. The urban is reframed by a language that emphasizes control nodes, data flows, memories, gateways, and interfaces. The Urban OS, as an operations center ... "would be like the arithmetic logic unit of the CPU, which does the adding and subtracting and multiplying". Like a person computer, this understanding of the urban is based on the idea of an internal processor (the intelligent operations center) working with external memory (analytics) to produce a visible outcome on a display (visualization on a screen or monitor)' (p. 49).
16 M. Christine Boyer, *CyberCities: Visual Perception in the Age of Electronic Communication* (New York: Princeton Architectural Press, 1996); Manuel Castells, *The Informational City: Information Technology, Economic Restructuring, and the Urban-Regional Process* (Oxford: Basel Blackwell, 1989); William Gibson, *Neuromancer* (New York: Ace Books, 1984); Germaine Halegoua, 'The Policy and Export of Ubiquitous Place: Investigating South Korean U-Cities', in: Marcus Foth et al. (eds.), *From Social Butterfly to Engaged Citizen: Urban Informatics, Social Media, Ubiquitous Computing, and Mobile Technology to Support Citizen Engagement* (Cambridge, MA: MIT Press, 2011), 315–334; Robert H. Kargon and Arthur P. Molella, *Invented Edens: Techno-Cities of the Twentieth Century* (Cambridge, MA: MIT Press, 2008); William J. Mitchell, *City of Bits: Space, Place, and the Infobahn* (Cambridge, MA: MIT Press, 1995).
17 Stephen Graham and Simon Marvin, *Telecommunications and the City: Electronic Spaces, Urban Places* (New York: Routledge, 1996); Light, *From Warfare to Welfare,* op. cit. (note 12); Shannon Mattern, 'Waves and Wires: Cities of Electric Sound', in: Shannon Mattern, *Code and Clay: 5000 Years of Urban Media* (Minneapolis: University of Minnesota Press, 2017); Mark Vallianatos, 'Uncovering the Early History of "Big Data" and "Smart City" in Los Angeles', *Boom California* (June 2015).
18 Matthew Gandy, 'Cyborg Urbanization: Complexity and Monstrosity in the Contemporary City', *International Journal of Urban and Regional Research* 29 (March 2005), 26–49; Peter Nientied, 'Metaphor and Urban Studies: A Crossover, Theory and a Case Study of SS Rotterdam', *City, Territory and Architecture* 3/21 (2016); William Solesbury, 'How Metaphors Help Us Understand Cities', *Geography* 99/3 (2014), 139–142; Tom Verebes, 'The Interactive Urban Model: Histories and Legacies Related to Prototyping the Twenty-First Century City', *Frontiers in Digital Humanities* 3 (February 2016).
19 See, for instance: Kevin Lynch, *Good City Form* (Cambridge, MA: MIT Press, 1981), 81–88.
20 Ibid., 89.
21 This paragraph is adapted from Shannon Mattern, 'Databodies in Codespace', *Places Journal* (April 2018).
22 On the historical relationship between urban planning and public health, see: Jon A. Peterson, 'The Impact of Sanitary Reform upon American Urban Planning, 1840–1890', *Journal of Social History* 13/1 (1979), 83–103; Jason Corburn, 'Reconnecting with Our Roots: American Urban Planning and Public Health in the Twenty-First Century', *Urban Affairs Review* 42/5 (2007); Jocelyn Pak Drummond, *Measuring and Mapping Relationships between Urban Environment and Urban Health: How New York City's Active Design Policies Can Be Targeted to Address the Obesity Epidemic* (master's thesis, Massachusetts Institute of Technology, 2013); Bonj Szczygiel and Robert Hewitt, 'Nineteenth-Century Medical Landscapes: John H. Rauch, Frederick Law Olmsted, and the Search for Salubrity', *Bulletin of the History of Medicine* 74/4 (2000), 708–734; and Sara Jensen Carr, *The Topography of Wellness: Health and the American Urban Landscape* (University of Virginia, forthcoming 2021).
23 Urban renewal in Europe after the First World War enabled modern architects to design new, hygienic forms of social housing, and many of those same architects also employed the tropes of modernism to design new sanatoriums, whose flat roofs, terraces, balconies and recliner chairs afforded

patients plenty of opportunities for open-air relaxation. Margaret Campbell, 'What Tuberculosis Did for Modernism: The Influence of a Curative Environment on Modernist Design and Architecture', *Medical History* 49/4 (2005), 463–488. See also: Giovanna Borasi and Mirko Zardini (eds.), *Imperfect Health: The Medicalization of Architecture* (Montreal: Canadian Centre for Architecture, Lars Muller, 2012). As the twentieth century proceeded, epidemiologists focused on germs and the biological causes of disease, while modernist architects turned towards formal concerns and rational master plans. Public health and urban planning drifted apart until the 1960s, when the environmental justice and community health centre movements brought them together again.
24 Jennifer Light, *The Nature of Cities: Ecological Visions and the American Urban Professions, 1920–1960* (Baltimore: Johns Hopkins University Press, 2009), 2.
25 Ibid., 10.
26 Leah Meisterlin, 'The City Is Not a Lab', *ARPA Journal* 1 (2014).
27 Among dozens of examples are: Stephanie Draper, 'Systems Thinking: Unlocking the Sustainable Development Goals', *Forum for the Future*, 29 September 2016; Ken Gibb, 'Systems-Thinking and Housing', *Catapult Future Cities*, 14 May 2019; Systems Studio, '100 Resilient Cities', https://100resilientcities.org/tools/systems-studio/.
28 Orit Halpern, *Beautiful Data: A History of Vision and Reason since 1945* (Durham, NC: Duke University Press, 2014); Light, *From Warfare to Welfare,* op. cit. (note 12).
29 See: Ranjodh Singh Dhaliwal, 'On Addressability, or What Even *Is* Computing?', Working Paper, 2020.
30 Hannah Knox, 'Cities and Organisation: The Information City and Urban Form', *Culture and Organization* 16/3 (2010), 187–188.
31 Lewis Mumford, *The City in History: Its Origins, Its Transformations, and Its Prospects* (New York: Harcourt, 1961), 344. See also: Shannon Mattern, 'Indexing the World of Tomorrow', *Places Journal* (February 2016).
32 See also: Friedrich A. Kittler, 'The City Is a Medium', trans. Matthew Griffin, *New Literary History* 27/4 (1996), 721–722.
33 Mattern, *Code and Clay*, op. cit. (note 17).
34 Mumford, *The City in History*, op. cit. (note 31), 569.
35 Marcus Foth's conception of 'urban informatics' is similarly capacious: it encompasses 'the collection, classification, storage, retrieval, and dissemination of recorded knowledge', either (1) in a city or (2) 'of, relating to, characteristic of, or constituting a city'. See: Marcus Foth (ed.), *Handbook of Research on Urban Informatics: The Practice and Promise of the Real-Time City* (Hershey, PA: Information Science Reference, 2009), xxiii. Such a definition acknowledges a wide variety of informational functions, contents and contexts. Yet his focus on *recorded* knowledge, and on informatics' reputation as a 'science' of data processing, still limit our understanding of the city's epistemological functions.
36 For more on the algorithm as a timely conceptual model, see: Massimo Mazzotti, 'Algorithmic Life', *Los Angeles Review of Books*, 22 January 2017.
37 For a survey of this work, see: Shannon Mattern, 'Cloud and Field', *Places Journal* (August 2016) and 'Infrastructural Tourism', *Places Journal* (July 2013). For prominent examples, see: Andrew Blum, *Tubes: A Journey to the Center of the Internet* (New York: HarperCollins, 2012), and the work of Ingrid Burrington and Mél Hogan.
38 Mattern, 'Cloud and Field', op. cit. (note 37). See also: Louise Amoore, 'Cloud Geographies: Computing, Data, Sovereignty', *Progress in Human Geography* (August 2016).
39 Ingrid Burrington, 'A Non-Exhaustive Taxonomy of Tools of Data-Driven Policing', *Urban Omnibus*, 19 June 2018.
40 James J. O'Toole, 'Back to the Future: Ernst Posner's *Archives in the Ancient World*', *American Archivist* 67 (2004), 161–175.
41 Alexandra Walsham, 'The Social History of the Archive: Record-Keeping in Early Modern Europe', *Past & Present* 230, Issue Supplement 11 (2016), 9–48.
42 Ann Stoler, *Against the Archival Grain: Epistemic Anxieties and Colonial Common Sense* (Princeton, NJ: Princeton University Press, 2010).
43 Shannon Mattern, 'Public In/Formation', *Places Journal* (November 2016).
44 Society of Architectural Historians Heritage Conservation Committee, 'Statement on the Removal of Monuments to the Confederacy from Public Spaces', 19 June 2020. See also: Michele H. Bogart, *Sculpture in Gotham: Art and Urban Renewal in New York City* (Chicago: University of Chicago Press, 2018); Catherine Clinton et al., *Confederate Statues and Memorialization* (Athens, GA: University of Georgia Press, 2019); Brett Devereaux, Twitter, 13 June 2020; William Sturkey, 'Beyond the Headlines: Confederate Monuments, Historical Memory & Free Speech', UNC Chapel Hill, 30 August 2017, https://www.youtube.com/watch?v=Ov_yL6kKcdM&t=5s; Eric R. Varner, *Mutilation and Transformation: Damnatio Memoriae and Roman Imperial Portraiture* (Boston: Brill, 2004); and the work of Monument Lab, https://monumentlab.com/.
45 'Re-visioning Black Urbanism', Goldsmiths, https://www.gold.ac.uk/cucr/research/revisioning-black-urbanism/. See also the work of urbanists, designers and planners Kristen Jeffers, Bryan C. Lee, Matthew Jordan Miller, Melvin Mitchell, Sue Mobley, Pete Saunders, Destiny Thomas, Amina Yasin and Sara Zewde.
46 Diana Taylor, *The Archive and the Repertoire: Performing Cultural Memory in the Americas* (Durham, NC: Duke University Press, 2003). See also: Maria Bonn, Lori Kendall and Jerome P. McDonough, 'Libraries and Archives and the Preservation of Intangible Cultural Heritage: Defining a Research Agenda', whitepaper, University of Illinois at Urbana-Champaign, 2017, https://www.ideals.illinois.edu/handle/2142/97228; Marisa J. Fuentes, *Dispossessed Lives: Enslaved Women, Violence, and the Archive* (Philadelphia: University of Pennsylvania Press, 2016); Saidiya Harman, 'Venus in Two Acts', *Small Axe* 26/12 (2008), 1–14.
47 Malcolm McCullough, *Ambient Commons: Attention in the Age of Embodied Information* (Cambridge, MA: MIT Press, 2013), 36, 42.
48 These ideas formed the premise behind the Urban Intelligence studio I taught at the New School in 2017 and 2018 (see: https://www.wordsinspace.net/urbanintel/spring2018/).
49 T.S. Eliot, *The Rock* (London: Faber & Faber, 1934).
50 Nikhil Sharma, 'The Origin of the "Data Information Knowledge Wisdom" (DIKW) Hierarchy', February 2008, https://www.researchgate.net/publication/292335202_The_Origin_of_Data_Information_Knowledge_Wisdom_DIKW_Hierarchy; David Weinberger, 'The Problem with the Data-Information-Knowledge-Wisdom Hierarchy', *Harvard Business Review*, 2 February 2010.
51 Frederick B. Thompson, 'The Organization Is the Information', *American Documentation* 19/3 (1968). See also: Marcia J. Bates, 'Information', in: Marcia J. Bates and Mary Niles Maac (eds.), *Encyclopedia of Library and Information Sciences*, 3rd ed. (New York: CRC Press, 2010), 2347–2360; and Rafael Capurro and Birger Hjørland, 'The Concept of Information', in: Blaise Cronin (ed.), *The Annual Review of Information Science and Technology*, vol. 37 (2003), 343–411.
52 Warren S. McCulloch and Walter H. Pitts, 'A Logical Calculus of the Ideas Immanent in Nervous Activity', *Bulletin of Mathematical Biophysics* 7 (1943), 115–133. See also: Gualtiero Piccinini, 'The First Computational Theory of Mind and Brain: A Close Look at McCulloch and Pitts's "Logical Calculus of Ideas Immanent in Nervous Activity"', *Synthese* 141 (2004), 175–325.
53 John von Neumann, *The Computer and the Brain* (New Haven, CT: Yale University Press, 1958).
54 Matthew Cobb, *The Idea of the Brain: The Past and Future of Neuroscience* (New York: Basic Books, 2020), x.
55 Robert J. Marks and Yuri Danilov, 'Is What We Know About the Brain All Wrong?', *Mind Matters*, 20 September 2019.
56 K.S. Lashley, quoted in: Magda B. Arnold, *Memory and the Brain* (Hillsdale, NJ: Lawrence Erlbaum Associates, 1984), 107.
57 Max Liboiron, 'The Power (Relations) of Citizen Science', *CLEAR*, 19 March 2019; Gleb Raygorodetsky, *The Archipelago of Hope: Wisdom and Resilience from the Edge of Climate Change* (New York: Pegasus, 2018); Paola Rosa-Aquino, 'To Share or Not to Share?', *Grist*, 21 November 2018.
58 Shannon Mattern, 'The Big Data of Ice, Rocks, Soils, and Sediments', *Places Journal* (November 2017).
59 Marcus Foth, Nancy Odendaal and Gregory N. Hearn, 'The View from Everywhere: Towards an Epistemology for Urbanites', in: *Proceedings of the 4th International Conference on Intellectual Capital, Knowledge Management and Organizational Learning* (Cape Town, South Africa, 2007).
60 Tom Verebes, 'The Interactive Urban Model: Histories and Legacies Related to Prototyping the Twenty-First-Century City', *Frontiers in Digital Humanities* 3 (February 2016).
61 See, for instance, the work of Michael Batty. My original 'The City Is Not a Computer' article, which was published in 2017, served as the namesake and a source of inspiration for a 2018 class at MIT's Department of Urban Studies and Planning, where designers and planners imagined how they might more critically integrate technology into their practice (Eric Huntley, 'A City Is Not a Computer: Histories and Theories of the Computational City', MIT Urban Planning, Fall 2018). The students' work was published at 'The Data Plays for a Year', 2018.
62 Taeyoon Choi, 'Hello, World!', *Avant.org*, 5 June 2017.

# ENGINEERE

ENGINEERED INEQUITY: ARE ROBOTS RACIST? First published in: Ruha Benjamin, *Race After Technology: Abolitionist Tools for the New Jim Code* (Cambridge UK: Polity, 2020).
Ruha Benjamin

*Welcome to the First International Beauty Contest Judged by Artificial Intelligence.*

So goes the cheery announcement for Beauty AI, an initiative developed by the Australian- and Hong Kong- based organization Youth Laboratories in conjunction with a number of companies who worked together to stage the first ever beauty contest judged by robots.[1] The venture involved a few seemingly straightforward steps:

1. Contestants download the Beauty AI app.
2. Contestants make a selfie.
3. Robot jury examines all the photos.
4. Robot jury chooses a king and a queen.
5. News spreads around the world.

As for the rules, participants were not allowed to wear makeup or glasses or to don a beard. Robot judges were programmed to assess contestants on the basis of wrinkles, face symmetry, skin colour, gender, age group, ethnicity and 'many other parameters'. Over 6,000 submissions from approximately 100 countries poured in. *What could possibly go wrong?*

On 2 August 2016, the creators of Beauty AI expressed dismay at the fact that 'the robots did not like people with dark skin'. All 44 winners across the various age groups except six were White, and 'only one finalist had visibly dark skin'.[2] The contest used what was considered at the time the most advanced machine-learning technology available. Called 'deep learning', the software is trained to code beauty using prelabelled images, then the images of contestants are judged against the algorithm's embedded preferences.[3] Beauty, in short, is in the trained eye of the algorithm.

As one report about the contest put it: 'The simplest explanation for biased algorithms is that the humans who create them have their own deeply entrenched biases. That means that despite perceptions that algorithms are somehow neutral and uniquely objective, they can often reproduce and amplify existing prejudices.'[4] Columbia University professor Bernard Harcourt remarked: 'The idea that you could come up with a culturally neutral, racially neutral conception of beauty is simply mind- boggling.' Beauty AI is a reminder, Harcourt notes, that humans are really doing the

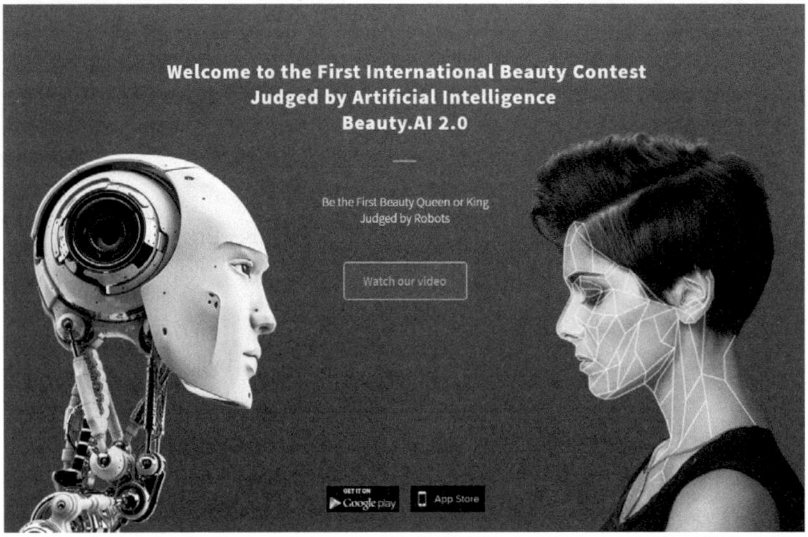


Fig.1 Beauty AI, Source: http://beauty.ai

thinking, even when 'we think it's neutral and scientific'.[5] And it is not just the human programmers' preference for Whiteness that is encoded, but the combined preferences of *all* the humans whose data are studied by machines as they learn to judge beauty and, as it turns out, *health*.

In addition to the skewed racial results, the framing of Beauty AI as a kind of preventative public health initiative raises the stakes considerably. The team of biogerontologists and data scientists working with Beauty AI explained that valuable information about people's health can be gleaned by 'just processing their photos' and that, ultimately, the hope is to 'find effective ways to slow down ageing and help people look healthy and beautiful'.[6] Given the overwhelming Whiteness of the winners and the conflation of socially biased notions of beauty and health, darker people are implicitly coded as unhealthy and unfit—assumptions that are at the heart of scientific racism and eugenic ideology and policies.

Deep learning is a subfield of machine learning in which 'depth' refers to the layers of abstraction that a computer program makes, learning more 'complicated concepts by building them out of simpler ones'.[7] With Beauty AI, deep learning was applied to image recognition; but it is also a method used for speech recognition, natural language processing, video game and board game programs, and even medical diagnosis. Social media filtering is the most common example of deep learning at work, as when Facebook auto-tags your photos with friends' names, or apps that decide which news and advertisements to show you to increase the chances that you'll click. Within machine learning there is a distinction between 'supervised' and 'unsupervised' learning. Beauty AI was supervised, because the images used as training data were prelabelled, whereas unsupervised deep learning uses data with very few labels. Mark Zuckerberg refers to deep learning as 'the theory of the mind ... How do we model—in machines—what human users are interested in and are going to do?'[8] But the question for us is, is there only *one* theory of the mind, and *whose mind* is it modelled on?

It may be tempting to write off Beauty AI as an inane experiment or harmless vanity project, an unfortunate glitch in the otherwise neutral development of technology for the common good. But, as explored in the pages ahead, such a conclusion is naive at best. Robots exemplify how race is a form of technology itself, as the algorithmic judgments of Beauty AI extend well beyond adjudicating attractiveness and into questions of health, intelligence, criminality, employment and many other fields in which innovative techniques give rise to new-fangled forms of racial discrimination. Almost every day a new headline sounds the alarm, alerting us to the New Jim Code:

'Some algorithms are racist'
'We have a problem: Racist and sexist robots'
'Robots aren't sexist and racist, you are'
'Robotic racists: AI technologies could inherit their creators' biases.'

Racist robots, as I invoke them here, represent a much broader process: social bias embedded in technical artefacts, the allure of objectivity without public accountability. Race as a form of technology—the sorting, establishment and enforcement of racial hierarchies with real consequences—is embodied in robots, which are often presented as simultaneously akin to humans but different and at times superior in terms of efficiency and regulation of bias. Yet the way robots can be racist often remains a mystery or is purposefully hidden from public view.

Consider that machine-learning systems, in particular, allow officials to outsource decisions that are (or should be) the purview of democratic oversight. Even when public agencies are employing such systems, private companies are the ones developing them, thereby acting like political entities but with none of the checks and balances. They are, in the words of one observer, 'governing without a mandate', which means that people whose lives are being shaped in ever more consequential ways by automated decisions have very little say in how they are governed.[9]

For example, in *Automated Inequality* Virginia Eubanks (2018) documents the steady incorporation of predictive analytics by US social welfare agencies. Among other promises, automated decisions aim to mitigate fraud by depersonalizing the process and by determining who is eligible for benefits.[10] But, as she documents, these technical fixes, often promoted as benefiting society, end up hurting the most vulnerable, sometimes with deadly results. Her point is not that human caseworkers are less biased than machines—there are, after all, numerous studies showing how caseworkers actively discriminate against racialized groups while aiding White applicants deemed more deserving.[11] Rather, as Eubanks emphasizes, automated welfare decisions are not magically fairer than their human counterparts. Discrimination is displaced and accountability is outsourced in this post-democratic approach to governing social life.[12]

So, how do we rethink our relationship to technology? The answer partly lies in how we think about race itself and specifically the issues of intentionality and visibility.

## I Tinker, Therefore I Am

Humans are toolmakers. And robots, we might say, are humanity's finest handiwork. In popular culture, robots are typically portrayed as humanoids, more efficient and less sentimental than *Homo sapiens*. At times, robots are depicted as having human-like struggles, wrestling with emotions and an awakening consciousness that blurs the line between maker and made. Studies about how humans perceive robots indicate that, when that line becomes too blurred, it tends to freak people out. The technical term for it is the 'uncanny valley'—which indicates the dip in empathy and increase in revulsion that people experience when a robot appears to be too much like us.[13]

Robots are a diverse lot, with as many types as there are tasks to complete and desires to be met: domestic robots, military and police robots, sex robots, therapeutic robots—and more. A robot is any machine that can perform a task, simple or complex, directed by humans or programmed to operate automatically. The most advanced are smart machines designed to learn from and adapt to their environments, created to become independent of their makers. We might like to think that robotic concerns are a modern phenomenon,[14] but our fascination with automata goes back to the Middle Ages, if not before.[15]

In *An Anthropology of Robots and AI*, Kathleen Richardson observes that the robot has 'historically been a way to talk about dehumanization' and, I would add, *not* talk about racialization.[16] The etymology of the word robot is Czech; it comes from a word for 'compulsory service', itself drawn from the Slav *robota* (servitude, hardship).[17] So yes, people have used robots to express anxieties over annihilation, including over the massive threat of war machines. But robots also convey an ongoing agitation about human domination over other humans![18]

The first cultural representation that employed the word robot was a 1920 play by a Czech writer whose machine was a factory worker of limited consciousness.[19] Social domination characterized the cultural laboratory in which robots were originally imagined. And, technically, *people* were the first robots. Consider media studies scholar Anna Everett's earliest experiences using a computer:



> In powering up my PC, I am confronted with the DOS-based text that gave me pause ... 'Pri. Master Disk, Pri. Slave Disk, Sec. Master, Sec. Slave.' Programmed here is a virtual hierarchy organizing my computer's software operations ... I often wondered why the programmers chose such signifiers that hark back to our nation's ignominious past ... And even though I resisted the presumption of a racial affront or intentionality in such a peculiar deployment of the slave and master coupling, its choice as a signifier of the computer's operations nonetheless struck me.[20]

Similarly, a 1957 article in *Mechanix Illustrated*, a popular 'how-to-do' magazine that ran from 1928 to 2001, predicted that, by 1965: 'Slavery will be back! We'll all have personal slaves again ... [who will] dress you, comb your hair and serve meals in a jiffy. Don't be alarmed. We mean robot "slaves".'[21]

It goes without saying that readers, so casually hailed as 'we', are not the descendants of those whom Lincoln freed. This fact alone offers a glimpse into the implicit Whiteness of early tech culture. We cannot assume that the hierarchical values and desires that are projected onto 'we'—*We, the People* with inalienable rights and not *You, the Enslaved* who serve us meals—are simply a thing of the past.

Coincidentally, on my way to give a talk—mostly to science, technology, engineering and mathematics (STEM) students at Harvey Mudd College—that I had planned to kick off with this

*The robots are coming! When they do, you'll command a host of push-button servants.*

**By O. O. Binder**

**You'll Own**

IN 1863, Abe Lincoln freed the slaves. But by 1965, slavery will be back! We'll all have personal slaves again, only this time we won't fight a Civil War over them. Slavery will be here to stay.

Don't be alarmed. We mean robot "slaves." Let's take a peek into the future

Robots will dress you, comb your hair and serve meals in a jiffy.



62 *Mechanix Illustrated*

Fig. 2 Robot Slaves, Source: Binder 1957

*Mechanix* ad, I passed two men in the airport restaurant and overheard one say to the other: 'I just want someone I can push around...' So simple yet so profound in articulating a dominant and dominating *theory of power* that many more people feel emboldened to state, unvarnished, in the age of Trump. *Push around*? I wondered, in the context of work or dating or any number of interactions. The slavebot, it seems, has a ready market!

For those of us who believe in a more egalitarian notion of power, of collective empowerment without domination, how we imagine our relation to robots offers a mirror for thinking through and against race as technology.

It turns out that the disposability of robots and the denigration of racialized populations go hand in hand. We can see this when police officers use 'throwbots'—'a lightweight, ruggedized platform that can literally be thrown into position, then remotely controlled from a position of safety'—to collect video and audio surveillance for use by officers. In the words of a member of one of these tactical teams: 'The most significant advantage of the throwable robot is that it "allows them [*sc.* the officers] to own the real estate with their eyes, before they pay for it with their bodies".'[22] Robots are not the only ones sacrificed on the altar of public safety. So too are the many *Black* victims whose very bodies become the real estate that police officers own in their trigger-happy quest to keep the peace. The intertwining history of machines and slaves, in short, is not simply the stuff of fluff magazine articles.[23]

While many dystopic predictions signal a worry that humans may one day be enslaved by machines, the current reality is that the tech labour force is already deeply unequal across racial and gender lines. Although not the same as the structure of enslavement that serves as an analogy for unfreedom, Silicon Valley's hierarchy consists of the highest-paid creatives and entrepreneurs, who are comprised of White men and a few White women, and the lowest-paid manual laborers—'those cleaning their offices and assembling circuit boards', in other words 'immigrants and outsourced labor, often women living in the global south', who usually perform this kind of work.[24] The 'diasporic diversity' embodied by the South Asian and Asian American tech workforce does not challenge this hierarchy, because they continue to be viewed as a 'new digital "different caste"'. As Nakamura notes: 'No amount of work can make them part of the digital economy as "entrepreneurs" or the "new economic men".'[25] Racism, in this way, is a technology that is 'built into the tech industry'.[26] But how does racism 'get inside' and operate through new forms of technology?

To the extent that machine learning relies on large, 'naturally occurring' datasets that are rife with racial (and economic and gendered) biases, the raw data that robots are using to learn and make decisions about the world reflect deeply ingrained cultural prejudices and structural hierarchies.[27] Reflecting on the connection between workforce diversity and skewed datasets, one tech company representative noted that 'if the training data is produced by a racist society, it won't matter who is on the team, but the people who are affected should also be on the team'.[28] As machines become more 'intelligent', that is, as they learn to think more like humans, they are likely to become more racist. But this is not inevitable, so long as we begin to take seriously and address the matter of how racism structures the social and technical components of design.

## Raising Robots

So, are robots racist? Not if by 'racism' we only mean white hoods and racial slurs.[29] Too often people assume that racism and other forms of bias must be triggered by an *explicit* intent to harm; for example, linguist John McWhorter argued in *Time* magazine that: 'Machines cannot, themselves, be racists. Even equipped with artificial intelligence, they have neither brains nor intention.'[30] But this assumes that self-conscious intention is what makes something racist. Those working in the belly of the tech industry know that this conflation will not hold up to public scrutiny. As one Google representative lamented: 'Rather than treating malfunctioning algorithms as malfunctioning machines ("classification errors"), we are increasingly treating tech like asshole humans.' He went on to propose that 'we [programmers] need to stop the machine from behaving like a jerk because it can look like it is being offensive on purpose'.[31] If machines are programmed to carry out tasks, both they and their designers are guided by some purpose, that is to say, intention. And in the face of discriminatory effects, if those with the power to design differently choose business as usual, then they are perpetuating a racist system whether or not they are card-carrying members of their local chapter of Black Lives Matter.

Robots are not sentient beings, sure, but racism flourishes well beyond hate-filled hearts.[32] An indifferent insurance adjuster who uses the even more disinterested metric of a credit score to make a seemingly detached calculation may perpetuate historical forms of racism by plugging numbers in, recording risk scores and 'just doing her job'. Thinking with Baldwin, someone who insists on his own racial innocence despite all evidence to the contrary 'turns himself into a monster'.[33] No malice needed, no N-word required, just lack of concern for how the past shapes the present—and, in this case, the US government's explicit intention to concentrate wealth in the hands of White Americans, in the form of housing and economic policies.[34] Detachment in the face of this history ensures its ongoing codification. Let us not forget that databases, just like courtrooms, banks, and emergency rooms, do not contain organic brains. Yet legal codes, financial practices and medical care often produce deeply racist outcomes.

The intention to harm or exclude may guide some technical design decisions. Yet even when they do, these motivations often stand in tension with aims framed more benevolently. Even police robots who can use lethal force while protecting officers from harm are clothed in the rhetoric of public safety.[35] This is why we must separate 'intentionality' from its strictly negative connotation in the context of racist practices, and examine how aiming to 'do good' can very well coexist with forms of malice and neglect.[36] In fact a do-gooding ethos often serves as a moral cover for harmful decisions. Still, the view that ill intent is always a feature of racism is common: 'No one at Google giggled while intentionally programming its software to mislabel black people.'[37] Here McWhorter is referring to photo-tagging software that classified dark-skinned users as 'gorillas'. Having discovered no bogeyman behind the screen, he dismisses the idea of 'racist technology' because that implies 'designers and the people who hire them are therefore "racists"'. But this expectation of individual intent to harm as evidence of racism is one that scholars of race have long rejected.[38]

We could expect a Black programmer, immersed as she is in the same systems of racial meaning and economic expediency as the rest of her co-workers, to code software in a way that perpetuates racist stereotypes. Or, even if she is aware and desires to intervene, will she be able to exercise the power to do so? Indeed, by focusing mainly on individuals' identities and overlooking the norms and structures of the tech industry, many diversity initiatives offer little more than cosmetic change, demographic percentages on a company pie chart, concealing rather than undoing the racist status quo.[39]

So, can robots—and, by extension, other types of technologies—be racist? Of course they can. Robots, designed in a world drenched in racism, will find it nearly impossible to stay dry. To a certain extent, they learn to speak the coded language of their human parents—not only programmers but all of us online who contribute to 'naturally occurring' datasets on which AI learn. Just like diverse programmers, Black and Latinx police officers are known to engage in racial profiling alongside their White colleagues, though they are also the target of harassment in a way their White counterparts are not.[40] One's individual racial identity offers no sure-fire insulation from the prevailing ideologies.[41] There is no need to identify 'giggling programmers' self-consciously seeking to denigrate one particular group as evidence of discriminatory design. Instead, so much of what is routine, reasonable, intuitive and codified reproduces unjust social arrangements, without ever burning a cross to shine light on the problem.[42]

A representative of Microsoft likened the care they must exercise when they create and sell predictive algorithms to their customers with 'giving a puppy to a three-year-old. You can't just deploy it and leave it alone because it will decay over time.'[43] Likewise, describing the many controversies that surround AI, a Google representative said: 'We are in the uncomfortable birthing stage of artificial intelligence.'[44] Zeros and ones, if we are not careful, could deepen the divides between haves and have-nots, between the deserving and the undeserving—rusty value judgments embedded in shiny new systems.

Interestingly, the MIT data scientists interviewed by anthropologist Kathleen Richardson ...

> ... were conscious of race, class and gender, and none wanted to reproduce these normative stereotypes in the robots they created ... [They] avoided racially marking the 'skin' of their creations ... preferred to keep their machines genderless, and did not speak in class-marked categories of their robots as 'servants' or 'workers', but companions, friends and children.[45]

Richardson contrasts her findings to that of anthropologist Stefan Helmreich, whose pioneering study of artificial life in the 1990s depicts researchers as 'ignorant of normative models of sex, race, gender and class that are refigured in the computer simulations of artificial life'.[46] But perhaps the contrast is overdrawn, given that colourblind, gender-neutral and class-avoidant approaches to tech development are another avenue for coding inequity. If data scientists do indeed treat their robots like children, as Richardson describes, then I propose a race-conscious approach to parenting artificial life—one that does not feign colour-blindness. But where should we start?

## Automating Anti-Blackness

As it happens, the term 'stereotype' offers a useful entry point for thinking about the default settings of technology and society. It first referred to a practice in the printing trade whereby a solid plate called a 'stereo' (from the ancient Greek adjective *stereos*, 'firm', 'solid') was used to make copies. The duplicate was called a 'stereotype'.[47] The term evolved; in 1850 it designated an 'image perpetuated without change' and in 1922 was taken up in its contemporary iteration, to refer to shorthand attributes and beliefs about different groups. The etymology of this term, which is so prominent in everyday conceptions of racism, urges a more sustained investigation of the interconnections between technical and social systems.

To be sure, the explicit codification of racial stereotypes in computer systems is only one form of discriminatory design. Employers resort to credit scores to decide whether to hire someone, companies use algorithms to tailor online advertisements to prospective customers, judges employ automated risk assessment tools to make sentencing and parole decisions, and public health officials apply digital surveillance techniques to decide which city blocks to focus medical resources. Such programs are able to sift and sort a much larger set of data than their human counterparts, but they may also reproduce long-standing forms of structural inequality and colourblind racism.

And these default settings, once fashioned, take on a life of their own, projecting an allure of objectivity that makes it difficult to hold anyone accountable.[48] Paradoxically, automation is often presented as a solution to human bias—a way to avoid the pitfalls of prejudicial thinking by making decisions on the basis of objective calculations and scores. So, to understand racist robots, we must focus less on their intended uses and more on their actions. Sociologist of technology Zeynep Tufekci describes algorithms as 'computational agents who are not alive, but who act in the world'.[49] In a different vein, philosopher Donna Haraway's 1991 classic *Simians, Cyborgs and Women* narrates the blurred boundary between organisms and machines, describing how 'myth and tool mutually constitute each other'.[50] She describes technologies as 'frozen moments' that allow us to observe otherwise 'fluid social interactions' at work. These 'formalizations' are also instruments that enforce meaning—including, I would add, racialized meanings—and thus help construct the social world.[51] Biased bots and all their coded cousins could also help subvert the status quo by exposing and authenticating the existence of systemic inequality and thus by holding up a 'black mirror' to society,[52] challenging us humans to come to grips with our deeply held cultural and institutionalized biases.[53]

Consider the simple corrections of our computer systems, where words that signal undue privilege are not legible. The red line tells us that only one of these phenomena, underserved and overserved, is legitimate while the other is a mistake, a myth (Fig. 3).

But power is, if anything, relational. If someone is experiencing the underside of an unjust system, others, then, are experiencing its upside. If employers are passing up your job application because they associate negative qualities with your name, then there are more jobs available for more appealing candidates. If, however, we do not have a word to describe these excess jobs, power dynamics are harder to discuss, much less intervene in. If you try this exercise today, your spellcheck is likely to recognize both words, which reminds us that it is possible to change technical systems so that they do not obscure or distort our understanding and experience of social systems. And, while this is a relatively simple update, we must make the same demand of more complex forms of coded inequity and tune into the socially proscribed forms of (in)visibility that structure their design.

If we look strictly at the technical features of, say, automated soap dispensers and predictive crime algorithms, we may be tempted to home in on their differences. When we consider the stakes, too, we might dismiss the former as relatively harmless, and even a distraction from the dangers posed by the latter. But rather than starting with these distinctions, perhaps there is something to be gained by putting them in the same frame to tease out possible relationships. For instance, the very idea of hygiene—cleaning one's hands and 'cleaning up' a neighbourhood—echoes a racialized vocabulary. Like the Beauty AI competition, many advertisements for soap conflate darker skin tones with unattractiveness and more specifically with dirtiness, as did an ad from the 1940s where a White child turns to a Black child and asks: 'Why doesn't your mama wash you with fairy soap?' Or another one, from 2017, where a Black woman changes into a White woman after using Dove soap. The idea of hygiene, in other words, has been consistently racialized, all the way from marketing to public policy. In fact the most common euphemism for eugenics was 'racial hygiene': ridding the body politic of unwanted populations would be akin to ridding the body of unwanted germs. Nowadays we often associate racial hygienists with the Nazi holocaust, but many early proponents were the American progressives who understood eugenics to work as a social uplift and a form of Americanization. The ancient Greek etymon, *eugeneia* (ευ' γένεια), meant 'good birth', and this etymological association should remind us how promises of goodness often hide harmful practices. As Margaret Atwood writes:

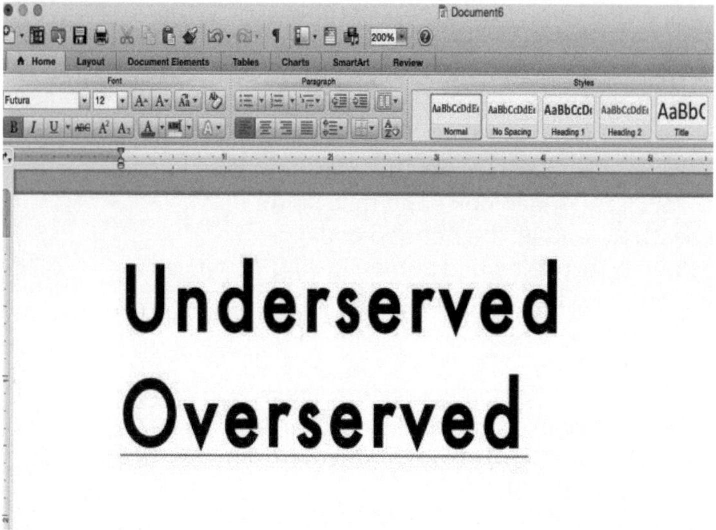


Fig.3 Overserved

'Better never means better for everyone ... It always means worse, for some.'

Take a seemingly mundane tool for enforcing segregation—separate water fountains—which is now an iconic symbol for the larger system of Jim Crow. In isolation from the broader context of racial classification and political oppression, a 'coloured' water fountain could be considered trivial, though in many cases the path from segregated public facilities to routine public lynching was not very long. Similarly, it is tempting to view a 'Whites only' soap dispenser as a trivial inconvenience. In a viral video of two individuals, White and Black, who show that their hotel soap dispenser does not work for the latter, they are giggling as they expose the problem. But when we situate in a broader racial context what appears to be an innocent oversight, the path from restroom to courtroom might be shorter than we expect.

That said, there is a straightforward explanation when it comes to the soap dispenser: near infrared technology requires light to bounce back from the user and activate the sensor, so skin with more melanin, absorbing as it does more light, does not trigger the sensor. But this strictly technical account says nothing about why this particular sensor mechanism was used, whether there are other options that recognize a broader spectrum of skin tones, and how this problem was overlooked during development and testing, well before the dispenser was installed. Like segregated water fountains of a previous era, the discriminatory soap dispenser offers a window onto a wider social terrain. As the soap dispenser is, technically, a robot, this discussion helps us consider the racism of robots and the social world in which they are designed.

For instance, we might reflect upon the fact that the infrared technology of an automated soap dispenser treats certain skin tones as normative and upon the reason why this technology renders Black people invisible when they hope to be seen, while other technologies, for example facial recognition for police surveillance, make them hypervisible when they seek privacy. When we draw different technologies into the same frame, the distinction between 'trivial' and 'consequential' breaks down and we can begin to understand how Blackness can be both marginal and focal to tech development. For this reason I suggest that we hold off on drawing too many bright lines—good versus bad, intended versus unwitting, trivial versus consequential. Sara Wachter-Boettcher, the author of *Technically Wrong*, puts it thus: 'If tech companies can't get the basics right ... why should we trust them to provide solutions to massive societal problems?'[54] The issue is not simply that innovation and inequity can go hand in hand but that a view of technology as value-free means that we are less likely to question the New Jim Code in the same way we would the unjust laws of a previous era, assuming in the process that our hands are clean.

## Engineered Inequity

In one of my favourite episodes of the TV show *Black Mirror*, we enter a world structured by an elaborate social credit system that shapes every encounter, from buying a coffee to getting a home loan. Every interaction ends with people awarding points to one another through an app on their phones; but not all the points are created equal. Titled 'Nosedive', the episode follows the emotional and social spiral of the main protagonist, Lacie, as she pursues the higher rank she needs in order to qualify for an apartment in a fancy new housing development. When Lacie goes to meet with a points coach to find out her options, he tells her that the only way to increase her rank in such a short time is to get 'up votes from quality people. Impress those upscale folks, you'll gain velocity on your arc and there's your boost.' Lacie's routine of exchanging five stars with service workers and other 'mid- to low-range folks' won't cut it if she wants to improve her score quickly. As the title of the series suggests, *Black Mirror* offers a vivid refection on the social dimensions of technology—where we *are* and where we might be going with just a few more clicks in the same direction. And, although the racialized dimensions are not often made very explicit, there is a scene towards the beginning of the episode when Lacie notices all her co-workers conspiring to purposely lower the ranking of a Black colleague and forcing him into a subservient position as he tries to win back their esteem ... an explicit illustration of the New Jim Code.

When it comes to engineered inequity, there are many different types of 'social credit' programs in various phases of prototype and implementation that are used for scoring and ranking populations in ways that reproduce and even amplify existing social hierarchies. Many of these come wrapped in the packaging of progress. And, while the idiom of the New Jim Code draws on the history of racial domination in the United States as a touchstone for technologically mediated injustice, our focus must necessarily reach beyond national borders and trouble the notion that racial discrimination is isolated and limited to one country, when a whole host of cross-cutting social ideologies make that impossible.

Already being implemented, China's social credit system is an exemplar of explicit ranking with far-reaching consequences. What's more, *Black Mirror* is referenced in many of the news reports of China's experiment, which started in 2014, with the State Council announcing its plans to develop a way to score the trustworthiness of citizens. The government system, which will require mandatory enrolment starting from 2020, builds on rating schemes currently used by private companies.

Using proprietary algorithms, these apps track not only financial history, for instance whether someone pays his bills on time or repays her loans, but also many other variables, such as one's educational, work and criminal history. As they track all one's purchases, donations and leisure activities, something like too much time spent playing video games marks the person as 'idle' (for which points may be docked), whereas an activity like buying diapers suggests that one is 'responsible'. As one observer put it: 'The system not only investigates behaviour—it shapes it. It "nudges" citizens away from purchases and behaviours the government does not like.'[55] Most alarmingly (as this relates directly to the New Jim Code), residents of China's Xinjiang, a predominantly Muslim province, are already being forced to download an app that aims to track 'terrorist and illegal content'.

Lest we be tempted to think that engineered inequity is a problem 'over there', just recall Donald Trump's idea to register all Muslims in the United States on an electronic database—not to mention companies like Facebook, Google and Instagram, which already collect the type of data employed in China's social credit system. Facebook has even patented a scoring system, though it hedges when asked whether it will ever develop it further. Even as distinct histories, politics and social hierarchies shape the specific convergence of innovation and inequity in different contexts, it is common to observe, across this variation, a similar deployment of buzzwords, platitudes and promises.

What sets China apart (for now) is that all those tracked behaviours are already being rated and folded into a 'citizen score' that opens or shuts doors, depending on one's ranking.[56] People are given low marks for political misdeeds such as 'spreading rumours' about government officials, for financial misdeeds such as failing to pay a court fine, or social misdeeds such as spending too much time playing video games. A low score brings on a number of penalties and restrictions, barring people from opportunities such as a job or a mortgage and prohibiting certain purchases, for example plane tickets or train passes.[57] The chief executive of one of the companies that pioneered the scoring system says that it 'will ensure that the bad people in society don't have a place to go, while good people can move freely and without obstruction'.[58]

Indeed, it is not only the desire to move freely, but all the additional privileges that come with a higher score that make it so alluring: faster service, VIP access, no deposits on rentals and hotels—not to mention the admiration of friends and colleagues. Like so many other technological lures, systems that seem to objectively rank people on the basis of merit and things we like, such as trustworthiness, invoke 'efficiency' and 'progress' as the lingua franca of innovation. China's policy states: 'It will forge a public opinion environment where keeping trust is glorious. It will strengthen sincerity in government affairs, commercial sincerity, social sincerity and the construction of judicial credibility.'[59] In fact, higher scores have become a new status symbol, even as low scorers are a digital underclass who may, we are told, have an opportunity to climb their way out of the algorithmic gutter.

Even the quality of people in one's network can affect your score —a bizarre scenario that has found its way onto TV shows like *Black Mirror* and *Community*, where even the most fleeting interpersonal interactions produce individual star ratings, thumbs up and down, giving rise to digital elites and subordinates. As Zeynep Tufekci explains, the ubiquitous incitement to 'like' content on Facebook is designed to accommodate the desires of marketers and works against the interests of protesters, who want to express dissent by 'disliking' particular content.[60] And, no matter how arbitrary or silly the credit (see 'meow meow beenz' in the TV series *Community*), precisely because people and the state invest it with import, the system carries serious consequences for one's quality of life, until finally the pursuit of status spins out of control.

The phenomenon of measuring individuals not only by their behaviour but by their networks takes the concept of social capital to a whole new level. In her work on marketplace lenders, sociologist Tamara K. Nopper considers how these companies help produce and rely on what she calls *digital character*—a 'profile assessed to make inferences regarding character in terms of credibility, reliability, industriousness, responsibility, morality, and relationship choices'.[61] Automated social credit systems make a broader principle of merit-based systems clear: scores assess a person's ability to conform to established definitions of good behaviour and valued sociality rather than measuring any intrinsic quality. More importantly, the ideological commitments of dominant groups typically determine what gets awarded credit in the first place, automating social reproduction. This implicates not only race and ethnicity; depending on the fault lines of a given society, merit systems also codify class, caste, sex, gender, religion and disability oppression (among other factors). The point is that multiple axes of domination typically converge in a single code.

Take the credit associated with the aforementioned categories of playing video games and buying diapers. There are many ways to parse the values embedded in the distinction between the 'idle' and the 'responsible' citizen so that it lowers the scores of gamers and increases the scores of diaper changers. There is the ableist logic, which labels people who spend a lot of time at home as 'unproductive', whether they play video games or deal with a chronic illness; the conflation of economic productivity and upright citizenship is ubiquitous across many societies.

Consider, too, how gender norms are encoded in the value accorded to buying diapers, together with the presumption that parenthood varnishes (and, by extension, childlessness tarnishes) one's character. But one may wonder about the consequences of purchasing too many diapers. Does reproductive excess lower one's credit? Do assumptions about sex and morality, often fashioned by racist and classist views, shape the interpretation of having children and of purchasing diapers? In the United States, for instance, one could imagine the eugenic sensibility that stigmatizes Black women's fertility and celebrates White women's fecundity getting codified through a system that awards points for diapers purchased in suburban zip codes and deducts points for the same item when purchased in not yet gentrified parts of the city—the geography of social worth serving as a proxy for gendered racism and the New Jim Code. In these various scenarios, top-down reproductive policies could give way to a social credit system in which the consequences of low scores are so far-reaching that they could serve as a veritable digital birth control.

In a particularly poignant exchange toward the end of the 'Nosedive' episode, Lacie is hitchhiking her way to win the approval of an elite group of acquaintances; and motorists repeatedly pass her by on account of her low status. Even though she knows the reason for being disregarded, when a truck driver of even lower rank kindly offers to give her a ride, Lacie looks down her nose at the woman ('nosedive' indeed). She soon learns that the driver has purposefully opted out of the coercive point system and, as they make small talk, the trucker says that people assume that, with such a low rank, she must be an 'antisocial maniac'. Lacie reassures the woman by saying you 'seem normal'. Finally, the trucker wonders about Lacie's fate: 'I mean you're a 2.8 but you don't *look* 2.8.' This moment is illuminating as to how abstract quantification gets embodied—that the difference between a 2.8 and a 4.0 kind of person should be self-evident and readable on the (sur)face. This is a key feature of racialization: we take arbitrary qualities (say, social score, or skin colour), imbue them with cultural importance, and then act as if they reflected natural qualities in people (and differences between them) that should be obvious just by looking at someone.[62]

In this way speculative fiction offers us a canvas for thinking about the racial vision that we take for granted in our day-to-day lives. The White protagonist, in this case, is barred from housing, transportation and relationships—a fictional experience that mirrors the forms of ethno-racial exclusions that many groups have actually experienced; and Lacie's low status, just like that of her real-life counterparts, is attributed to some intrinsic quality of her person rather than to the coded inequity that structures her social universe. The app, in this story, builds upon an already existing racial arithmetic, expanding the terms of exclusion to those whose Whiteness once sheltered them from harm. This is the subtext of so much science fiction: the anxiety that, if 'we' keep going down this ruinous road, then *we might be next*.

Ultimately the danger of the New Jim Code positioning is that existing social biases are reinforced—yes. But new methods of social control are produced as well. Does this mean that every form of technological prediction or personalization has racist effects? Not necessarily. It means that, whenever we hear the promises of tech being extolled, our antennae should pop up to question what all that hype of 'better, faster, fairer' might be hiding and making us ignore. And, when bias and inequity come to light, 'lack of intention' to harm is not a viable alibi. One cannot reap the reward when things go right but downplay responsibility when they go wrong.

Notes

1 Visit Beauty.AI First Beauty Contest Judged by Robots, at http://beauty.ai.
2 2 Jordan Pearson, 'Why an AI-judged Beauty Contest Picked Nearly All White Winners', *Motherboard*, 5 September 2016.
3 Ibid.
4 Sam Levin, 'A Beauty Contest Was Judged By AI and the Robots Didn't Like Dark Skin', *The Guardian*, 8 September 2016.
5 Both Harcourt quotations are from: ibid.
6 See: http://beauty.ai.
7 See: https://machinelearningmastery.com/what-is-deep-learning.
8 Cade Metz, 'Facebook's Deep Learning Guru Reveals the Secret of AI', *Wired*, 12 December 2013.
9 Field note, Jack Clark's Keynote Address at the Princeton University AI and Ethics Conference, 10 March 2018.
10 The flip side of personalization is what Eubanks refers to as an 'empathy override'. Virginia Eubanks, *Automating Inequality: How High-Tech Tools Profile, Police, and Punish the Poor* (New York: St. Martin's Press, 2018); see also: Alyssa Edes, '"Automating Inequality": Algorithms in Public Services Often Fail the Most Vulnerable', *NPR*, 19 February 2018.
11 Cybelle Fox, *Three Worlds of Relief: Race, Immigration, and the American Welfare State from the Progressive Era to the New Deal* (Princeton, NJ: Princeton University Press, 2012).
12 'Homelessness Is Not a Systems Engineering Problem, It's a Carpentry Problem.' Eubanks, *Automating Inequality*, op. cit. (note 10), 125.
13 The term 'uncanny valley' was coined by Masahiro Mori in 1970 and translated into English by Reichardt. Jasia Reichardt, *Robots: Fact, Fiction, and Prediction* (Middlesex: Penguin Books,1978).
14 But it is worth keeping in mind that many things dubbed 'AI' today are, basically, just statistical predictions rebranded in the age of big data—an artificial makeover that engenders more trust as a result. This point was made by Arvind Narayanan in response to a Microsoft case study at a workshop sponsored by the Princeton University Center for Human Values and Center for Informational Technology Policy, 6 October 2017.
15 Elly R. Truitt, *Medieval Robots: Mechanism, Magic, Nature, and Art* (Philadelphia: University of Pennsylvania Press, 2015).
16 Sarah Richardson and Hallam Stevens (eds.), *Postgenomics: Perspectives on Biology after the Genome* (Durham, NC: Duke University Press, 2015), 5.
17 Ibid., 2.
18 As Imani Perry explains: 'Mary Shelley's Frankenstein provided a literary example of the domestic anxiety regarding slavery and colonialism that resulted from this structure of relations. ... Frankenstein's monster represented the fear of the monstrous products that threatened to flow from the peculiar institutions. The novel lends itself to being read as a response to slave revolts across the Atlantic world. But it can also be read as simply part of anxiety attendant to a brutal and intimate domination, one in which the impenetrability of the enslaved was already threatening.' Imani Perry, *Vexy Thing: On Gender and Liberation* (Durham, NC: Duke University Press, 2018), 49.
19 Richardson and Stevens, *Postgenomics*, op. cit. (note 16), 2.
20 Anna Everett, *Digital Diaspora: A Race for Cyberspace* (Albany, NY: SUNY Press, 2009), 1.
21 O.O. Binder, 'You'll Own "Slaves" by 1965', *Mechanix Illustrated* (January 1957).
22 These passages come from a PoliceOne report that cautions us: 'As wonderful an asset as they are, they cannot provide a complete picture. The camera eye can only see so much, and there are many critical elements of information that may go undiscovered or unrecognized ... Throwable robots provide such an advance in situational awareness that it can be easy to forget that our under- standing of the situation is still incomplete', visit: https://www.policeone.com/police-products/police-technology/robots/articles/320406006–5-tactical-considerations-for-throwable-robot-deployment.
23 Amelie O. Rorty, 'Slaves and Machines', *Analysis* 22 (1962), 118–120.
24 Jessie Daniels, 'My Brain-Database Doesn't See Skin Color: Colorblind Racism in the Tech Industry and in Theorizing the Web', *American Behavioral Scientist* 59/11 (2015), 1379; see also: Marion Crain et al., *Invisible Labor: Hidden Work in the Contemporary World* (Berkeley: University of California Press, 2016); Radhika Gajjala, *Cyber Selves: Feminist Ethnographies of South Asian Women* (Walnut Creek, CA: Alta Mira Press, 2004); K.J. Hossfeld, '"Their Logic against Them": Contradictions in Sex, Race, and Class in Silicon Valley', in: Alondra Nelson, Thuy Linh N. Tu and Alicia Hedlam Hines (eds.), *Technicolor: Race, Technology and Everyday Life* (New York: NYU Press, 1990), 34–63; Stephen J. Pitti, *The Devil in Silicon Valley: Northern California, Race, and Mexican Americans* (Princeton, NJ: Princeton University Press, 2004); Johanna Shih, 'Circumventing Discrimination: Gender and Ethnic Strategies in Silicon Valley', *Gender & Society* 20/2 (2006), 177–206.
25 Lisa Nakamura, *Cybertypes: Race, Ethnicity, and Identity on the Internet* (London: Routledge, 2002), 24.
26 Jessie Daniels, 'Race and Racism in Internet Studies: A Review and Critique', *New Media & Society* 15/5 (2012), 679.
27 Safiya Noble and Brendesha Tynes (eds.), *The Intersectional Internet: Race, Sex, Class and Culture Online* (New York: Peter Lang, 2016).
28 Field note from the Princeton University Center for Human Values and Center for Informational Technology Policy Workshop, 6 October 2017.
29 The notion of 'racist robots' is typically employed in popular discourse around AI. I use it as a rhetorical device to open up a discussion about a range of contemporary technologies, most of which are not human-like automata of the kind depicted in films and novels. They include forms of automation integrated in everyday life, like soap dispensers and search engines, bureaucratic interventions that seek to make work more efficient, as in policing and healthcare, and fantastical innovations first imagined in science fiction, such as self-driving cars and crime prediction techniques.
30 John McWhorter, '"Racist" Technology Is a Bug—Not a Crime', *Time*, 12 September 2016.
31 Field note from the Princeton University Center for Human Values and Center for Informational Technology Policy Workshop, 6 October 2017.
32 The famed android Lieutenant Commander Data of the hit series Star Trek understood well the distinction between inputs and outputs, intent and action. When a roughish captain of a small cargo ship inquired whether Data had ever experienced love, Data responded: 'The act or the emotion?' And when the captain replied that they're both the same, Data rejoined: 'I believe that statement to be inaccurate, sir.' Just as loving behavior does not require gushing Valentine's Day sentiment, so too can discriminatory action be fueled by indifference and disregard, and even by good intention, more than by flaming hatred.
33 James Baldwin, *James Baldwin: Collected Essays* (New York: Library of America, 1998), 129.
34 See: https://www.nclc.org/images/pdf/credit_discrimina tion/InsuranceScoring-WhitePaper.pdf.
35 Policeone.com, at https://www.policeone.com/police-products/police-technology/robots.
36 This is brought to life in the 2016 HBO series Silicon Valley, which follows a young Steve Jobs-type of character, in a parody of the tech industry. In a segment at TechCrunch, a conference where start-up companies present their proof of concept to attract venture capital investment, one presenter after another exclaims 'we're making the world a better place' with each new product that also claims to 'revolutionize' some corner of the industry. See: https://longreads.com/2016/06/13/silicon-valley-masterfully-skewers-tech-culture.
37 McWhorter, '"Racist" Technology Is a Bug', op. cit. (note 30).
38 Sociologist Eduardo Bonilla-Silva argues that 'if racism is systemic, this view of "good" and "bad" whites distorts reality'. Eduardo Bonilla-Silva, *Racism without Racists: Colorblind Racism and the Persistence of Racial Inequality in the United States* (New York: Rowman & Littlefield, 2006), 132). He quotes Albert Memmi saying: 'There is a strange enigma associated with the problem of racism. No one, or almost no one, wishes to see themselves as racist; still, racism persists, real and tenacious.' Ibid., 1.
39 Grace Dobush, 'White Men Dominate Silicon Valley Not by Accident, But by Design', *Quartz*, 16 March 2016.
40 Perry explains how racial surveillance does not require a 'bogeyman behind the curtain; it is a practice that emerges from our history, conflicts, the interests of capital, and political expediency in the nation and the world... Nowhere is the diffuse and individuated nature of this practice more apparent than in the fact that overpolicing is not limited to White officers but is instead systemic.' Imani Perry, *More Beautiful and More Terrible: The Embrace and Transcendence of Racial Inequality in the United States* (New York: NYU Press, 2011), 105.
41 Calling for a post-intentional analysis of racism, Perry argues that intent is not a good measure of discrimination because it 'creates a line of distinction between "racist" and "acceptable" that is deceptively clear in the midst of a landscape that is, generally

speaking, quite unclear about what racism and racial bias are, who [or what] is engaging in racist behaviors, and how they are doing so'. Ibid., 21.

42 Zach Schonbrun, 'Tarnished by Charlottesville, Tiki Torch Company tries to move On', *The New York Times*, 20 August 2017.

43 Field note from the Princeton University Center for Human Values and Center for Informational Technology Policy Workshop, 6 October 2017.

44 Ibid.

45 Richardson and Stevens, *Postgenomics*, op. cit. (note 16), 12.

46 Ibid.; see also: Stefan Helmreich, *Silicon Second Nature: Culturing Artificial Life in a Digital World* (Berkeley: University of California Press, 1998).

47 See: s.v. 'stereotype' in the *Online Etymology Dictionary* at https://www.etymonline.com/word/stereotype.

48 'It is to say, though, that all those inhabiting subject positions of racial power and domination—notably those who are racially White in its various formulations in different racially articulated societies—project and extend racist socialities by default. But the default is not the only position to occupy or in which to invest. One remains with the default because it is given, the easier to inhabit, the sociality of thoughtlessness.' David T. Goldberg, *Are We All Postracial Yet?* (Cambridge: Polity, 2015), 159–160.

49 Zeynep Tufekci, 'Algorithmic Harms beyond Facebook and Google: Emergent Challenges of Computational Agency Symposium Essays', *Colorado Technology Law* 13 (2015), 203–218.

50 Donna Haraway, *Simians, Cyborgs and Women: The Reinvention of Nature* (New York: Routledge, 1991), 164.

51 Ibid.

52 This potential explains the name of the provocative TV series *Black Mirror*.

53 According to Feagin and Elias, systemic racism refers to 'the foundational, large-scale and inescapable hierarchical system of US racial oppression devised and maintained by whites and directed at people of colour ... [It] is foundational to and engineered into its major institutions and organizations.' Joe Feagin and Sean Elias, 'Rethinking Racial Formation Theory: A Systemic Racism Critique', *Ethnic and Racial Studies* 36/6 (2013), 936.

54 Sara Wachter-Boettcher, *Technically Wrong: Sexist Apps, Biased Algorithms, and Other Threats of Toxic Tech* (New York: W.W. Norton, 2017), 200. On the same page, the author also argues: 'We'll only be successful in ridding tech of excesses and oversights if we first embrace a new way of seeing the digital tools we rely on—not as a wonder, or even as a villain, but rather as a series of choices that designers and technologists have made. Many of them small: what a button says, where a data set comes from. But each of these choices reinforces beliefs about the world, and the people in it.'

55 Rachel Botsman, 'Big Data Meets Big Brother as China Moves to Rate Its Citizens,' *Wired UK*, 21 October 2017.

56 Clinton Nguyen, 'China Might Use Data to Create a Score for Each Citizen Based on How Trustworthy They Are', *Business Insider*, 26 October 2016.

57 David Z. Morris, 'China Will Block Travel for Those with Bad "Social Credit"', *Fortune*, 18 March 2018.

58 State Council ,'Planning Outline for the Construction of a Social Credit System (2014–2020)', China Copyright and Media, 2014, https://chinacopyrightandmedia.wordpress.com/2014/06/14/planning-outline-for-the-construction-of-a-social-credit-system-2014-2020.

59 Ibid.

60 Zeynep Tufekci, *Twitter and Tear Gas: The Power and Fragility of Networked Protest* (New Haven, CT: Yale University Press, 2017), 128.

61 Tamara K. Nopper, 'Digital Character in "the Scored Society": FICO, Social Networks, and Competing Measurements of Creditworthiness', in: Ruha Benjamin (ed.), *Captivating Technology: Race, Carceral Technoscience, and Liberatory Imagination in Everyday Life* (Durham, NC: Duke University Press, 2019), 170.

62 Ian Hacking, 'Kinds of People: Moving Targets', *Proceedings of the British Academy*, vol. 151: 2006 Lectures (Oxford: Oxford University Press, 2007), 285–317.

# FROM DATA

## FROM DATA CENTRE TO DATA FOREST

Marina Otero Verzier

Imagine walking through a park. A park that is actually a library. Where every flower, plant and shrub is a data archive. As you sit down on a bench and touch a leaf with your portable reader, you start listening to the Rolling Stones, the music coming directly from the information stored in the plant's DNA. Later, amid the greenery, you read a novel or watch a documentary.[1]

### Cartesian Enclosures

Envisioning futures is not as easy as it might look. The 'defuturing' processes (as coined by design theorist Tony Fry) and the slow cancellation of the future (as argued by author Mark Fisher) of capitalist patriarchal modernity have created an unliveable world for humans and non-humans.[2] Despite the fact that other paradigms are urgently needed, however, too many of us have lost our ability to imagine alternatives. We are subjected to an appalling sense of exhaustion and finitude and too often driven by inertia.

That is not the case of Karin Ljubič Fister, a doctor and genetic researcher from the University of Maribor (Slovenia), whose curiosity for storing binary data in the DNA of plants and seeds engendered a radical vision for data centres. According to Fister, by containing information—be it an image, a song, a book, a drawing—in every cell of the host organism, billions of gigabytes of data could be archived for millennia and possibly beyond—and without the human. Despite unavoidable controversies surrounding biotechnology and genetic-engineering experiments, and the difficulties of materializing this alternative data storage, Fister's project allows architecture to be imagined differently. In particular, it opens up the possibility of a future for the architecture of data centres that goes beyond Cartesian logic.

As a philosophical and scientific system founded by René Descartes (1596–1650), Cartesianism developed around the prerogative of 'mind over matter'. By refusing to attribute a soul—reason—to animals and plants, Cartesians rendered humans unique beings, separated from the rest of nature and occupying a superior position in relation to others.[3] The human, and particularly the Western white man, was then constructed as a universal, rational subject, entitled to dominate the landscape, resources and other beings. This categorization of beings, and the compartmentalization and instrumentalization of human and non-human relations, subsequently enabled the discrimination of entities, bodies and identities for the benefit of that particular human.

Enmeshed in the space we continue to inhabit, Cartesian dualisms are at the core of the discipline of architecture. They define what I call the 'Cartesian enclosure', a controlled space, shaped by technological rationales, coalescing in unassuming architectures organized according to borders that neglect the surrounding environment. Supporting Western man's exceptionality, these architectures follow economic efficacy rationales and extractivist logics, often at the expense of ethical and ecological awareness.[4]

Data centres are one of the most recent exponents of the 'Cartesian enclosure'. A central piece of today's political, cultural, socioeconomic system, data centres mediate the creation, storage and transmission of digital records across physical networks. Their functioning, however, involves vast energy and water consumption, land occupation, large global carbon emissions, and is reliant on extractivist practices.[5] In Fister's imagination, though, data centres transfigure into a spatial and material entanglement difficult to describe under dual, Cartesian categories: a more-than-human architecture, non-contained by walls, non-dependent on fossil fuels, efficacy or productivity, and not propelled by the punitive work of 'the other'.

## DNA Data Storage

Even if easily categorized as science fiction, the plant-DNA architecture described by Fister is an alternative worth pursuing to tackle the unprecedented worldwide demand for data storage and its ecological footprint. As humans continue to generate and consume more and more data, the digital-data production is outpacing the scalability of today's storage solutions. According to International Data Corporation (IDC), data generated globally will grow by 23 per cent annually from 2020 to 2025, reaching 180 zettabytes in 2025.[6] The developments in AI, the Internet of Things and the Metaverse, which demand vast amounts of data to be stored for extended periods, will only exacerbate this condition. Gartner, in its report 'Market Trends: Evolving Enterprise Data Requirements', refers to a coming Zone of Potential Insufficiency and estimates that hyperscale vendor data-storage demand, which grew by nearly 35 per cent CARG (Compound Annual Growth Rate) from 2013 to 2019, could potentially soar to 50 per cent CARG from 2020 through 2030.[7]

Following these estimates, growth appears to be the only available option, even in the face of climate catastrophe. Growth, these reports strategically insist, is inevitable. And by disregarding the possibility of data production and consumption behaviours that reduce dependence on data centres, these statistics push governments and the industry to find new technological fixes rather than conceive other forms of living. In this context, data centre architecture has turned into a critical site of investment and innovation. Unsurprisingly, the largest commercial real estate services company in the world, CBRE, has recently launched a data centre division, while architecture and engineering firms such as Corgan, HDR, Gensler, AECOM and Arup are also expanding their operations.[8]

And yet, the architectural solutions for reinventing these facilities fall short to the point that the Metaverse cannot be delivered without a planetary meltdown. While most digital infrastructure providers experience difficulties coping with demand, environmental targets, governmental bans and regulations, industry conglomerates are taking the opportunity to invest in stable and scalable data-storing mediums not dependent on copious energy, water and land consumption. Molecular and biological computation are among the thriving methods praised as a solution to the shortcomings of current storage models. Their implementation is opening previously inconceivable (and convoluted) futures that reach beyond the existing architecture of data centres.

In Cambridge, Massachusetts, the Whitesides Research Group is experimenting with fluorescent dye storage.[9] Microsoft's Optics for the Cloud group at the Microsoft Research Lab in the UK recently launched Project HSD, which focuses on hologram data storage, and Project Silica, which studies the capacity of crystals to deliver long-term archival storage.[10] Companies, including Illumina, Microsoft and Twist, have created the DNA Data Storage Alliance to develop DNA-based data storage solutions in California. They propose DNA data as a long-term, reliable and ecological alternative to tape drives or HDDs.[11] Professor George Church is carrying out similar work at Harvard's Wyss Institute in Cambridge.[12]

In 2013, building on these studies yet embracing a different ethos, Karin Fister and Iztok Fister Jr conducted tests using plant-based DNA as a storage medium. Their experiments instigated the image of the data forest we walked through at the beginning of this essay. As Karin explained in an interview with *New Scientist*, the idea was born out of frustration with the continuous lack of disc space on her computer and the determination to find an alternative system.[13] Aware of the compatibility of genetic programming and binary code, she wondered if plant DNA could also be used as a storage medium. Together with colleagues at the university, Fister inserted a basic computer program called Hello World into a tobacco plant's DNA. The sequence of 0s and 1s was translated into artificial DNA by turning 00 into A, 10 into C, 01 into G and 11 into T. Upon synthesizing the resulting DNA sequence, the team transferred the DNA into a bacterium with which they infected one of the plant's leaves. They later took a fragment of the infected leaf, planted it, and grew a tobacco plant whose leaves, seeds and future offspring contained the Hello World program encoded in their DNA.

The research success encouraged Karin Fister and Iztok Fister Jr to publish the paper 'How to Store Wikipedia into a Forest Tree: Initial Idea' (2014). By storing information in the DNA of plants and seeds, the Fisters proposed creating off-the-grid, living Wiki forests that capture CO2 and produce oxygen.[14] The project had revolutionary potential. Yet, while the encoding process was successful, the data stored in the DNA was read-only, preventing users from editing entries.[15] Proving the system was better suited to archiving did not impede the Fisters' ambition. Rather the opposite. 'All of the archives in the world could be stored in one box of seeds,' Karin stressed.[16] And, just like that, by scaling up the operation from a box of seeds to a collection of them (similar to the Svalbard Global Seed Vault), the Fisters envisioned no less than an architecture in which to keep all the world's information.[17]

Still, understanding seeds as storing drives and plants as data archives implied the possibility of data retrieval. According to Fister, the extraction of the plant's DNA and sequence could be carried out using standard methods and specialized botanical equipment, although it would involve the destruction of some leaf material. As DNA encoding and retrieval technologies evolve, Fister anticipated, they could decrease costs and production time and avoid the living organism's damage during the process. In the meantime, until such technologies are widely available, the seeds-drive could be frozen and have their germination ability preserved for millennia.

### Beyond Technological Fixes

Fister's work has important implications for the relations between data storage and the environment. Whereas their latest proposal still considers architecture as a stable container—a seed vault—it unleashes the architectural imagination, making it possible to conceive of alternative spatial, non-Cartesian worldmaking models. There are troubles in paradise, however. Whereas DNA data storage solutions may challenge aspects of the Cartesian enclosure, the system still requires the modification of living organisms. Commercially driven applications often treat plants and, in particular, their DNA as a resource to be extracted and unproblematically subjected to human utilitarian needs, triggering ethical concerns about the exploitation of beings on the grounds of human's benefit and exceptionalism (which, as we have seen, lies at the core of Cartesianism).

At the same time, the narratives surrounding this technology's market application assume the inevitability of growth and the role of the planet as a resource to meet human's compulsive and devastating desires. They propose, as researchers and scholars Madeleine Mendell, Mél Hogan and Deb Verhoeven have argued, a frictionless storage medium for a limitless accumulation, possibly beyond the human: while most forms of life on the planet may be wiped out due to climate catastrophe, human information will survive.[18]

Rather than attempting to address both the need for storage and the environmental collapse by adopting alternative data-production and storage practices, scientists invest in technological fixes imbricated in apocalyptic thinking.[19] Fixes such as genomics research and synthetic DNA that, as Hogan reminds us, depend on a profit-driven system and a global communications infrastructure and computational power reliant on exploitative logics: cheap labour, extraction of natural resources, and proprietary software and hardware. Driven by a fear of extinction and loss, these scientific, market-driven remedies hold an empty promise of keeping information unaffected by the passing of time. But they fail to acknowledge that any archival holding is continually renewed through practices of maintenance, that meanings are transformed as they are continually performed and that remembering simultaneously concerns knowledge generation and its loss.

By contrast, non-industry examples conceived by Karin Ljubic Fister and others offer generative ideas based on scientific realism and speculative design that account for the environmental impact of DNA-based infrastructures, but also the poetics and politics of storing and preserving.[20] They prompt world makings that understand data storages as archives enmeshed in states of proliferation and decay. Preservation processes that recognize memory not as what is contained and stored, but as a dynamic practice that involves acts of remembering and forgetting.[21] Through states of gradual decomposition and recomposition, what is archived generates future meanings that exceed the intentions and interpretations of those who deposited and contained it.

The data garden is one of these futures. Conceived by Grow Your Own Cloud (GYOC), a duo formed by Monika Seyfried and Cyrus Clarke, the data garden transcends the notion of a repository of seeds and proposes a dynamic system.[22] One made of plants carrying our more precious memories. One that turns plant caring in our domestic spaces, at a community flower shop, in a public garden or a forest, into a collective and more-than-human memory-making practice. And, perhaps more importantly, one that challenges architecture's Cartesian postulates and their imperative of rationalization. Instead of the conventional stable form of the black box storage, the resulting data garden is an information ecosystem that transforms solar energy into living matter, captures CO2 and creates oxygen. This architectural assemblage of space, data and organisms, grows unpredictable environments, structures and relations that surpass discrete forms of being, their meanings and knowledge.

In considering plant DNA data storage as a generative idea, however, ethical questions arise. What is at stake when we encode human data in the DNA of a plant? How can an ethical human-vegetal engagement be established that also speaks to climate justice? And which theoretical framework would help define the resulting knowledge from the melding of human and plant thinking?

## Plant Intelligence and Ethical Conundrums

As professor of philosophy Michael Marder argues, research on plant intelligence carried out over the past three decades questioned traditional understandings of vegetal life and, with them, previously held moral stances.[23] Critical plant studies have unleashed theoretical arguments and experimental evidence on the state of plant existence that challenges Cartesian postulates, destabilizes Western philosophy's treatment of plants and supports the idea of plant intelligence.

Marder is one of those arguing for the recognition of the multidimensional intelligence of plants. The imageless and non-representational material memories stored in the plant's cells are, according to him, constituents of the 'non-conscious intentionality'.[24] A kind of 'thinking before thinking', evident in plants striving towards the sun, their roots seeking nutrients in the soil and other actions that demonstrate their creative interaction with their environment. This intentionality in plants demands human's acceptance of 'the ontological fact that they exist neither for human enjoyment and consumption, nor for the sake of anything or anyone'. An acknowledgment that could bring about a renewed and reciprocal relation between humans and plants. A relation receptive to plant's otherness and, as such, not organized around the benefit of the human nor the calculative rationality of capitalism, but based on solidarity and cohabitation.[25]

In the context of the data forest and data garden, to counter extractivist logics implies, following

Marder's argument, resisting its conceptualization around instrumental rationality and, instead, establishing a human-plant creative relation of knowledge co-production and communication. Rather than controlling and dominating plant behaviour to preserve information and treating plants as resources to be exploited, humans should respect their ontological freedom and let vegetal beings be. This 'letting be' implies preserving their wilderness—vegetal wilderness as a category constructed in opposition to plant cultivation and genetic engineering for the sake of efficiency and profit.[26] For it is precisely the appeal of monoculture (and the selection of plant species on the basis of their DNA-storing capacities) for optimizing the system that the project ethos contests. In addition, 'letting be' also demands preserving plants' existential potentialities, including their 'reproductive capacity and, hence, futurity'.[27] Whereas the deployment of a plant's genetic modification has been a common design practice resulting in patented seeds with a very limited lifespan—ensuring continuous demand and hence revenues for the corporations holding their patents—the data forest and data garden aim to support non-proprietary futures for plants and other beings on the planet.

Having reached this point, we should also scrutinize the possible effects of encoding data on plants. A site of material memory, the body of plants is a receptive medium that channels energy and registers an incredible biological hospitality.[28] It is, as is visible in the rings of a tree trunk, an embodied spatiotemporal archive of their vegetal memory and the physical stimuli that have affected them.[29] For the subjectivity of plants is distributed throughout their bodies, from the roots to the leaves. Might the encoded data therefore radically transform their way of being?

Plants have, in fact, an extraordinary capacity for regeneration and for making use of that which otherwise could be a threat. The fact that they exhibit 'growth by increments, quantitative additions, or reiterations of already existing parts' makes plants resilient against external factors and alterations, but will inevitably also affect the stability of the information stored.[30] While these processes of information altering render plant DNA storage unsecure for commercial enterprises that prefer, instead, to focus on ungerminated seeds as storage medium, the relevance of the forest data relies precisely on coming to terms with the prospect of plants exceeding the frameworks we set for them. That is, recognizing and celebrating their autonomy and embracing a human-plant communication beyond commodification and the mere circulation of information through digital bytes.

As plants germinate, grow, multiply and even repair their DNA without human aid, encoded human data would not stay unchanged. Errors and mutations would alter the code over time, leaving room for unexpected developments—entanglements of matter and data that resist utilitarian motivations and surpass and exceed human intentions and memories. Even more, by resorting to the Socratic analogy between bodily generation and the birthing of ideas, it is possible to argue, as Marder does, that these developments and material reproduction happening in the plant's body are a prototype of thought. An intelligence that transcends human's narrow conceptions of thinking and produces new knowledges and meaningful material enmeshments. Resulting vibrant entanglements and, why not, aberrations, transcend any purpose apparent to humans without necessarily making them meaningless or irrelevant.

In the data forest, plants won't only keep, but also rewrite human knowledge, opening cracks in the Cartesian logic and its historical living beings' divide.[31]

## The Annihilation of the Line

What, we should ask, does this mean for the discourse and practice of architecture? A grid of computer server racks, data halls and high-density PODs generally organize the architecture of the data centre. Systems controlling temperature, humidity and dust ensure their smooth operation. Fibreoptic cables connect data centres with users at the speed of light across the entangled geographies of mining sites, factories, ports, switch-points, mobile telephone towers and the space of everyday life.[32] Yet, despite their enmeshment with multiple sites, scales and times, most data centres are conceived as rectangular black boxes.

Whereas the environment is in a constant state of emergence, the discipline of architecture has been largely founded on the model of solids and the Cartesian paradigm. The Western architectural canon is organized according to solids geometry and the Cartesian grid. The latter, a coordinate system and seemingly neutral method for categorization based on X, Y and Z axes, and one of Descartes's most important legacies, enabled the rationalization of space in order to visualize, calculate, draw, optimize, replicate, standardize and, ultimately, control it.[33] The Cartesian grid organizes the Cartesian space, where humans position themselves as separate from the environment, often in a hierarchical, antagonistic position. A position from which to seek their survival and dominion over the medium at any cost, even at the expense of the entire planet.

If the discipline of architecture asserts the separation of human and non-human, plant-thinking implies, in contrast, an entangled, embedded, non-appropriative, non-dominant relation between living organisms and the environment. Plants are, according to Marder, 'guarantors of environmental justice', as 'the vegetal *it thinks* will moderate the lethal tendencies of the human *I think*'. Plant-thinking disregards metaphysical distinctions between plants, animals and human beings, and goes beyond the fictitious enclosure of a reified and self-sufficient identity. In doing so, plant-thinking helps to imagine a data centre that is more than a repository of information. It allows us to see it as an architecture of evolving forms of materiality in time and space, where plants are agents responsible for material changes (including modification, decomposition, disintegration), where not all information is extractable textual or graphic human knowledge, where loss is an inevitable condition for transformation and forgetting is inseparable from remembering.

Marder's unpacking of plant-thinking shows the way for architecture to be conceived otherwise. So do Marisol de la Cadena's theories on indigenous political strategies. Equally not bound by binary logic and Cartesian categories, De la Cadena's work invites us to slow down our practice of knowing the world and the organisms that constitute it,[34] to embrace worlds embodying immeasurability and mutual difference. By suspending the compartmentalization of the world and exceeding the categories that we mobilize to know it, De la Cadena explains, we could instigate worlding practices and futures based on mutual care, empathy, life-in-common. In embracing the spaces and ways of being of the 'between', and the 'be with', plants are plants *but not only*. Humans are humans *but not only*.

Marder and De la Cadena's theories help to muddle the lines that have long fortified the Cartesian divide; lines blurred in the architectures proposed by Fister and the GYOC that spark the imagination of what the architecture of a data centre, a library, an archive, a museum or a school could be. Despite many unsolved questions regarding the data forest and the set of rituals and power relations that might take place there (ownership, political context, status and provenance of the encoded information, conditions under which the information is accessed and decoded, etcetera) it provides an image and metaphor of a non-Cartesian architectural space. That is, an architecture inseparable from its environment, where the building and the documents stored are entangled in a spatial system that germinates, blooms, flourishes. A contingent, temporary assemblage of living matter. A place shared by all living beings and not defined by distinctions, categories and hierarchies, but by trans-species interactions. A space of inhabitation and knowledge-sharing not based on containment, not defined by walls. A spatial system that grows slowly and opens previously inconceivable futures. That is orchestrated to serve the many and is the result of a multifarious agency instead of a discrete set of designers. That exists in continuity with the environment. That is both alive and gives life. That makes breath possible. An architecture that breathes and is breathed. That changes our affective relation to others and the world.

Imagine walking through a forest. A forest that is actually a library, a school, a museum. Where every tree, flower, plant and shrub is an archive and an architecture of generative flourishing. As you stop amid the greenery and touch a leaf, you attune to previously unimageable worlds. Worlds far away from the myths of immortality. Far from the defuturing effects of the Cartesian enclosure, and the false promises of endless growth.

Notes

1 Sean O'Neill, 'Imagine Storing All the World's Archives in a Box of Seeds', *New Scientist* (13 January 2016), https://www.newscientist.com/article/mg22930560-400-imagine-storing-all-the-worlds-archives-in-a-box-of-seeds/#ixzz7PVJsrePH.
2 Tony Fry, *Defuturing: A New Design Philosophy* (London: Bloomsbury Visual Arts, 2020). See also: Arturo Escobar, *Designs for the Pluriverse: Radical Interdependence, Autonomy, and the Making of Words* (Durham, NC: Duke University Press, 2018), 16–17; philosopher Franco 'Bifo' Berardi coined the phrase 'the slow cancellation of the future' in his book *After the Future* (2011) to describe the postmodern condition. Cultural theorist Mark Fisher mobilized this notion to explain the deflation of expectations as a result of neoliberal capitalism and the inability of contemporary cultural production to came up with anything genuinely new, a symptom of the impossibility of imagining a better future or a world beyond our own.
3 Cartesianism, its imperative of rationalization and theory of the animal-machine, drove the mechanical age and the formation of Western capitalism. Influenced by the technical creations of the early seventeenth century, its founder René Descartes (1596–1650) referred to machines as models to explain the functioning of organisms in what came to be known as the Cartesian theory of the animal-machine. Departing from the observation of the parallels between animal movements and automatic mechanical movements (early machines were necessarily powered by humans or animals), Cartesians attempted to explain physical and biological phenomena solely by technical models. See also: Georges Canguillhem, 'Machine and Organism', translated by Mark Cohen and Randall Cherry, in: Jonathan Crary and Sanford Kwinter (eds.), *Incorporations* (New York: Zone Books, 1992), 52.
4 See Marina Otero Verzier, 'Cartesian Enclosures', *New Geographies* 12, 'Commons', edited by Mojdeh Mahdavi and Liang Wang (March 2022), 39–57.
5 Data Centres and Data Transmission Network, Tracking report, *International Energy Agency website* (November 2021), https://www.iea.org/reports/data-centres-and-data-transmission-networks, accessed 18 January 2022.
6 IDC, Worldwide Global StorageSphere Forecast, 2021–2025: To Save or Not to Save Data, That Is the Question, IDC Doc #US47509621 (March 2021).
7 Gartner Market Trends: Evolving Enterprise Data Requirements—How Much Is Not Enough? (ID G00727554).
8 Data Center Solutions, *CBRE website*, https://www.cbre.com/services/property-types/data-center, accessed 6 February 2022; see also: Peter Judge, 'CBRE Launches Brokerage for Data Center Investors', *Data Center Dynamics* (1 April 2019), https://www.datacenterdynamics.com/en/news/cbre-launches-brokerage-for-data-center-investors/, accessed 16 December 2022.
9 Amit A. Nagarkar et al., 'Storing and Reading Information in Mixtures of Fluorescent Molecules', *ACS Central Science* (13 October 2021).
10 'Project HSD: Holographic Storage Device for the Cloud', *Microsoft.com*, https://www.microsoft.com/en-us/research/project/hsd/, accessed 2 February 2022; 'Project Silica', *Microsoft.com*, https://www.microsoft.com/en-us/research/project/project-silica/, accessed 2 February 2022.
11 'Preserving our Digital Legacy: An Introduction to DNA Data Storage', first whitepaper published by The DNA Data Storage Alliance, *DNA Data Storage Alliance website*, https://dnastoragealliance.org/dev/wp-content/uploads/2021/06/DNA-Data-Storage-Alliance-An-Introduction-to-DNA-Data-Storage.pdf, accessed 3 January 2022.
12 'DNA Data Storage: Integrated Information Storage Technology for Writing Large Amounts of Digital Information in DNA Using an Enzyme-Driven, Sustainable, Low-Cost Approach', *Wyss Institute Harvard website*, https://wyss.harvard.edu/technology/DNA-data-storage, accessed 2 February 2002.
13 O'Neill, 'Imagine storing', op. cit. (note 1).
14 Karin Fister and Iztok Fister Jr, 'How to Store Wikipedia into a Forest Tree: Initial Idea', paper first delivered at the Conference Multimedia, Scientific Information and Visualization for Information Systems and Metrics (MSIVISM 2014), Maribor, Slovenia, 29–31January 2014, http://www.iztok-jr-fister.eu/static/publications/35.pdf, accessed 5 December 2021.
15 Ibid.
16 O'Neill, 'Imagine storing', op. cit. (note 1).
17 Ibid.
18 Madeleine Mendell, Mél Hogan and Deb Verhoeven, 'Matters (and Metaphors) of Life and Death: How DNA Storage Doubles Back on Its Promise to the World', *The Canadian Geographer* IF1.5 (11 January 2022), 46.
19 Mél Hogan and Deb Verhoeven, 'Sustainable DNA: In Conversation', in: *Right Research: Modelling Sustainable Research Practices in the Anthropocene* (Cambridge: Open Book Publishers, 2021), http://books.openedition.org/obp/20563.
20 Ibid.
21 See: Laurajane Smith, *Uses of Heritage* (London/New York: Routledge, 2006).
22 Grow Your Own Cloud website, https://growyourown.cloud, accessed 20 December 2021.
23 Michael Marder, 'Is It Ethical to Eat Plants?', *parallax* 19/1 (2013), 29.
24 Michael Marder, 'What Is Plant-Thinking?', *Klesis, Revue Philosophique* 25, 'Philosophies de la nature' (2013), 125.
25 Marder, 'Is It Ethical', op cit. (note 23), 31.
26 Ibid., 34.
27 Ibid.
28 Ibid., 33.
29 Marder, 'What Is Plant-Thinking?', op. cit. (note 24), 126; see also: Marder, 'Is It Ethical', op. cit. (note 23), 33.
30 Marder, 'Is It Ethical', op. cit. (note 23), 36.
31 See: Marina Otero Verzier, 'Cartesian Enclosures: From Grid to Cloud', *Archined*, 'The Persistence of Questioning' (December 2021), https://www.archined.nl/2021/12/cartesian-enclosures-from-grid-to-cloud/.
32 Analysing the material implications of data infrastructures was one of the aims of the architecture studio ADS8: Data Matter: Digital Networks, Data Centres & Posthuman Institutions, led by Kamil Hilmi Dalkir, Marina Otero Verzier and Ippolito Pestellini Laparelli at the Royal College of Art in London, https://www.rca.ac.uk/study/schools/school-of-architecture/architecture/ads-themes-2019–20/ads8-data-matter-digital-networks-data-centres-posthuman-institutions/.
33 Descartes introduced the Cartesian Grid in *Discourse on Method* and more precisely in one of its three appendices titled *La Géométrie*, published in 1637. The system makes it possible to specify the position of any point or object on a surface using two intersecting axes as measuring guides, and to exactly duplicate geometric figures.
34 See: Marisol de la Cadena, *Earth Beings: Ecologies of Practice across Andean Worlds* (Durham, NC: Duke University Press, 2015).

# VIBRANT DA

## VIBRANT DATA: THE SEWER AS INFORMATION INFRASTRUCTURE

Sabrina Chou

(An Archaeology)

This is a transcription of social movement and the transmission of bodies, complete and incomplete. It is a record of circulations past and present. It moves at the horizon between political techniques of objective totalization and technologies of subjective individualization, and their expressions through code and protocol on the one hand, and desire and consumption on the other.[1] A peristaltic—and thus moving, changing—image of the social, through the sewer.

This is a reading of excrement through theories of information.[2] It is a materialization of flows of information and social circulations of belief and desire. To excavate the sewer as a site of vibrant[3] data is not only to contend with millennia of civilizations built and eroded, or with the management of bodily fluids, disease and populations, but it is also to confront a symbolic structure of the state form,[4] and the convergence in contemporaneity of a high-speed networked global economy with the slow-moving sludge of bodily waste. It is an archaeology that peers into the depths of our own ex-embodied excreta. But rather than suspending excrement in a realm of abject self-alienation,[5] this study recuperates our shit to reveal the glimmers of its potency in historical and contemporary fertilization and medicinal practices, as well as its potential as a source of collectivized data. Reading shit in this way has implications for an embodied (meta)physical approach to data as a social substance. This is not to say that we must instrumentalize and capitalize on even those residues of excrement that are passively produced by the shared labour that is our digestion.[6] Instead, what we trace here is a 'zone of irreducible indistinction', indeed, a zone of biopolitics that can—and must—be extricated from the realm of semiotic biopower and quantifiable code.[7]

By bringing together the recent use of sewage surveillance for public health and Covid-19 monitoring with the installation of fibreoptic data cables in existing sewers, I propose a reconsideration of the metaphysics of data that gestures towards a communal and biosocial understanding of information. The vibrant data of the sewer might revitalize the notion of the multitude via Gabriel Tarde's concept of the microsocial, tracing the flows of the social through the dynamic material flows of the sewer and the coded informational flows of the fibreoptic cable. In Tarde's theory, the binary of individual versus collective is set aside for the concept of the microsocial, in which social flows—beliefs and desires—are in motion, aggregating into social forms like institutions.[8] The notion of vibrant data might also reveal a semantic valence of biopolitics, one that can offer a mode of engaging information that refuses immediate reduction to positions of greater or lesser value and instead attends to the contextual, relational aspects of transmission and communication. Can we treat the use and study of data not as individualized points of consumer preferences or overcoded collective masses, but instead as sets of social flows that produce and allow for multiple and overlapping meanings?

In foregrounding the sewer as an infrastructure for the circulation of information, perhaps we can better understand collective pictures of local, communal microbiometrics, their correlations with ecological factors and their implications for public health. By acknowledging the vibrancy of data and its relational status, perhaps we can also shift our notions of public and collective, of life and matter, of how the social forms and flows—even in the bowels of our cities.

## Viral Data: Biopolitics of the Sewer

In *History of Shit*, Dominique Laporte describes the institution of protocols around hygiene, including the handling of all manner of waste and the building of cesspools in the year 1539 in France. In the same year, these hygienic codes collided with an edict establishing the official language of France, thus unifying and reinforcing the power of the monarchy. Together, these protocols, where shit and language coincide, can be read as a registration of the idea of the nation-state.[9] Here, base bodily function meets with the establishment of the institution of language—which might be considered the basis for and which makes possible political life.[10] Giorgio Agamben describes biopolitics as the 'zone of irreducible indistinction', where bare life and political life, where inside and outside, can no longer be separated. He calls for a new politics to be invented.[11] Why wallow in the sewage of our already existing present? The sewer is a site for the collision of bare life and political life. What has historically been an infrastructural necessity has become once again an index of the biopolitical.

The Covid-19 pandemic has created new global circumstances that have demanded large-scale responses and technological developments at a previously untested scale and speed. The pandemic has brought with it new governmental protocols, social behaviours and economic problems. One of the innovative by-products of the pandemic has been the initiation of sewage surveillance, which studies and monitors the prevalence of the SARS-CoV-2 virus and the RNA of specific virus strains in sewage.[12] For example: the National Institute for Public Health and the Environment of the Netherlands first monitored sewage from Schiphol Airport in mid-February 2020 and has now partnered with the 21 regional water boards of the country to monitor all of the more than 300 sewage treatment plants in the Netherlands for SARS-CoV-2 virus particles.[13] In the United States, the Centers for Disease Control and Prevention and the US Department of Health and Human Services initiated the National Wastewater Surveillance System to understand the prevalence of SARS-CoV-2 in local communities.[14] In the UK, the Health Security Agency has started testing for SARS-CoV-2 RNA in the Environmental Monitoring for Health Projection wastewater programme, taking samples from sewage treatment works and sewer network sites.[15]

Sewage surveillance can provide a collective picture of the scale of infection in a community or communities, even when virus symptoms are not present or when testing is unavailable or not used. Sewage surveillance can thus provide local health administrators and citizens with early warnings of possible spikes in infection localized in particular communities, prior to detection through testing or clinical presentation. Importantly, sewage surveillance is not dependent on behavioural compliance or changes in policy. It is based on the bare, material fact of regular bodily functions. These sewage surveillance programmes have potential uses in other public health initiatives—for studying populations and their health, or for looking at broader ecosystems and watersheds and the effects of climate change on them. For instance, sewage and wastewater surveillance has been used to monitor other viruses, such as noroviruses, hepatitis E and polio; to examine antibiotic resistant bacteria or antimicrobial resistance; to monitor spatial and temporal trends of recreational, illicit and pharmaceutical drug use on local and international scales; and to monitor the presence and possible ecological effects of microplastics in wastewater.[16] Collecting excrement and studying it has become a source of vital information about local public health.

Shit itself is vital. Humans have long used the excrement of both humans and animals as fertilizer. Consider, for example, the pigeon towers of Isfahan, which were built solely to house pigeons and their excrement to be used as manure.[17] More recent initiatives have returned to ancient practices of converting human faeces into fertilizer to deal with waste accumulation while nourishing poor soil, or to process excrement for use as fuel after removing water and sanitizing it.[18] We use excrement now not just to fertilize land or to burn as fuel, but also to support life and living. Recent research into the gut microbiome and faecal microbiota transplants has made this more urgently evident. Jane Bennet has written about tracing chemical exuberances through bodies and sites, and Michelle Murphy about how microbiome research reveals that humans are collectivities, rather than discrete, contained bodies.[19] Studying the microbiome has made clear that our bodies are not singular, but instead always already irreducible and ungraspable multiplicities. Microbiomes are not bounded within the confines of our bodies, but instead spill over and out, commingling to produce new ecosystems, relations and assemblages. But while these studies are applicable on the level of broader populations, the use of faecal matter in faecal microbiota transplants is still individualized, on a one-to-one, bowel-to-bowel basis. The *United European Gastroenterology Journal* specifically states that:

> Pooling (mixing) of multiple donor faeces during processing is not recommended because: (a) it hampers the traceability of the faecal preparation to the individual donor; (b) the risk of transmissible disease may be increased; and (c) the principle of transfusing a balanced microbiota preparation may be lost.[20]

Ancient historical practices, while perhaps failing to meet present-day medical and hygienic standards, might point to more collective and communal uses of faecal treatments. For instance, even as early as the fourth century CE, ancient Chinese medical texts refer to cures for the treatment of food poisoning, severe diarrhoea and otherwise incurable conditions with various faecal or fermented faecal treatments administered by mouth and otherwise, including the sealing of liquorice root in bamboo that was left in a village latrine for a year, then retrieved and ground up for poultices or other therapies.[21] Such approaches demonstrate the basis for a social and local use of faecal matter, for a socially and locally situated understanding of the internal microbiome and its co-constitution with, by and of local, social, bodies.

As individuals, we are necessarily and urgently situated in an environment, within an ecological assemblage of other humans, of networked communication, of macro-watersheds and of the microbial matter that co-constitutes our bodies. Locality matters here. Methodologies can be translated and scaled up or down, but it matters *whose* shit is mixing *where*. The vitality of our sewage, which is teeming with microscopic life—with the diversity of microbiotic bacteria and viral RNA, next to traces of environmental chemical leakages and the residues of microplastics that have also journeyed to and subsequently through our bodies—points much more to *multitude* and crowd than to the sewer's ancient architectural symbolic status of *civilization*. Michel Serres has written about the crowd and its solidification into institutions or foundations: 'The crowd is fluid. An institution stable, is solid. The foundation solidifies the crowd.'[22] Attending to the biopolitics of sewage and its data re-liquifies such solidified crowds into fluidities, into flows, into multitudes.[23] These are crowds and pluralities that exist in a state of multiplicity and motion, where they are not unified or homogenized and thus are not (yet) objectified into a people or a collective subject of the state. In *The Grammar of the Multitude*, Paolo Virno writes about Spinoza's concept of the multitude, wherein 'the *multitudo* indicates a plurality which persists as such in the public scene, in collective action, in the handling of communal affairs, without converging into a One, without evaporating within a centripetal form of motion.'[24] A centripetal form of motion focuses on a singular, central point, bringing everything together into a simultaneity, at the same speed and in the same direction. But the multitude flows freely, retaining a plurality of possible directions. Virno describes multitude as 'the form of social and political existence for the many, seen as being many'.[25] The multitude is the crowd before it becomes a people, an addressee or a targeted collectivity. It is this 'many, seen as being many' that is in dynamic motion in the circulating excrement—and chemical, biological and ecological information—that flows through our sewers. The sewer, then, or the lower stratum of the built and inhabited environment, is where the material substance of the multitude flows, alongside its potency as a streaming source of viral, vital data.

If the biopolitics of the sewer speaks multitudes, what else can we find as we trace the moving contents of these social infrastructures?

## Digital Waste: The Sewer in the Information Age

Excavating sewers reveals much about the societies that built them. They are archaeological evidence of feats of engineering and can trace histories of urban development and morphology.

The contents of sewers reveal the social histories of a particular group of inhabitants, as much about cultural dietary practices as about differences in socioeconomic class.[26] In 'Deep Time of Media Infrastructure', Shannon Mattern describes how our infrastructures—and in particular, infrastructures of communication—are layered over time, rather than merely replaced or phased out.[27] The sewer presents itself as such an infrastructure that layers large-scale hygienic utilities with communications infrastructure: next to excrement, sewers also host fibreoptic cables that are installed there to avoid the expense, disruption and sometimes destruction of digging new tunnels under cities. Like excrement, which has long been sequestered underground, data cables are not ordinarily visible or present. Nicole Starosielski locates the invisibility of cable systems within 'a broader social tendency to overlook the distribution of modern communications in favour of the more visible processes of production and consumption' rather than 'any intentional desire to obscure cable systems'.[28] Making visible such communication distribution systems might be akin to attending to the concealed bodily processes of digestion and excretion that have long been ignored or overlooked in light of production and consumption.[29]

A closer look at the layered infrastructures of the sewer and the data cable shows a pragmatic coupling of the two. The use of existing infrastructure like sewers and pipes for the installation of fibreoptic cables offers financial benefits, as well as the possibility of monitoring the—often aging—infrastructures for leaks during installation.[30] The first instance of installing fibreoptic cables in existing sewers was in the United Kingdom in 1984, by the Water Research Centre.[31] Since then, different techniques have been implemented globally for installing fibreoptic cables in sewers. While early methods used robots and drill and dowel techniques, other methods include the implementation of cabling trays, or the installation of lining systems that simultaneously fortify deteriorated sewers and provide protective housing for cabling.[32] Fibreoptic cables continue to be installed in sewer lines, especially in ongoing efforts to rapidly extend broadband capacity to more and more users, particularly those outside of more densely inhabited metropolitan areas.[33] The sewer has thus become part of the conduit system of networked infrastructures that produce and circulate social media, platform-based economies and new financial instruments.

This convergence of the network and the sewer—or, the network *in* the sewer—brings the material and immaterial multitude together. The mingling of bodily and disembodied activity also muddles the already blurred boundaries of public and private. Tung-Hui Hu compares the privatization of the Internet following its initial design and usage via time-sharing to the privatization of hygiene via the sewer system: 'Centuries before computers were invented, sewers kept each household's private business private even as it extended the armature of the state into individual homes.'[34] While the sewer was designed to keep the waste of individuals private, opening up the sewer as a vital source of social data might be able to de-privatize human functions both excretal and digital.

Subterranean fibreoptic cables carry new products of biopower: digital waste and data collected from our content production and consumption; violent, digital pollution that is cleared by the invisible labour of content moderators; and platform-mediated, network-based economic exchanges. The data cable is a part of a larger network of the Internet of things through which soft pollution—disembodied waste and its biopolitical affects—circulates. In the sewer, we come to find the circulation of multiple streams of excreted and expelled vital data. While there are myriad ways to analyse data and its social effects, one mode of analysis might point to Gabriel Tarde's early description of social flows as *beliefs* and *desires*.[35] In fact, these elements are perhaps some of the most valuable data currently collected and analysed. Marketing to beliefs allows for the shaping of political views, opinions and even electoral outcomes, and marketing to desires generates advertising revenues, optimized product development research and increased sales.[36] Might it be possible that beliefs and desires are some of the most instrumentalized aspects of digital activities? The profiles that are created from the traces of our digital activities—the digital traces of our beliefs and desires—'produce estimating patterns and anticipate potentialities: consumption preferences, economic value, behavioural inclinations, professional capacities, virtual diseases, political preferences.'[37]

The content and composition of our collective excrement can also be studied in terms of their material manifestation of desire and belief. The consumption of food, pharmaceutical and recreational drugs, and other substances are often tied to our corporeal desires and structures of belief, whether rooted in religious observance or within a body of sociomedical knowledge. Agamben writes about the term *diet*, wherein maintenance of life coincides with a regulatory and formal process. The ancient medical term *diaita,* he writes, 'designates the regime of life, the "diet" of an individual or a group, understood as the harmonic proportion between food (*sitos*) and physical exercise or labour (*ponos*)', while also being the term for 'arbitration that decides a suit not according to the letter of the law but according to circumstances and equity'.[38] The term has further developed to mean 'a political assembly with decision-making power'.[39] As both 'mode of life' and the 'governance and regime of life', *diet* thus regulates the body at the intersection of the biological axis and the political axis of life and living.[40] Along these lines, we find that code and protocol, expressed through both digital activity and dietary practice, circulate the social through the sewer. What are the biopolitical implications of these sociopolitical flows, if we are not to be completely and immediately reduced and subsumed to mechanisms of capitalism as units or profiles of consumer and consumption? How might the multitude and the beliefs and desires—the excremental traces of data and digestion—that move and make the social offer another perspective, another possibility for life and living?

The confluence of digital and dietary waste in the sewer points to a composite notion of sociality. Rather than thinking through the binary of individuals versus collective mass, it might be helpful to turn to Fernanda Bruno and Pablo Manolo Rodriguez's theorization of a 'complex dividual-individual composition' that is constituted via 'biotechnologies, digital culture, and financial capitalism' and its appearance via biogenetics, digital profiles, mobile applications, wearable technology or financial subjects.[41] According to anthropological perspectives, and as described by Marilyn Strathern in her discussion of Melanesian persons and personhood, the dividual contains 'a generalized sociality within. Indeed, persons are frequently constructed as the plural and composite site of the relationships that produce them.'[42] In the sewer we find a social manifestation of the plural and composite site of relations, made up of not only the biomass of persons, divided and 'disseminated in a "biological matter" related to, but not contained in, this individuality', or the 'traces about our lifestyles that feed huge and valuable databases' that together make up digital profiles, but also the material and immaterial expressions of our ongoing relationships.[43]

This site of relations is where, as described by Rosi Braidotti, a 'missing and virtual' people can be 'actualized and assembled'.[44] This is not the people in opposition to the multitude, created as a manageable mono-ubject of the state. Rather, it is 'the result of a praxis, a collective engagement to produce different assemblages'.[45] In the sewer we find that although 'we are not one and the same ... we can interact together'.[46] And in encountering the sewer's disparate traces of forms of living, from data to bodily waste, we might be able to view life and living as 'on-going flows and transformations of forces'.[47]

To better approach the fluid dynamics of the crowded multitude streaming through the sewer,[48] we might turn to the notion of *flow*.

## Vibrant Flows: Circulating the Social

Viewing shit as data—and data next to shit—allows for a consideration of the vital materiality and movement of both digital data and physical excrement. Barbara Orland conceptualizes the history of knowledge of material, writing that '*flow*, or current, describes the directed kinetic movement of either a fluid continuum of a substance or a multitude of similar singularities'.[49] Within the sewer, the fluid continuum of excretal sludge and the multitude of digital dividuals flow together.[50] This course of movement is the terrain of biopolitics that Eugene Thacker describes in which a decomposing, disintegrated, diseased body politic is addressed by forms of governance that regulate 'networks, flows, and circulations'.[51]

Here we turn to Gabriel Tarde's concept of microsociology in which 'the social ... is a circulating fluid',[52] or the movement of flows of 'belief or desire'.[53] Tarde looks at diverse, small social interactions as the basis for analysing the social, rather than generalizing about collective social wholes, institutions or processes.[54] For Tarde, the 'distinction between the social and the individual loses all meaning since flows are neither attributable to individuals nor overcodable by collective signifiers'.[55] These flows are 'created, exhausted, or transformed, added to one another, subtracted, or combined'.[56] In Tarde's microsociology, a social unit can be understood as an 'ensemble, compound, or configuration of previously disperse flows of desires and beliefs'.[57] These compositions are subject to change, increasing or decreasing depending on what is added to or subtracted from the ongoing current.[58] Yet while the social and individual are not the primary figures in Tarde's microsociology, they can still be 'understood as open ensembles of immanent, contingent, and partial relationships of beliefs and desires in continuous change'.[59] It is such an open assemblage of immanence, contingency and partiality that we find in the swirling multitude of the sewer, and in its continuous, changing movement.

If prior practices of biopolitics have been cached in the realm of valuation, classification, quantification and regulation—practices related to signification and the semiotic—then perhaps the notion of shit as data and the vital materiality of data can shift these practices into a semantic realm, into meanings that shift and move, just as our data and excrement do. The notion of vibrant data attempts to resist the semiotic, financialized arrest of data by grounding information *through* materiality and its social flows. N. Kathleen Hayle discusses the disembodiment of information during the Macy Conferences on Cybernetics, when Norbert Wiener and Claude Shannon established a stable conception of information that was divorced from meaning and instead focused on the transmission of messages, excluding consideration of their context or the observer. For Shannon, 'defining information as a probability function was a strategic choice that enabled him to bracket semantics'.[60] What would it mean to return semantics to information? How might we allow semantics to enter into the circulation of data and shit that flush through the sewers, so that the meeting of sewage and language does not merely reproduce Laporte's recollection of the institution of centralized state power?

The semantic, according to linguist Émile Benveniste, opens up a notion of language that is a site of transformations, language as something that moves, rather than merely a linguistic system that rigidly upholds signs.[61] A reconsideration of the semantic aspect of information revisits the establishment of cybernetics and the construction of information as disembodied and dematerialized to retrace an alternate route in which information is necessarily dependent on and subject to structural, social and material conditions. The information in our data and biomass is necessarily relational, as are our biogenetics: 'Like the dividual self in an anthropological sense, genomic data gain their meaning and utility in reference to the wider genomic cohort.'[62] The sewer as an information infrastructure allows for a relational and semantic approach to individual and dividual excrement and data. For in studying sewage, it is impossible to identify individual sources. Instead, that information must be considered in the context of the other matter that flows with it, without being coded back to original addresses. If we take a similar approach and study collected sets of digital data rather than data attributed to identified individuals, perhaps we can view data in light of its relations, as sets of changing and growing social flows.[63] In this sense, data—and its meanings—are not objectified or reified, but rather, they remain in motion, as sites for transformations.

A semantic approach to the flow of information—both digital and excretal—does not instigate the creation, maintenance and subjugation of sets of individualized subjectivities, nor does it actuate the overcoded collective subjectivities that can become tools of market research or optimized product development. Rather, this is a social subjectivity that is contingent. It produces flows of information that necessarily depend on senders, observers and environs—and these not merely as positions of value. It reconfigures information as 'compositions of the multiple collective flows' of the microsocial.[64] The excess traces of our digital and bodily activities, data and excrement manifest this informational flow that is 'the social tissue [that] always both composes and overflows the social organs'.[65]

In allowing our digital and material biosocial substances to overflow, we might arrive at a notion of grotesque data by way of Mikhail Bahktin's analysis of Rabelais's writing and world. In Rabelais's writing, 'the body and bodily life have here a cosmic and at the same time an all-people's character'.[66] The body is not individualized, and the 'material bodily principle is contained not in the biological individual ... but in the people, a people who are continually growing and renewed'.[67] This, writes Bakhtin, is why 'all that is bodily becomes grandiose, exaggerated, immeasurable', why the overflowing, overgrowing body becomes grotesque.[68] The sewer circulates the social through the continually changing flow of bodily excrement and digital waste. It is the site for the circulation of embodied information and the materialization of data as a social substance. Reading data through the vital contents of the sewer might allow the multitude to overflow both individualization and mass collectivities, to instead circulate through multiple meanings and relational compositions. These social flows might enable the creation of alternate biopolitics, of a biopolitics that escapes individual capture and collective overcoding, or subjective individualization and objective totalization.

Grotesque social overflows of the body are sites of becoming, and sites of the potency of life. The sewer is the space that hosts this immense potentiality.

Notes

1 Giorgio Agamben, *Homo Sacer*, trans. Daniel Heller-Roazen (Stanford: Stanford University Press, 1998), 5.
2 For example, Eric Hayot discusses the conflation of 'information' with 'communication' by Claude Shannon and Warren Weaver in *The Mathematical Theory of Communication* as a paradigm shift in the theorization of information. In modern history, information has come to mean 'that which has no fixed form, but can be passed on from one person to another, or one medium to another, without either being significantly altered in itself, or altering that which it touches'. Eric Hayot, introduction to: Eric Hayot, Anatoly Detwyler and Lea Pao (eds.), *Information: A Reader* (New York: Columbia University Press, 2022), 6–7.
3 Or, in line with Jane Bennett's analysis, the vital materiality of things and substances and their capacity as forces or agents. Jane Bennett, *Vibrant Matter: A Political Ecology of Things* (Durham, NC: Duke University Press, 2010), 113–119.
4 Georges Bataille, 'Architecture', in: Neil Leach (ed.), *Rethinking Architecture* (London: Routledge, 1997), 19–20.
5 See: Julia Kristeva and Leon S. Roudiez, *Powers of Horror: An Essay on Abjection* (New York: Columbia University Press, 1982). Kristeva elaborates on the abject as something other than the self that is, for instance, unclean or unhealthy, that which 'disturbs identity, system, order. What does not respect borders, positions, rules. The in-between, the ambiguous, the composite.' Within this plane of analysis, 'excrement and its equivalents (decay, infection, disease, corpse, etc.) stand for the danger to identity that comes from without.' Ibid., 4, 71.
6 See, for instance: Sabrina Chou, 'Live Tenders: An Incomplete Theory of Social Digestion', *Thresholds* 48 (2020), 176–185, for a possible notion of the political implications of collective digestion.
7 Agamben, *Homo Sacer*, op. cit. (note 1), 9.
8 Gilles Deleuze and Félix Guattari, *A Thousand Plateaus: Capitalism and Schizophrenia*, trans. Brian Massumi (Minneapolis: University of Minnesota Press, 1987), 219; see also: Gabriel Tarde, *Monadology and Sociology,* ed. and trans. Theo Lorenc (Melbourne: re.press, 2012).
9 Dominique Laporte, *History of Shit*, trans. Nadia Benabid and Rodolphe el-Khoury (Cambridge, MA: MIT Press, 2000), 2–9.
10 Hannah Arendt describes speech and action as what allows humans to appear to one another, to relate to one another, and as what constitutes the space of the *polis*, rather than any physical location. Hannah Arendt, *The Human Condition* (Chicago: University of Chicago Press, 1958), 174–88, 196–99.
11 Agamben, *Homo Sacer*, op. cit. (note 1), 9.
12 Tomáš Mackul'ak et al., 'Wastewater-Based Epidemiology as an Early Warning System of the Spreading of SARS-CoV-2 and Its Mutations in the Population', *International Journal of Environmental Research and Public Health* 18/11 (2021), 5629.
13 'Coronavirus Monitoring in Sewage Research', Rijksinstituut voor Volksgezondheid en Milieu (Dutch National Institute for Public Health and the Environment), https://www.rivm.nl/en/covid-19/sewage, last modified 12 September 2022.
14 'National Wastewater Surveillance System', Center for Disease Control and Prevention, https://www.cdc.gov/healthywater/surveillance/wastewater-surveillance/wastewater-surveillance.html, last modified 21 March 2022.
15 'Wastewater Testing Coverage Data for the Environmental Monitoring for Health Protection (EMHP) Programme', UK Health Security Agency, https://www.gov.uk/government/publications/wastewater-testing-coverage-data-for-19-may-2021-emhp-programme/wastewater-testing-coverage-data-for-the-environmental-monitoring-for-health-protection-emhp-programme, last modified 10 June 2021.
16 Mackul'ak et al., 'Wastewater-Based Epidemiology', op. cit. (note12), 5629; Kevin V. Thomas et al., 'Comparing Illicit Drug Use in 19 European Cities Through Sewage Analysis', *Science of the Total Environment* 432 (2012), 432–439; Jing Sun et al., 'Microplastics in Wastewater Treatment Plants: Detection, Occurrence and Removal', *Water Research* 152 (2019), 21–37.
17 Michael Hensel et al., 'Towards an Architectural History of Performance', *Architectural Design* 82/3 (2012), 26–37; Elisabeth Beazley, 'The Pigeon Towers of Isfahan', *Iran* 4 (1995), 105–109.
18 Chelsea Wald, 'The New Economy of Excrement', *Nature* 549 (2017), 146–148.
19 Bennett, *Vibrant Matter*, op. cit. (note 3), 113–119; Michelle Murphy, 'Against Population, Towards Alterlife', in: Adele E. Clarke and Donna Haraway (eds.), *Making Kin Not Population* (Chicago: Prickly Paradigm Press, 2018), 115. For a discussion of the multiplicity of bodies in relation to excrement and microbiota, see also: Anna Wolodzko, 'Bodies Within Affect' (PhD dissertation, University of Leiden, 2018), 145, 162–165.
20 Josbert J. Keller et al., 'A Standardised Model for Stool Banking for Faecal Microbiota Transplantation: A Consensus Report from a Multidisciplinary UEG Working Group', *United European Gastroenterology Journal* 9 (2021), 229–247.
21 See: Pieter de Groot, 'Fecal Microbiota Transplantation in Metabolic Syndrome: History, Present and Future', *Gut Microbes* 8/3 (2017), 253–267; 王国华, '最好的藥' ['The best medicine'], 視野 [*Horizons*] 13 (2014), 21.
22 Michel Serres, *Rome: The First Book of Foundations,* trans. Randolph Burks (London: Bloomsbury, 2015), 201.
23 Recent studies have framed the digital crowd as a 'form of contemporary collective life for practicing symbolic politics' that also extend 'the crowd's affective and cognitive attributes, well beyond its duration of physical gathering and actions'. Hazem Ziada, 'The Digital Crowd', *Architecture and Culture* 8/3–4 (2020), 653–666.
24 Paolo Virno, *A Grammar of the Multitude: For an Analysis of Contemporary Forms of Life*, trans. Isabella Bertoletti, James Cascaito and Andrea Casson (Los Angeles, Semiotexte, 2004), 21–22.
25 Ibid., 21–22.
26 Erica Rowan, 'Bioarchaeological Preservation and Non-elite Diet in the Bay of Naples: An Analysis of the Food Remains from the Cardo V Sewer at the Roman Site of Herculaneum', *Environmental Archaeology: The Journal of Human Palaeoecology* 22/3 (2017), 318–336.
27 Shannon Mattern, 'Deep Time of Media Infrastructure', in: Lisa Park and Nicole Starosielski (eds.), *Signal Traffic: Critical Studies of Media Infrastructures* (Chicago: University of Illinois Press, 2015), 95–112.
28 Nicole Starosielski, *The Undersea Network* (Durham, NC: Duke University Press, 2015), 4.
29 See: http://www.fiberatlantic.com/submarinecablemap/ for a map of undersea cables.
30 Stefan Stanko and Ivana Mahrikova, 'Implementation of Fibre Optic Cables in Sewage System', in: Petr Hlavinek et al. (eds.), *Integrated Urban Water Resources Management* (Dordrecht: Springer, 2016), 171–180.
31 Sanjiv Gokhale, 'Deployment of Fibre Optic Networks through Underground Sewers in North America', *Journal Of Transportation Engineering* 132/8 (2006), 672–682.
32 Ibid., 673–677.
33 Ibid., 672.
34 Tung-Hui Hu, *A Prehistory of the Cloud* (Cambridge, MA: MIT Press, 2016), 37–71, 41.
35 Tarde, *Monadology and Sociology*, op. cit. (note 8), 16–18.
36 Shoshanna Zuboff, *The Age of Surveillance Capitalism* (New York: PublicAffairs, 2019), 37–51.
37 Fernanda Bruno and Pablo Manolo Rodriguez, 'The Dividual: Digital Practices and Biotechnologies', *Theory, Culture and Society* 39/3 (2021), 39–40.
38 Giorgo Agamben, *The Use of Bodies*, trans. Adam Kotsko (Redwood City, CA: Stanford University Press, 2016), 225–226.
39 Ibid., 225.
40 Ibid., 225–226.
41 Bruno and Rodriguez, 'The Dividual', op. cit. (note 37), 40.
42 Marilyn Strathern, *The Gender of the Gift: Problems with Women and Problems with Society in Melanesia* (Berkeley: University of California Press, 1990), 13.
43 Bruno and Rodriguez, 'The Dividual', op. cit. (note 37), 39–40.
44 Rosi Braidotti, *Posthuman Knowledge* (Cambridge: Polity Press, 2019), 101. Judith Butler writes about the potential in acts of assembly that are enactments of the gathering of bodies that allow for difference rather than mass homogeneity: '*... the assembly is already speaking before it utters any words* ... To act in concert does not mean to act in conformity; it may be that people are moving or speaking in several different directions at once, even at cross purposes.' Judith Butler, *Notes Toward a Performative Theory of Assembly* (Cambridge, MA: Harvard University Press, 2015), 156–157.
45 Ibid., 101.
46 Ibid., 101.
47 Ibid., 100. Here Braidotti is discussing Deleuze's assertion that life is not one system.
48 Indeed, the movement of crowds is studied using the physics of fluid dynamics. See, for example: Muhammad Umer Farooq et al., 'Motion Estimation of High Density Crowd Using Fluid Dynamics', *The Imaging Science Journal* 68/3 (2020), 141–155.
49 Barbara Orland. 'Matter in Flux: How to Study the Dynamic States of the Material World', in: Jennifer Teets (ed.), *Electric Brine* (Berlin: Archive Books, 2021), 43–63.

50 See Manuel Castells's discussion of the network as the 'space of flows'. Manuel Castells, *The Rise of the Network Society* (Oxford: Blackwell Publishers, 2000), 442. For the engineering of sewage flows, see: Thomas Barlow, *Hydraulics: Gauging of Sewage Flows, etc.: A Handbook of Rules and Tables for Engineers and Managers of Sewage Disposal Works* (London: Lockwood, 1926); Franz Tscheikner-Gratl et al., 'Sewer Asset Management: State of the Art and Research Needs', *Urban Water Journal* 16/9 (2019), 662–675.
51 Eugene Thacker, 'Necrologies; or, the Death of the Body Politic', in: Patricia Ticineto Clough and Craig Willse (eds.), *Beyond Biopolitics: Essays on the Governance of Life and Death* (Durham, NC: Duke University Press, 2011), 152.
52 Bruno Latour, *Reassembling the Social: An Introduction to Actor-Network-Theory* (Oxford: Oxford University Press, 2007), 13.
53 Deleuze and Guattari, *A Thousand Plateaus*, op. cit. (note 8), 219.
54 Sergio Tonkonoff, 'A New Social Physic: The Sociology of Gabriel Tarde and Its Legacy', *Current Sociology* 61/3 (2013), 267–282.
55 Deleuze and Guattari, *A Thousand Plateaus*, op. cit. (note 8), 219.
56 Ibid., 219.
57 Tonkonoff, 'A New Social Physic', op. cit. (note 54), 275.
58 Ibid., 275.
59 Ibid., 276.
60 N. Katherine Hayles, *How We Became Post-Human* (Chicago: The University of Chicago Press, 1999), 50–55.
61 Émile Benveniste, *Last Lectures: College de France 1968 and 1969*, ed. Jean-Claude Coquet and Irene Fenoglio, trans. John E. Joseph (Edinburgh: Edinburgh University Press, 2019), 125–127.
62 Ian McGonigle, 'Genomic Data and the Dividual Self', *Genetics Research* 101/12 (2019), 2.
63 For instance, Mélanie Dulong de Rosnay and Felix Stalder have written about the digital commons, wherein resources of 'data, information, culture, and knowledge ... are created and/or maintained online'. They posit the digital commons as 'a political institution and as a way to expand democratic participation beyond the framework of representative democracy, through self-governance, 'participation, flexibility, and collaboration throughout society'. Significantly, the notion of a data commons could provide resistance against the 'increasing centralisation and commodification of data in the hands of a small number of companies', to instead govern data as a commons. Yet, as De Rosnay and Stalder write, this commons is yet to be constituted, as it still lacks a 'conceptual and legal framework', and the differentiation between open data and members-based commons data. Mélanie Dulong de Rosnay and Felix Stalder, 'Digital Commons', *Internet Policy Review* 9/4 (2020), 2, 15, 16.
64 Tonkonoff, 'A New Social Physic', op. cit. (note 54), 279.
65 Ibid., 275.
66 Mikhail Bakhtin, *Rabelais and His World*, trans. Hélène Iswolsky (Bloomington: Indiana University Press, 1984), 19.
67 Ibid., 19.
68 Ibid., 19.

# MACHINES

## MACHINES IN LANDSCAPE/TERRITORIAL INSTRUMENTS

### On the Sociotechnical Production of Planetary Platforms

Ali Fard

#### Elusive by Design

In 2009, Kevin Kelly, one of the cofounders of WELL and the long-time executive editor of *Wired* magazine, asked his followers on the web to draw their interpretation of the Internet as a map.[1] Since then, the project has garnered more than 270 submissions on Kelly's Flickr page.[2] A cross-sectional reading of all submissions reveals the overwhelming abstracted conceptions of the Internet that dominate most of these maps. Projects like this point to the lack of clarity and articulation that prevails in popular imaginaries of large technical systems, and the duality at the core of how planetary technology platforms are produced. On the one hand, metaphors and marketing generate a sociotechnical imaginary of data platforms that intensifies myths of immateriality, inherent sustainability and technologically mediated societal progress. This sociotechnical imaginary actively undermines the material geographies, power relations and labour conditions that are at the core of platforms and their operations. On the other hand, the production of data and its mobility necessitate the spatial (re)production of operational landscapes of extraction, storage and processing, as well as infrastructural geographies of circulation that accommodate the continual expansion of platforms and the enclosure of resources at the core of twenty-first-century data capitalism. These infrastructural extensions in turn create an uneven geography of data that is essential to the production of platforms, but that has largely been black-boxed behind imagery and metaphor and has escaped critical attention in discussions of urbanization and technology.

To fully capture the intricacies of how planetary platforms are produced, we need to come to terms with this duality. Given the interwoven relationship between representation and materiality in the production of platforms, the framework of sociotechnical imaginaries presents a productive starting point to examine the coproduction of sociocultural practices and technological development. Emerging discourses in urban studies, such as discussions around planetary urbanization, that reengage with urbanization through a relational reading of both its concentrated and extended moments, also provide a helpful conceptual basis for examining the spatial impact and imprints of platforms beyond the city. Ultimately, a material rereading of the infrastructural landscapes of data, complemented by a critical understanding of their political imaginary, begins to complicate the smooth and frictionless reading of the operations of tech platforms and how they are produced.

## Capturing the Cloud

The notion of the 'cloud', which has been widely popularized since its emergence in the early 2000s, is a good example of how metaphors are operationalized in the sociotechnical imaginaries of data infrastructure. Clouds of the water vapor kind are naturally formless. They are in a state of constant flux, changing shape and direction based on temperature fluctuations and wind dynamics. Hence, to represent a cloud is to abstract its amorphous nature, to freeze its continually changing form in time and space, an abstraction that is necessary for its representation. The same logic of abstraction is at play in reference to the cloud of the digital kind. Emerging from the technical network drawings of the 1970s, the metaphor of the 'cloud' renders the complex ever-expanding global network of digital infrastructures as 'a single, virtual, object'.[3] A representation that is increasingly important to the marketing of the services and products that run on its critical infrastructure.

An important component of this abstraction is virtualization, which generates an analogue of computing resources with the aim of efficiently sharing them among multiple users at any given time. Evolving from the time-sharing of mainframe computers in the 1960s, virtualization enables the cloud 'to distribute, meter, and charge for these computing resources in highly granular and flexible ways'.[4] But the concept can also more generally be understood as 'a technique for turning real things into logical objects', hence abstracting their relationship to a material base.[5] The reconfigurative power of virtualization persists through the representation of the cloud. The very term 'the cloud' denotes a singularity that sweeps under the rug the multiplicities of network forms and the dynamically competitive world of cloud computing, which boasts a large variety of actors, services and user types. A single object is finite; it can be secured; it can be controlled. The structure of an object is clear; it can be understood and communicated. An object can be mastered and naturalized, hence becoming a tool of advancement. An object can be fetishized, hence marketed and sold. Not all of these things could be done if the cloud was represented as a complex, infinitely expandable collection of networks of varied devices, protocols and connections.

Virtualization, as both a technical process and a representational practice, has therefore enabled the imagination of a parallel, more data-driven, spatial condition to that of 'real' space. A Google image search for 'cloud computing' returns an array of similarly shaped, abstract and cartoonish depictions of clouds. The results reflect the dominance of the virtual icon in all aspects of technological representation, from marketing images to network diagrams. It also portrays a general ambivalence to what the cloud actually is or what it does. The icons and the images that stand in for the cloud instead accommodate the virtualization of the material basis upon which it is constructed: while they manage to capture the ease, the security and the familiarity of the operations of the cloud, they concurrently blur the actual effort, vulnerability, space and infrastructure of its delivery, as well as the extended material geography and the environmental degradation that it entails.

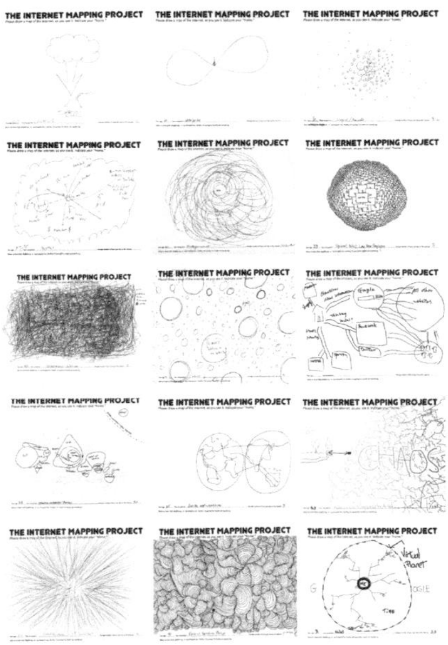


Fig. 1 A representative sample of submissions to Kelly's Internet Mapping Project. Images courtesy of Kevin Kelly on Flickr



In the absence of clear technical or easily digestible representations, the info-industrial complex has always relied on metaphors for marketing and generating a sociotechnical imaginary of its projects.[6] The sociotechnical 'co-production' of imaginaries is a particularly useful concept to the critical analysis of technology platforms.[7] Jasanoff elaborates on sociotechnical imaginaries as 'collectively held, institutionally stabilized, and publicly performed visions of desirable futures, animated by shared understandings of forms of social life and social order attainable through, and supportive of, advances in science and technology'.[8] And while sociotechnical imaginaries are often promoted through the state, as Sadowski and Bendor have shown in the context of 'smart' urbanism, the particularities of contemporary data technologies and their development and marketing situate corporations as the main actors in shaping and extending sociotechnical imaginaries of platforms. In this sense, 'large technology vendors set the tone, enroll other actors, and weave the narratives' that allow the tech sector's strategists, now turned urbanists, to strongly position themselves and their platforms as harbingers of a new paradigm of urbanism.[9]

## Urbanizing Platforms

In the emerging sociotechnical imaginary of tech platforms, the city is treated as a frontier of data-driven projects that attempt to deal with a variety of issues facing contemporary urbanization. Abstraction and metaphors abound within these imaginaries, which re-present the spatial projects of tech as part of a singular, clean, objective, efficient and transformative platform logic. This further imbues the urban operations of data platforms with techno-utopian ideologies of progress that attempt to justify the colonization of everyday life through technology as a necessary consequence of responding to the socioenvironmental challenges of urbanization.[10] Within these abstractions it is easy to forget that every click, every sensor, every screen and every AI platform in the urban model propagated by tech is ultimately linked up to clouds of data that

Fig. 2 An image search of 'cloud computing' returns a slew of abstracted cloud icons

depend on infrastructural landscapes and material geographies that stretch well beyond the city. Yet, the urban projects of tech platforms are portrayed as highly defined entities with clear boundaries that operate on the city as a wholly knowable, manageable and ultimately controllable archipelago in a sea of an undefined periphery.

In a process evocative of the undergrounding of urban systems at the turn of the twentieth century, the infrastructural landscapes of data production and technology development form the blurry background to the over-hyped operations of platforms in cities. The disappearance of the urban monuments of modernism from the cityscape, in response to the failure of the emancipatory project of modernity and the unfulfilled promise of delivering a better society, prompted the construction of new sociotechnical imaginaries built around notions of lightness, sustainability, convenience and effortlessness.[11] By the 1990s a new techno-utopian ideology was in full effect; this time driven by rapid advancements in digital information and communication technologies. As Graham elaborates, the mystique that was attached to electronic (and later digital) communication technologies between the 1960s and the 1990s was such that 'ICT-mediated shifts away from place and city-based lives were often uncritically assumed to automatically also involve shifts towards more democratic, egalitarian, decentralized and ecologically sensitive societies'.[12]

While these decentralized visions still animate the sociotechnical imaginaries of platforms, rapid urbanization and the densification of global population centres—with their massive cache of potential data—has meant the re-ascendance of the city as the site, and the target, of the urban operations of platforms. The emerging literature on platform urbanism is indicative of this city-centric, administratively bounded and anthropocentric focus.[13] Similarly, discussions of 'smart cities' are bogged down by a 'methodological cityism' that has limited a broader and more complex understanding of how urban technology regimes are constructed atop globally extended material geographies with inherent political, social and spatial intricacies.[14] And while the patina of immateriality is slowly wearing off, the hidden forms of tech platforms and the obscurity of the means of data production, mediated through abstractions and marketing, still privilege the city over the operational landscapes of data, where the primary production of platforms takes place. Even when the operational spaces of platforms are highlighted, their representation is folded into a narrative of technological progress and peaceful cohabitation of technology and environment, which attempt to iron out the social, environmental and political tensions that are at the core of the territorial project of planetary platforms.

## Machines in Landscape and Territorial Instruments

In 2012, Google released a series of photographs of its data centres, which up to that point had existed in a shroud of mystery. In addition to closeup shots of flickering lights and carefully lit rows of server stacks, the facilities were shown surrounded by mountains, rivers and wildlife. This idyllic imagery was very different from the screen-side representation of Google or the images of its campus in California that had dominated the representations of the company up to that point. Yet the juxtaposition of the massive machinery of data centres, and their backup generators and clouds of steam, with the surrounding sleepy countryside was not easy to digest. The sublime imagery simultaneously recalled the disruptive quality of technology in Leo Marx's *The Machine in the Garden*, and Hans Hollein's fantastical collages of technical objects in landscapes.[15] Whereas Marx's writing examined the relationship between culture and technology in nineteenth- and twentieth-century American literature, Hollein's collages belonged to the radical architecture movement of the 1960s, which included Archigram, Superstudio and Archizoom. The movement that emerged partly in response to post-Second World War modernist architecture, attempted to re-evaluate the relationship between society and technological changes brought on by mass communication and information technologies, particularly the telephone and the computer.[16]

Hollein's collages from this period, which included surreal juxtapositions of rural landscapes with aircraft carriers or technical objects like sparkplugs and theodolites, attempted to create a more appropriate (architectural) imagery of its time. This was not the first and would not be the last time that architects would mobilize technical objects of their time as foundational drivers of a 'new' architecture. But while Hollein's collages were meant to be provocative by playing up the uncomfortable juxtaposition of technology and landscape, Google's images attempt to strike a harmonious tone between data infrastructure and nature. In this sense, 'where the internet lives' is presented matter-of-factly as a part of—and perhaps positively contributing to—the natural environment.

But the technological sublime of these images goes well beyond pacifying the environmental impact of platforms. In the context of the competitive dynamics between major tech platforms, representing infrastructure and technical apparatus enables a staking of ground within corporate territory as platforms extend their territorial claim and operations. As Holt and Vonderau have argued, the infrastructural politics that is played out through this imagery is 'about the hypervisibility created around some of an infrastructure's component parts, all while most of the relations it engenders and the rationality embodied in its overall system sink deeply in obscurity'.[17] The obscurity of their means of production

Fig. 3 Google's Council Bluffs data centre in Iowa during wildflower bloom and visited by deer. Images courtesy of Google.



and the power relations enacted through them enables platforms to produce an abstraction that is essential for their representation and marketing, and hence their territorial expansion.

This points to an important contradiction between abstraction, metaphors and the material operations of platforms. While technology platforms have introduced new dynamism and hypermobility to information, capital and social interactions, this fluidity is continually materialized in the very physical infrastructure of connectivity that enables the operations of platforms. This 'spatial fixity' is at the core of capitalism's tendency towards geographic and territorial expansion.[18] Central to this expansion is the construction of massive infrastructural networks and capital investment into pipes, cables, roads, ports and railways that underlie and connect concentrated moments of agglomeration, a process that 'grounds' these infrastructural networks in the very space and time they are trying to overcome. Consequently, the 'annihilation of space through time' (that is, the creation of frictionless, efficient conduits of capitalist production and circulation) necessitates investments into physical infrastructures of transportation and communication.

As mediators between the contemporary information society and its data, tech platforms are entangled in a similar process. The production and circulation of data entails the spatial (re)production of operational landscapes of extraction, storage and processing, as well as infrastructural geographies of circulation that are necessary for the continual expansion of platforms. The processes at the core of platform capitalism demand continual territorial expansion and competitive enclosure of increasingly sparse resources.[19] Google, for example, has spent more than $30 billion on its global cloud infrastructure.[20] Amazon has spent $35 billion on data centres in Northern Virginia alone.[21] Microsoft, Amazon, Google and

Fig. 4 Hans Hollein's Aircraft Carrier City in Landscape (1964). © Private archive Hollein; digital Image © The Museum of Modern Art/Licensed by SCALA / Art Resource, NY



Facebook have invested so heavily in undersea fibreoptic lines that their private networks' bandwidth capacity now completely overshadows the backbone of the Internet.[22] All these developments point to the importance of infrastructural landscapes and material support systems to the expansionist operations of platforms. It is these operational geographies that platforms leverage to exert their considerable power in cities. Recent discussions in urban studies have productively positioned concentrated moments of urbanization in cities in direct relationship to material and infrastructural extensions beyond population centres.[23] Elaborating on these extended 'operational landscapes' of platforms would allow for a more nuanced reading of the spaces, scales and more-than-human geographies that are not captured in the current city-centric discussions of platform urbanism and the surficial reading of the 'countryside' as a categorically separate part of contemporary urbanization.[24]

## Excavating the Infrastructural Palimpsest of Platforms

The grounding of the cloud entails a spatial logic based on a geopolitical cocktail of economic, political and environmental conditions.[25] In locating their nodes, platforms have followed in the footprints of previous rounds of capitalist spatial production. Data centres are often located on top of the ghosts of the manufacturing economy or within the spatial vestiges of the military industrial complex of the Cold War.[26] The infrastructural and security surplus built into these leftover spaces means significant savings for data-centre companies. This problematizes the idealized spatial imaginary of contemporary technology platforms. Whereas lightness and revolutionary spatial dynamism tend to characterize data technology and its spatial footprints, in reality tech platforms are highly dependent on previous rounds of infrastructural development and the established processes of enclosure and privatization at the core of neoliberal urbanization.

The conceptual framework of palimpsest can help articulate the sectional interrelationships and territorial interdependencies inherent to the sociotechnical production of planetary platforms. While the metaphor of the palimpsest has been used in architecture and urban design to describe the temporal layering and stratification of spatial interventions, in the discussion of platforms it can help explode the otherwise flat image of data technology. As Sébastien Marot has explained in relationship to contemporary landscapes, 'the palimpsest contributes a much needed dimension of materiality, givenness, gravity, inertia, and care' to the 'hyperlandscapes' of the contemporary urban condition.[27] A palimpsestuous reading of the production of planetary platforms integrates the otherwise hidden contribution of more-than-human geographies, environments and species, and reintroduces weight, depth and intricacy to the spatial discussion of technology platforms. The palimpsest as a conceptual framework for the analysis of the grounded spatial relationships of platforms is particularly helpful in the context of recent reconceptualization of planetary-scale computing, especially in the work of Benjamin Bratton and the notion of the Stack.[28] Whereas the Stack presents the planetary computing system as a series of interrelated but otherwise discrete layers, the palimpsest frames it as an inherently complex and intertwined condition. Through the palimpsest it is difficult to separate the layers from one another as they form a new whole. This ultimately strengthens the relational reading of spatial interdependencies at the core of the sociotechnical production of platforms.

## Furnishing the Cloud

To fully articulate the potential of this palimpsestuous reading we can go back to Google. The company's Lenoir data centre was among the first series of large purpose-built data centres that Google constructed in the early 2000s. Lenoir is a small industrial city of around 18,000 people and the seat of Caldwell County in western North Carolina. Lenoir's location in the foothills of the Blue Ridge Mountains ensures relatively cooler temperatures compared with areas in the eastern part of the state, which is very beneficial for the maintenance of the many servers in Google facilities. The area is not disaster prone, and the proximity of Lenoir to Charlotte, NC and other major East Coast population centres also makes the city and the region an attractive destination for data centres. The city's previous industrial history was also critical in Google's decision to locate there. By the early 2000s the city's many furniture factories were facing massive competition from their Asian counterparts. Between 1992 and 2012, furniture manufacturing in North Carolina declined 56 per cent and employment in manufacturing has also steadily fallen.[29] Between 2004 and 2007, Lenoir saw the closure of seven furniture factories and the loss of 2,100 jobs.[30] The closing of factories meant that the excess power capacity that had been built for manufacturing over the past 50 years became available. The local electricity utility, Duke Energy, was willing to offer this excess capacity at very low rates for industrial use.[31]

By late 2005 when Google came calling, Lenoir's city officials, led by the Caldwell County Economic Development Commission and supported by the state of North Carolina, were ready to offer the company a massive incentive package to locate its data centre in the city. Nanette Byrnes, writing for *BusinessWeek* in 2007, valued the incentives offered by the city, the county and the state at $211.7 million for Google's initial $600 million facility.[32] As part of the incentive package the state legislature enacted a 'Google-driven' bill that allowed for data centre-specific sales and use tax exemptions. The state legislature's efforts were pushed along by Google's threats to back out of Lenoir if the bill was not enacted quickly.[33] The legislation was passed in a month. The bill effectively exempts Google from paying sales taxes for the massive amounts of the electricity that its data

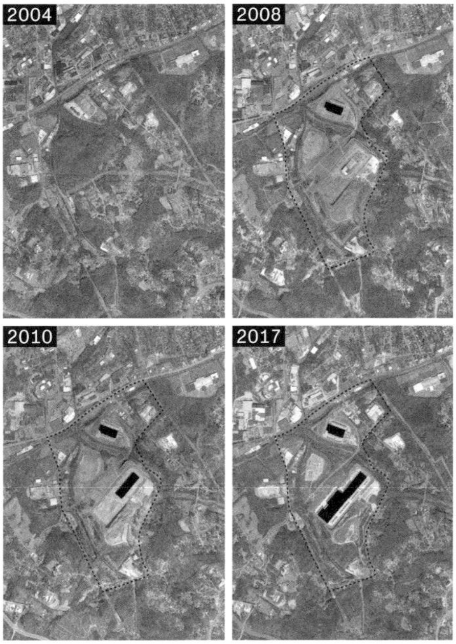


Fig. 5. Progressive development of Google's data campus in Lenoir, NC, 2007–2017

centres use. Servers, software and other equipment (comprising 'HVAC and mechanical systems, including chillers, cooling towers, air handlers, pumps and other capital equipment used for these purposes') would also be exempt.[34] For its part, the city of Lenoir offered a 100 per cent discount on local property taxes and 80 per cent off real estate taxes for 30 years. A substantial offer given Google's initial purchase of 216 acres of land. Since its initial 2007 project in Lenoir, Google has expanded its facilities and has doubled its investment to $1.2 billion.[35] In 2008 the company purchased an additional 60 acres, bringing its total to 276 acres of land.[36] This also means that the value of the incentive package offered by the city and the state has in turn increased well beyond the initial $211.7 million.

Beyond incentives, the heavy energy usage of data centres entangles these footprints of platforms with the geopolitics and the environmentally degradative extractive practices of energy production and transmission. Greenpeace has highlighted the heavy toll that all this increased energy consumption has on the environment. In Greenpeace's 2015 report, of the major tech companies that it analysed only two source more than 50 per cent of their electricity needs from clean energy sources. Additionally, energy from coal is still responsible for a considerable chunk of the energy used by data centres and platform companies.[37] Following the report and relying on their huge purchasing power in North Carolina, Google and Apple have successfully encouraged Duke Energy to start a renewable energy programme and to invest heavily in solar energy production in the region.[38] This is not a small shift for the United States' largest electricity utility. In 2015 only 2 per cent of the energy provided by Duke Energy was from renewable sources, while coal accounted for 32 per cent.[39] And even though the new investment presents a positive shift towards renewable energy production, it only accounts for a meagre 0.006 per cent (278 of 50,000 megawatts) of the capacity produced nationally by the company.[40] As one of the catalytic forces that brought Google and Apple to the state, Duke Energy is by North Carolina law the sole provider of power to customers. And the energy used in data centres is as clean or dirty as the

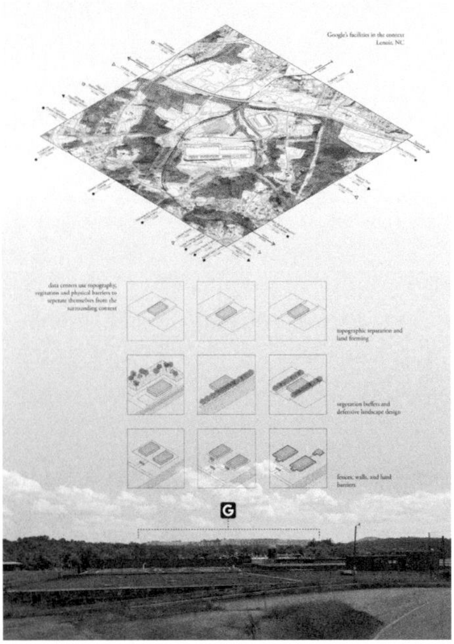


Fig. 6 Google's Lenoir facilities in context. Bottom: Google's data centres perched on top of the city in the midst of abandoned factories. Viewed from 467 Virginia Street SW. Middle: typical spatial strategies of data centres. Top: Lenoir data centres and their proximities and spatial dependencies.

local power grid that supplies it. One may question then whether the new push by Google for Duke to adopt renewable energy actually holds any merit or if it is solely cosmetic. In either case, the extractive logic of platforms makes sure that the decision ultimately comes down to economics. 'The price of renewable-power sources are far less volatile than power derived from commodities like coal, which is useful for things like long-term budgeting and cost-cutting.'[41] The decision is as much a business tactic as it may be a marketing one.

So, what is the actual value of having a data centre in your backyard? Initially the Lenoir project was touted for its economic regenerative potential for the city. Following years of decline, the city was ready to embrace a 'new economy', and to be part of the twenty-first century. As Byrnes wrote in 2007, Google was welcomed in Lenoir with open arms and was seen as 'a vital morale booster, if not a full replacement for the lost furniture factories'.[42] Although the construction of the data centres would mean more construction jobs initially, these are temporary employment opportunities. Data centres are known for their lean workforce. By 2015 Google employed 250 people in Lenoir, of which 150 were 'white-collar technical staff' who attend to the 50,000 servers of the facility. The remaining 100 employees were mostly security guards and HVAC technicians. Only about two thirds of the employees actually lived in Caldwell County.[43] Based on Google's own estimate, the company has created 340 jobs in all of North Carolina.[44] That number can hardly replace the more than 3,000 jobs that have been lost in Caldwell County since Google came to town.[45] Even spatially, a typical data centre's defensive strategies use topography, vegetation and hard barriers like fences and walls to cut it off from its immediate context while maintaining heavy dependence on regional infrastructures of energy, water and data.

Hence, the expansionist logic of platforms demands a complex set of geographic requirements and entails a series of spatial dependencies that together form the infrastructural palimpsest at the core of the sociotechnical production of planetary platforms. In Lenoir, the infrastructural surplus left over from many years of industrial development, inexpensive energy and real estate, proximity to population centres, political/economic lubrication through incentives and customized legislation, optimal climatic and environmental conditions, and, as importantly, the social and cultural conditioning of the local population through socially progressive imaginary propagating the 'transformative' role of twenty-first-century technology, all coalesce into a dense and complex sociotechnical palimpsest that cannot be easily decoded or simplified. Too often, however, narratives on platforms nullify this complexity by isolating and focusing solely on one of these factors, be it local economic generation or environmental impact. A palimpsestuous reading of these critical but latent dependencies allows for a richer, albeit a much more complex, relational understanding of the spatial imprints of planetary platforms.

## Complicating Things as Method

The above episodes help to illustrate the intertwined relationship between the sociocultural representation and the material geographies of data infrastructures. While at times each subverts or obfuscates the other, both are critical components of the sociotechnical production of planetary platforms. This helps extend the argument that, as opposed to 'accidental megastructures', technology platforms are highly designed sociotechnical entities.[46] Tech platforms combine ideology, myth and marketing to construct amorphous and all-encompassing narratives that socially and culturally condition their users, while actively hiding their dependence on the infrastructural landscapes, material geographies and operational extensions that increasingly encompass the globe. As the computational logic of digital platforms begins to dominate urban life, tech corporations leverage this planetary geography to exert ever more control over the organization of urban spaces and the people whose data they depend on. However, by externalizing these operational landscapes, tech platforms essentially evade responsibility and avert attention from the grounded dynamics and the geopolitical messiness of data technology. As shown through cases like Lenoir, the scrambling for resources and territory within the expansionist logic of platforms exacerbates their negative social, political and environmental impact.

As planetary platforms like Meta, Amazon and Alphabet come under increasing scrutiny, their sociopolitical impacts need to be contextualized within the extended geography of their operations, and the real geographic and environmental toll of their infrastructures. Excavating the tension between how these platforms are represented and the palimpsestuous spatial dependencies at the base of their construction can greatly contribute to our understanding of the actual impacts these platforms have on global processes of urbanization. And while initially this may complicate the smooth and effortless image of platforms, this complexity will ultimately be productive in generating a more intricate reading of the intertwined relationship between narrative and material conditioning of the ground as tech corporations' power and influence continues to expand.

Notes

1 The Whole Earth 'Lectronic Link, or the WELL, is one of the earliest forms of a virtual community that was launched in 1985.
2 The drawings have continued to be collected on the page since March 2009. The most recent entry, at the time of writing, was in January 2019. See: Kevin Kelly, 'Internet Mapping Project', Flickr, 2009, https://www.flickr.com/photos/kevinkelly/sets/72157613562011932/.
3 Tung-Hui Hu, *A Prehistory of the Cloud* (Cambridge, MA: MIT Press, 2015), X.
4 Jean-François Blanchette, 'From Digital Exceptionalism to Computational Scarcity', in: Ali Fard and Taraneh Meshkani (eds.), *Geographies of Information (NG7)* (Cambridge, MA: Harvard University Press, 2015), 26.
5 Hu, *A Prehistory of the Cloud*, op. cit. (note 3), X.
6 Shannon Mattern, 'A City Is Not a Computer', in: Shannon Mattern, *A City Is Not a Computer: Other Urban Intelligences* (Princeton, NJ: Princeton University Press, 2021), 51–72.
7 Sheila Jasanoff, *States of Knowledge: The Co-Production of Science and Social Order* (New York: Routledge, 2004).
8 Sheila Jasanoff, 'Future Imperfect: Science, Technology, and the Imaginations of Modernity', in: Sheila Jasanoff and Sang-Hyun Kim (eds.), *Dreamscapes of Modernity: Sociotechnical Imaginaries and the Fabrication of Power* (Chicago/London: The University of Chicago Press, 2015), 4.
9 Jathan Sadowski and Roy Bendor, 'Selling Smartness: Corporate Narratives and the Smart City as a Sociotechnical Imaginary', *Science, Technology, & Human Values* 44/3 (2019), 543.
10 Adam Greenfield, *Radical Technologies: The Design of Everyday Life* (London/New York: Verso, 2017).
11 See: Stephen Graham and Simon Marvin, *Splintering Urbanism: Networked Infrastructures, Technological Mobilities and the Urban Condition* (London/New York: Routledge, 2001); Maria Kaika and Erik Swyngedouw, 'Fetishizing the Modern City: The Phantasmagoria of Urban Technological Networks', *International Journal of Urban and Regional Research* 24/1 (2000), 120–138.
12 Stephen Graham, *The Cybercities Reader* (London/New York: Routledge, 2004), 8.
13 Peter Mörtenböck and Helge Mooshammer (eds.), *Platform Urbanism and Its Discontents* (Rotterdam: nai010, 2021); Sarah Barns, *Platform Urbanism* (Singapore: Springer Singapore, 2020).
14 Hillary Angelo and David Wachsmuth, 'Urbanizing Urban Political Ecology: A Critique of Methodological Cityism', *International Journal of Urban and Regional Research* 39/1 (2015), 16–27; Ali Fard, 'Cloudy Landscapes: On the Extended Geography of Smart Urbanism', *Telematics and Informatics* 55 (2020), 101450.
15 Leo Marx, *The Machine in the Garden: Technology and the Pastoral Ideal in America* (Oxford: Oxford University Press, 1964).
16 Simon Sadler, *Archigram: Architecture without Architecture* (Cambridge, MA: MIT Press, 2005).
17 Jennifer Holt and Patrick Vonderau, '"Where the Internet Lives": Data Centers as Cloud Infrastructure', in: Lisa Parks and Nicole Starosielski (eds.), *Signal Traffic: Critical Studies of Media Infrastructures* (Chicago: University of Illinois Press, 2015), 80.
18 David Harvey, 'Globalization and the "Spatial Fix"', *Geographische Revue* 3/2 (2001), 23–30.
19 Nick Srnicek, *Platform Capitalism* (Cambridge: Polity Press, 2017).
20 Eric Schmidt, the Chairman of Alphabet, mentioned the $30 billion investment in Google's infrastructural expansion at Google Cloud's Next conference in San Francisco on 8 March 2017. See: 'Google Cloud Next, 17—Day 1 Keynote', YouTube video, 2:04:35, posted by 'Google Cloud', 8 March 2017, https://youtu.be/j_K1YoMHpbk. See also: João Marques Lima, 'Google's Cloud CAPEX Hits $30bn with More to Come as Giant Plans 10 New Data Centres', *Data Economy* (March 2017), https://data-economy.com/googles-cloud-capex-hits-30bn-come-giant-plans-10-new-data-centres/.
21 Rich Miller, 'AWS Has Spent $35 Billion on Its Northern Virginia Data Centers', *Data Center Frontier* (2021), https://datacenterfrontier.com/aws-has-spent-35-billion-on-its-northern-virginia-data-centers/.
22 Alan Mauldin, 'Rising Tide: Content Providers' Investment in Submarine Cables Continues', *TeleGeography Blog* (2016), https://blog.telegeography.com/rising-tide-content-providers-investment-in-submarine-cables-continues.
23 Neil Brenner and Christian Schmid, 'The "Urban Age" in Question', *International Journal of Urban and Regional Research* 38/3 (2014), 731–755; Neil Brenner, 'Theses on Urbanization', *Public Culture* 25/1 (2013), 85–114; Neil Brenner and Christian Schmid, 'Planetary Urbanisation', in: Matthew Gandy (ed.), *Urban Constellations* (Berlin: Jovis, 2011), 10–13.
24 Neil Brenner and Nikos Katsikis, 'Operational Landscapes: Hinterlands of the Capitalocene', *Architectural Design* 90/1 (2020), 22–31.
25 Thomas Freeman, 'What's Important in the Data Center Location Decision?', *Area Development* (2011), https://www.areadevelopment.com/siteSelection/may2011/data-center-location-decision-factors2011–62626727.shtml; Paul T. Jaeger et al., 'Where Is the Cloud? Geography, Economics, Environment, and Jurisdiction in Cloud Computing', *First Monday* 14/5 (2009).
26 Sebastian Moss, 'Children of the Cold War', *Data Centre Dynamics* (2018), https://www.datacenterdynamics.com/analysis/children-of-the-cold-war/; Cade Metz, 'Google Reincarnates Dead Paper Mill as Data Center of Future', *Wired* (2012), https://www.wired.com/2012/01/google-finland/.
27 Sébastien Marot, 'Envisioning Hyperlandscapes', *Harvard Design Magazine* 36 (2013), 94.
28 Benjamin Bratton, *The Stack: On Software and Sovereignty* (Cambridge, MA: MIT Press, 2015).
29 Duke University Center for Globalization Governance & Competitiveness, 'Furniture', North Carolina in the global Economy, n.d., http://www.ncglobaleconomy.com/furniture/overview.shtml.
30 Nanette Byrnes, 'The High Cost of Wooing Google', *BusinessWeek* (July 2007), 51.
31 Rich Miller, 'The Apple-Google Data Center Corridor', *Data Center Knowledge*, 4 August 2009, https://www.datacenterknowledge.com/archives/2009/08/04/the-apple-google-data-center-corridor/.
32 Byrnes, 'The High Cost of Wooing Google', op. cit. (note 30), 52.
33 Ibid., 54.
34 'Data Centers Sales & Use Tax Exemptions', The Economic Development Partnership of North Carolina, n.d., https://edpnc.com/incentives/datacenter-sales-and-use-tax-exemption/.
35 Rich Miller, 'Google Expands in North Carolina, Will Boost Renewables', *Data Center Knowledge*, 19 April 2013, https://www.datacenterknowledge.com/archives/2013/04/19/google-expands-in-north-carolina-will-boost-renewables/.
36 Rich Miller, 'Google Buys More Lands at Lenoir Data Center', *Data Center Knowledge*, 15 December 2008, https://www.datacenterknowledge.com/archives/2008/12/15/google-buys-more-land-at-lenoir-data-center/.
37 Gary Cook, *Clicking Clean: A Guide To Building the Green Internet* (Washington, DC: Greanpeace, 2015), 8, http://www.greenpeace.org/usa/wp-content/uploads/legacy/Global/usa/planet3/PDFs/2015ClickingClean.pdf.
esearchGate (2013)Electricity consumption
38 'Duke Energy Commits $500 Million to North Carolina Solar Power Expansion', *Duke Energy*, 15 September 2014, https://news.duke-energy.com/releases/duke-energy-commits-500-million-to-north-carolina-solar-power-expansion; Bruce Henderson, 'Duke Energy to Spend $500 Million on N.C. Solar', *The Charlotte Observer*, 15 September 2014, http://www.charlotteobserver.com/news/business/article9165593.html.
39 Duke Energy Carolinas, 'Integrated Resource Plan (Annual Report)' (2014), https://starw1.ncuc.gov/NCUC/ViewFile.aspx?Id=c3c5cbb5-51f2-423a-9dfc-a43ec559d307.
40 Marti Maguire, 'Duke Energy to Invest $500 Mln in North Carolina Solar Power', Reuters, 15 September 2014, https://www.reuters.com/article/usa-duke-energy-solar-idUSL1N0RG2D520140915.
41 Ingrid Burrington, 'The Environmental Toll of a Netflix Binge', *The Atlantic*, 16 December 2015, https://www.theatlantic.com/technology/archive/2015/12/there-are-no-clean-clouds/420744/.
42 Byrnes, 'The High Cost of Wooing Google', op. cit. (note 30), 52.
43 Wes Mayberry, 'Google Data Center in NC Sets Good Example for Potential Impact of Jackson County Site', *The Daily Sentinel*, 26 June 2015, http://jcsentinel.com/feature_story/article_aa7eb456-1c4b-11e5-bd1f-b39a1408974a.html.
44 Google, 'Google Economic Impact: North Carolina' (2021), https://economicimpact.google.com/state/nc/.
45 Based on annual average of monthly employment levels for a given year, 2007–2021. Data from: United States Department of Labor, 'Quarterly Census of Employment and Wages, 2007–2022', Bureau of Labor Statistics, n.d., https://www.bls.gov/data/.
46 Bratton, *The Stack*, op. cit. (note 28).

# THIS WAY: E

## THIS WAY: EXPRESSIVE SELF-TRACKING AS CREATIVE PRACTICE

Joost Grootens

In the early 1960s, Suriname-born Dutch conceptual artist Stanley Brouwn (1935–2017) approached random pedestrians and asked them to draw directions to a particular place on a piece of paper. The resulting maps consist of sometimes clumsily drawn loops, lines, circles, arrows, crosses and street names. Subsequently, Brouwn stamped the text 'This Way Brouwn' on the drawings and exhibited them as his works of art.[1]

In the above description of the work of art *This Way Brouwn*, several themes emerge that I will address in this article about self-tracking as creative practice: navigation, drawing, mapmaking, expression, appropriation and the role of non-specialists. I will explore how evolving digital technologies influence how we use, measure and display our surroundings, and draw comparisons between Brouwn's work of art and a map drawn some 60 years later by a Dutch cyclist, to analyse the technological, economic, political, sociological and theoretical dimension of fitness apps and GPS navigation. The article will focus on the material side of these technologies and the visualization practices they enable.

### Self-Tracking as a Creative Practice

Self-tracking is the recording and monitoring of specific features of one's own life. It is also called lifelogging, personal analytics, personal informatics and the quantified self.[2] These practices go back to the keeping of a diary, but have changed with the dispersion of mobile digital devices. Activity tracking or fitness tracking is a specific kind of self-tracking. Using the Global Positioning System (GPS) and other functionalities of mobile and smart devices, the position and duration of sport activities are recorded as well as the condition of the body in terms of heart rate and blood pressure. Fitness apps keep track of the recorded data and combine these to give users insight and comparison through infographics and maps.

Writing a text or drawing an image making use of satellite navigation is called GPS drawing. Individuals run, cycle, swim, drive, boat or fly a preplanned route to create a large-scale image. Using smart devices with GPS functionality, and software such as fitness apps, to capture the drawing in a GPS exchange format (GPX), the image is then visualized as a line on a map. Other names for GPS drawing are GPS art and Strava art, the latter named after the widely used fitness app and social network for athletes Strava, about which more later. Below, I will describe a specific GPS drawing and identify and untangle the various aspects of the map.

### Giro d'Italia

In the spring of 2020, at the height of the first wave of the Covid-19 pandemic, René Koppert (1960) gets on his racing bike to cycle through the south of the Netherlands. The carefully planned trip is recorded on his Garmin GPS device. The resulting drawing is a map of Italy, approximately at a scale of 1:50, projected onto the landscape of the province of North Brabant. Koppert, a former professional cyclist who raced in Dutch and Italian cycling teams in the 1980s, is based in the Netherlands but has children living in Italy.[3] In the spring of 2020, the world was in the grip of the corona virus. Italy was particularly hard hit during the first months of the pandemic. National borders were closed, events were cancelled, or postponed. The Giro d'Italia, the annual three-week Italian cycling race usually held in May, was postponed to October. After having cycled the contour of Italy, Koppert sends a screenshot of his GPS drawing to his children in Italy by WhatsApp.

In addition to the closed borders that prevented him from visiting his children, and the cancelled cycling race in which he once participated, Koppert indicates that he was also inspired by the Strava Art movement to create his map of Italy.[4] According to Koppert, GPS drawing gained popularity during the corona pandemic when it was no longer possible to exercise in groups. Making a drawing through a sporting activity and sharing it online became a substitute social activity.

Koppert designs the route of Italy's contour by opening the Strava and Google Maps websites side by side on his computer, the first zoomed in on the map of the south of the Netherlands, and the latter on the map of Italy. On the map in Strava, Koppert looks for parts that correspond in form to the characteristic outline of Italy. He starts by looking for the shape of the heel of the boot of Italy, the regions of Puglia, Basilicata and Calabria, and finds it in the road pattern around the towns of Heeswijk Dinther, Veghel and Schijndel. The drawing that is thus created is sent as a GPX-file to the Garmin GPS device on his bicycle that transforms the drawing into an itinerary. While cycling, Koppert adjusts the route, removing illogical parts and avoiding traffic lights. The line that Koppert cycles is recorded in the Strava app and opened again on his computer, after which the route is further refined, optimized and brought closer to the shape of the map of Italy. This goes back and forth five or six times between designing and cycling until the map is done.

The final route, with the shape of the outline of Italy, is about 100 km long. The northern Italian border is formed by the Maas River, the Zuid-Willemsvaart forms the Adriatic coast, the surroundings of the village of Liempde pass for Sicily, the Tyrrhenian coast is formed by the road between Schijndel and Vught, while the city of 's-Hertogenbosch forms the north of Italy, from the provinces of Liguria and Emilia Romagna upwards. Koppert's map is missing Sardinia, but this will be added later by others as the map makes its way on social media after being shared on Strava by former pro cyclist Erik Dekker[5] and reported in a local newspaper.[6]

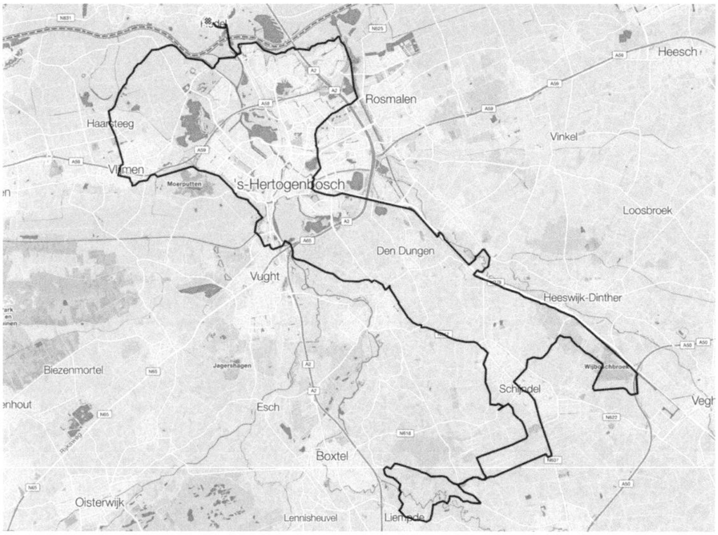


Fig. 1 René Koppert, *Giro d'Italia in The Netherlands*, Map, published on Strava, 2020.



## Satellite Navigation

The technology underlying GPS drawings, such as that of René Koppert, is satellite navigation. A network of satellites and ground stations provide users with a portable receiver with accurate information about their location—anywhere, anytime and in all weather conditions. Today there are several global navigation satellite systems: GPS (United States), GLONASS (Russia), BeiDou (China) and Galileo (European Union). GPS, the oldest of the networks and the first to open its signal to all users, consists of 32 satellites circling the Earth in six orbits at an altitude of 20,200 km.[7] The satellites carry atomic clocks that are synchronized with each other and with ground stations. A GPS receiver monitors multiple satellites and calculates the exact position by measuring the time it takes to receive the different signals. A minimum of four satellites must be in 'view' of the receiver to calculate its three-dimensional location (latitude, longitude and altitude).

GPS was initially developed as a military technology, but its signal was gradually made available for civilian use as well. Releasing the signal to all sparked a commercial GPS equipment and service industry. The early GPS receivers were large, heavy and expensive. Take, for instance, the first portable consumer GPS device, the NAV 1000, manufactured by Magellan from 1989 and widely used in the first Gulf War (1990–1991). The NAV 1000 measured 22.2 × 8.9 × 6.3 cm, weighed almost 700 grams, and had a retail price of $3,000. Over time, GPS receivers became smaller, lighter, cheaper and more widely available. The Garmin Edge 530, a device similar to the one with which René Koppert made his drawing of Italy, is one tenth of the dimensions, weight and price of the NAV 1000: measuring 5.8 × 11.4 × 1.9 cm, weighing 75.8 grams and costing just under $300. Over time, GPS tracking became a feature on other devices as well, such as digital cameras, adding location coordinates to digital photos, cell phones, smartphones, where GPS tracking is a standard feature, and wearable technology such as fitness trackers and smartwatches.

While GPS technology offers individuals possibilities to compile the 'intimate particularity' of their 'personal plots', satellite navigation also provides opportunities for commercial enterprises and the state, in their quest for 'total knowledge'.[8] 'Surveillance capitalism'[9] and 'liquid surveillance'[10] are terms used to describe these monitoring practices in which the user's behavioural data is the source of new economic value, respectively the ubiquitous technologies and apps that constantly check, monitor, test, assess, value and judge us.

## Strava

To design his drawing of Italy and later share it with others as itinerary, René Koppert uses the Strava app. Strava is a social network for runners, cyclists and triathletes.[11] With the Strava website and mobile app, users can track their activities and share and compare their performance with other Strava users. The app, used by both professional and amateur athletes, is free but offers paid subscriptions with additional features.

Founded in San Francisco in 2009, Strava is one of the last independent fitness trackers. Other trackers are developed or purchased by technology companies such as Apple (Apple Health), Fitbit (Fitstar), Google (Google Fit), Microsoft (MSN Health & Fitness), Nokia (Nokia Sports Tracker) and Samsung (Samsung Health), or sportswear companies such as Adidas (Runtastic), Asics (RunKeeper), Nike (Nike Training Club) and Under Armor (Endomondo, MyFitnessPal, MapMyFitness). For technology companies, it is lucrative to collect behavioural data from their users as they can link this data to saleable advertising space. For sportswear companies, it is lucrative to know where which sports are practiced, by whom and in what way. As Kevin Plan, CEO of Under Armour, said in an interview with *The Financial Times* in 2015 after his company purchased three fitness apps: 'We now have the world's largest digital health and fitness community. We believe ultimately this will help us sell more shirts and shoes, reach more athletes, and make them better.'[12]

Strava is not yet profitable and has raised venture capital to compensate its expenses.[13] The company adopts a different economic strategy than its competitors with roots in technology and sportswear. In 2013, Strava published a Global Heatmap showing the aggregated and anonymized activities of its users. The map sparked the interest of city planners and transport services, for whom it provided a unique insight into cyclists' movements through their cities. Traffic movement recording is a common method used by planners, but Strava's dataset provided a unique opportunity to access many movements, at different times of the day, including routes and speed. In 2014, this culminated in the creation of a new company: Strava Metro sells Strava users' data to work with cities to improve infrastructure for cyclists and pedestrians.[14] With the commercial venture Strava Metro in mind, Strava's Global Heatmap is more than a map that allows users to 'discover new places to be active',[15] above all it presents a showcase of the technical and commercial capabilities of the company and the data it can offer.

René Koppert uses the Strava Global Heatmap for sports activities. In his case, if he goes to unfamiliar places, he uses the Heatmap as a source of information to discover popular bicycle routes. On the other hand, for local cycling tours, he uses the Heatmap to help him avoid roads frequently used by other athletes.

Not only does Strava passively track data for Strava Metro, it also actively encourages users to collect data outside of their leisure activities and share this information. In 2016, Strava invited cyclists to participate in the Global Bike to Work Day Challenge. It also launched the #CommutesCount campaign. A short animation accompanying this includes the message: 'Traveling with Strava makes cities better.'[16] While Strava encourages its users to collect and share data, users have the possibility to opt out. The app's default setting, however, is to accept sharing of user data and users have to make a real effort to disable tracking in the app's settings—which most do not.

The Strava Global Heatmap became controversial in January 2018, when a military analyst discovered that the map displays the location of secret military bases.[17] Ironically, GPS—the technology that lies at the basis of fitness apps like Strava—was originally a closed-off military technology that was later made available for civilian use, partly to support the developing US commercial GPS equipment and service industry. After the discovery of the sensitive information about the locations of secret military bases, the discussions in the news media focussed on the privacy of users of social media and the power of technology companies like Strava, Facebook and Google to track our behaviour and use it to control our lives.

## Expressive Self-Tracking

Not all self-tracking practices are taken up voluntarily. Deborah Lupton, professor of sociology at the University of New South Wales, Sydney, identifies five different modes: private, pushed, communal, imposed and exploited self-tracking.[18] In the case of the Strava app, a variety of these types can be observed. Many individual Strava users will use the self-surveillance app to collect information about themselves, raise self-awareness and optimize their lives.

Since Strava is both a fitness tracking app and a social medium, the focus of the tracking goes beyond the individual user. Users are part of a community of trackers. The platform enables the comparing and sharing of data with other members. Another form of communality is the message Strava puts forward in its Commutes Count campaign. In an animation for the 2017 campaign, users are persuaded to take responsibility for their city and take part in an event: 'Every time you commute on Strava you create anonymous data. The data show urban planners how to improve your city. Commuting by bike is good for you anyway. Now it's good for your city too.'[19]

As mentioned above, the Strava Global Heatmap revealed the locations of secret military bases because the personnel on those bases were recording their activities. It is ironic that some military personnel were given the devices to track their activities by the army itself. In 2013 the US Army provided some of its personnel with fitness trackers to promote a healthier lifestyle.[20] Lupton calls this mode of self-tracking, where the initial incentive does not come from the user, pushed self-tracking.[21] Or even imposed self-tracking if the benefit is only to the advantage of others than the user.[22]

Exploited self-tracking are those practices where personal data of users are repurposed for the (commercial) benefit of others. This is the case when Strava Metro sells the behavioural data of users to cities. One could argue that in the anonymized aggregated data it is not possible to find the data patterns of individual users. However, hacking practices and instances where datasets have been combined to uncover secrets show that the boundary between personal small data and anonymous big data is blurry at best, and non-existent at worst.

Where should we situate René Koppert's self-tracking practice in Lupton's categorization? The recording of leisure activities to track and improve his physical well-being could be classified as private self-tracking. Since Koppert uses Strava and shares the activities with other users of the fitness app, it can also be classified as communal self-tracking. Although Koppert does not have an open profile where his activities can be accessed by everyone, he does share his activities with a small group of followers that he has given permission to do so. If Koppert does not feel the pressure of fellow athletes to keep track of his data, then his practice cannot be classified as pushed self-tracking. And if Koppert has not given Strava permission to use his data, then exploited self-tracking is not the correct classification either.

Koppert's GPS drawing fits into a larger group of practices aimed at expressive self-tracking. There are specialized practices like that of artist Jeremy Wood, who has been using GPS technology to create drawings since 2000.[23] But also practices of non-specialists who express their ideas through GPS drawing in their leisure and fitness activities. The American Claire Pisano, for instance, publishes on Instagram under the moniker @ dick_run_claire GPS drawings of penises projected on urban cityscapes that she draws through running. Her drawings are both an expression of fitness culture, the position of Western women in the twenty-first century, and contemporary US urbanity.[24] The website strav.art lists more than 3,000 GPS-drawings categorized by subject in categories such as Elephants, Food & Drink, Plants, Sport and Words & Numbers.[25] The Geography section lists about 75 drawings of the contours of countries and continents, including nine maps of Italy, among which that of René Koppert.[26] Another category on the strav.art website is Burbing, drawing all of the roads of a suburb.[27]

Lupton's categorization feels limited in situating GPS drawing practices. The public nature of the drawings shared on social media, either by the original creator or by followers who recreate the drawing, extends so much further than the term community suggests in communal self-tracking. Koppert's Italy drawing also received attention outside the group of cyclists he is part of when it was shared online and reported in a newspaper.[28] In the case of GPS drawings, there is also the aspect that these self-tracking practices aim to create an image, or write text, that expresses a thought or feeling intended by the creator to be seen by others. For these practices, I would suggest extending Lupton's classification with the category expressive self-tracking.

## Post-Representational Cartography

In the text above I have addressed Koppert's practice but not yet his map. It could be argued that the contour of Italy that Koppert cycled is not a map but an illustration, a picture. This reading ignores that the drawing is a 'diagrammatic representation of a piece of land'[29] and that the image, and digital GPX file behind the drawing, can be used as an itinerary or to navigate an environment.

With the mere red line on the Open Street Map base map, Koppert's contour of Italy perhaps lacks the visual qualities of the rich cartographic heritage. It is more the action that made the line, and the skill with which it was created, that are appealing. Understanding and appreciating Koppert's map requires an approach that places less emphasis on what the map shows visually and more on the process behind it.

To situate René Koppert's Italy contour drawing as a map, I propose to use a post-representational approach. Post-representational cartography, a term coined by British-American geographer John Pickles, understands a map as a process rather than an object.[30] According to this approach, maps are never fully formed and their work is never complete, they are in a constant state of becoming. A map is constantly being made and remade, every time a user engages with it. According to this approach to cartography, the binary division between making and using no longer applies.[31] Since Koppert first shared his map on Strava in 2020, the route has been cycled 650 times.[32]

When approached as a process, it is interesting to consider at what moment the production of a map ends.[33] Is it when it is conceptualized, when it is embedded in other content on a page, when it is loaded on the screen of a digital device, when it is read, or maybe never? Similar considerations can be made when contemplating the use of maps. When is it first used? During the process of creation, the moment when the mapmaker sees the whole through the fragments? Or, in the case of Koppert's map, the consideration is whether the map was created when it was designed on the computer, or when the route was cycled? This also applies to the use of the map, was Koppert the first user when he saw the contour line appear on the map in Strava, when the map was shared as an image via WhatsApp, or when the route was cycled by himself, or others?

As the term 'post' in post-representational cartography suggests, it is a phase in the thinking about maps. Earlier stages are representational and more-than-representational cartography. Representational[34] or cognitive cartography,[35] developed after the Second World War, has as its premise that the world can be objectively known and truthfully mapped. Cartography develops by using controlled scientific experiments to improve the way a map communicates.

At the end of the 1980s, ideas from poststructuralist thinking, social constructionism and actor-network theory caused a shift in the thinking about maps. In more-than-representational[36] or critical cartography,[37] maps are seen as the product of power, but also as producing power themselves. Mapping is not objective or neutral but charged with power, where knowledge is not revealed but created.

The early 2000s saw a third shift in thinking about maps. In post-representational cartography, the fundamental status of a map is questioned. Maps are seen as a process and never fully formed. In previous ways of thinking the map was regarded as diverse, rhetorical, relational and complex, but nevertheless a stable product. The terms ontological and ontogenic have been used to describe the difference between post-representational thinking about mapmaking and the two earlier approaches.[38] Ontological refers to how things are, ontogenetic to how things become. The map has an ontologically safe basis in representational and more-than-representational cartography: a map is a map, and it is constant. Post-representational cartography questions that ontological basis and looks at the entire mapmaking process. This approach, which examines the entire process of production and use, and all its overlap and intermediate forms, makes it particularly suitable for considering maps in the digital age, in which the distinction between making and using is sometimes vague and often non-existent.

The three different approaches to cartography described above are based on distinct epistemological methods. Representational cartography uses quantitative methods from cognitive psychology. Methods from textual and linguistic deconstruction are used in more-than-representational cartography. The process-based approach to post-representational cartography uses a variety of methods, including genealogies, ethnographies, ethnomethodology, participant observation, observant participation and deconstruction, to capture the full scope of how a map is created and produced, and how social, embodied, political and economic relations play a role in this messy process.

## This Way

There are clear parallels between Stanley Brouwn's work of art *This Way Brouwn* and the GPS drawing of the contour of Italy that René Koppert created. Both are maps that involve non-specialist mapmakers. Both projects are intended to be shared with a larger audience, one in an art exhibition, the other on social media. Where the projects differ is the authorship and degree of completeness. Brouwn's work has a clear author: Brouwn himself. From stamping the work, giving it a title with his name in it, and by displaying it in a context that matches his position as an artist, it becomes clear that he is the creator of the map, even if he did not put the lines and street names on the sheet of paper. The work can be repeated, and Brouwn has done that several times, but after it is stamped and exhibited it is complete. With Koppert's map, the issue is more ambiguous: there is no final version of the map, the map is recreated every time someone views the image or travels the route. This also makes its authorship less clear. Not only because there are several GPS drawings of Italy in circulation, but mainly because the map is an ongoing process and not a finished product.

This makes Brouwn and Koppert's projects typical of the technological periods in which they were made: the analogue era in which Brouwn worked, and the digital era in which Koppert created his map. Ambiguity of authorship and degree of completeness are aspects of both maps. But where Brouwn's work is limited to questioning these aspects, with Koppert's map the ambiguity about the creation and status of the final object is fundamental and total because the map is an open process without a clear beginning or end. That is not to say that Koppert's map has only advantages over Brouwn's analogous version. The openness of process and the ambiguity between making and using has downsides. The digital tool will eventually limit and constrain the user.

American social scientist and professor emirata at Harvard Business School Shoshanna Zuboff formulated three laws of surveillance.[39] First, everything that can be automated will be automated. Second, everything that can be informed will be informed. And third, every digital application that can be used for surveillance and control will be used for surveillance and control, irrespective of its original intention. It is especially this third law that comes into play in the practices of GPS drawing where technology companies share apps and services with users in return for behavioural data. Knowledge about where and when athletes do which activity is profitable when you sell advertising space or sneakers, and this explains why tech and sportswear companies are providing fitness tracking apps to users for free or for a small fee. It is not yet fully clear how venture capital-funded Strava will exploit user data. The app with which Koppert made his Italy drawing might be acquired by a tech or sportswear giant, or by a completely different type of company with more problematic intentions than selling ads or sportswear.

More generally, the use of mobile devices is vulnerable to all kinds of surveillance practices. Documents leaked in 2014 by former National Security Agency contractor Edward Snowden show that US and UK intelligence agencies intercept and store Google Maps searches on smartphones with the location information from where the search was made. This was so successful that a 2008 document points out that it basically means that anyone who uses Google Maps on a smartphone is working in support of the intelligence services.[40]

In addition to these surveillance issues, users of self-tracking tools only have freedom within the limits set by technology companies. There are several examples of tech companies choosing to restrict the capabilities of their products for economic reasons or under pressure from states.[41] Every tool reflects the choices of its maker and their thoughts about how it should be used. The more extensive the tool, the more drastic the curation of its functionalities and thus limiting the possible outputs. This calls for awareness and for developing strategies to undermine and hack the digital tools that we use in our lives. In that respect we can learn from Stanley Brouwn who, by not picking up the pencil himself, was able to circumvent the conventions of his field, and thereby hack the work of art as format.

The powerful technology companies monopolize and centralize the technologies that enable us to find our way, to express ourselves by recording our activities and to share these with others. Self-tracking technologies control how we view the world and the data we thus generate is exploited as raw material for economic gain. By being aware and developing subversive strategies, we can determine our own ways, otherwise we will become casualties of the tools that we can no longer do without.

Notes

1 For more information on the work of Stanley Brouwn, see: Antje von Graevenitz, '"We Walk on the Planet Earth": The Artist as a Pedestrian: The Work of Stanley Brouwn,' translated by Ruth Koenig. *Dutch Art and Architecture* (June 1977), https://www.impossibleobjectsmarfa.com/ fragments/stanley-brouwn, accessed 12 June 2022.
2 Deborah Lupton, 'Self-Tracking Modes: Reflexive Self-Monitoring and Data Practices', paper for the 'Imminent Citizenships: Personhood and Identity Politics in the Informatic Age' workshop, 27 August 2014, ANU, Canberra, https://www.academia.edu/76659435/Self_Tracking_Modes_Reflexive_Self_Monitoring_and_Data_Practices, accessed 12 June 2022.
3 René Koppert cycled for TI-Raleigh (1982), Termolan-Gali (1983) and Santini-Conti-Galli (1984), and was the winner of the prologues in Tour de Romandie and Critérium du Dauphiné Libéré and a stage in the Tour of Germany.
4 Interview with René Koppert on 15 April 2022.
5 Erik Dekker (1970), professional cyclist from 1992 to 2006, winner of four stages in the Tour de France, Amstel Gold Race, Clásica San Sebastián, Paris-Tours and Tirreno-Adriatico.
6 Ties Cleven, 'Oud-prof Erik Dekker en zoon David rijden Giro d'Italia, maar dan in Brabant', *Brabants Dagblad*, 6 May 2020, https://www.bd.nl/brabant/oud-prof-erik-dekker-en-zoon-david-rijden-giro-ditalia-maar-dan-in-brabant~ac80b4ae4/?referrer=https%3A%2F%2Fwww.google.com%2F, accessed 12 June 2022.06.06.2022.
7 For more information on GPS and satellite navigation, see: Joost Grootens, 'The Strava Global Heatmap', in: Joost Grootens, *Blind Maps and Blue Dots: The Blurring of the Producer-User Divide in the Production of Visual Information* (Zurich: Lars Müller, 2021), 96–119; and Laura Kurgan, *Close Up at a Distance: Mapping, Technology, and Politics* (New York: Zone Books, 2013).
8 Lisa Parks, 'Plotting the Personal: Global Positioning Satellites and Interactive Media', *Cultural Geographies* 8 (2001), 209–222.
9 Shoshana Zuboff, 'Google as a Fortune Teller: The Secrets of Surveillance Capitalism', *Frankfurter Allgemeine Zeitung*, 5 March 2016, http://www.faz.net/aktuell/feuilleton/debatten/the-digital-debate/shoshana-zuboff-secrets-of-surveillance-capitalism-14103616.html?printPagedArticle=true, accessed 2 April 2018.
10 David Lyon, 'Liquid Surveillance: The Contribution of Zygmunt Bauman to Surveillance Studies', *International Political Sociology* 4 (2010), 325.
11 Strava, 'About Us', https://www.strava.com/about, accessed 12 June 2022
12 Tim Bradshaw, 'Under Armour Snaps up Fitness Apps', *Financial Times*, 5 February 2015, https://www.ft.com/content/2eed0aac-acc7-11e4-beeb-00144feab7de, accessed 12 June 2022.
13 Joseph B. Lassiter III, William A. Sahlman and Sid Misra, 'Strava: Harvard Business School Case 814–055', February 2014 (Revised August 2016).
14 'Strava Metro', website, https://metro.strava.com, accessed 12 June 2022.
15 'Strava Global Heatmap', Strava, https://www.strava.com/heatmap, accessed 12 June 2022.
16 '#CommutesCount', Strava Metro, animation, 24 April 2016, https://www.youtube.com/watch?v=LYUmhlAwuRA, accessed 24 August 2019.
17 Alex Hern, 'Fitness Tracking App Gives away Location of Secret US Army Bases', *The Guardian*, 28 January 2018, https://www.theguardian.com/world/2018/jan/28/fitness-tracking-app-gives-away-location-of-secret-us-army-bases, accessed 12 June 2022.
18 Lupton, 'Self-Tracking Modes', op. cit. (note 2).
19 'Commutes Count', Strava, animation, 3 May 2017, https://www.youtube. com/watch?v=WcGoR5RWQNU, accessed 12 June 2022.
20 Kevin Lilley, '20,000 Soldiers Tapped for Army Fitness Program's 2nd Trial', *Army Times*, 27 July 2015, https://www.armytimes.com/news/your-army/2015/07/27/20000-soldiers-tapped-for-army-fitness-programs-2nd-trial/, accessed 12 June 2022.
21 Lupton, 'Self-Tracking Modes', op. cit. (note 2), 7.
22 Ibid., 9.
23 Janet Abrams and Peter Hall, 'Skywriting: Janet Abrams and Peter Hall Talk to Jeremy Wood', in: Janet Abrams and Peter Hall (eds.), *Else/Where: Mapping: New Cartographies of Networks and Territories* (Minneapolis: University of Minnesota Design Institute, 2005), 2–9 and 268–275.
24 'dick_run_claire', https://www.instagram.com/dick_run_claire, accessed 12 June 2022.
25 'Strav.art', https://www.strav.art, accessed 12 June 2022.
26 'Geography', https://www.strav.art/geography, accessed 12.06.2022, accessed 12 June 2022.
27 'The "Burbing" Phenomenon', *Bicycle-network*, 12 August 2020, https://www.bicyclenetwork.com.au/newsroom/2020/08/12/the-burbing-phenomenon, accessed 12 June 2022.
28 Cleven, 'Oud-prof Erik Dekker', op. cit. (note 6).
29 'Map', *Oxford English Dictionary*, https://www.lexico.com/definition/map, accessed 12 June 2022.
30 John Pickles, *A History of Spaces: Cartographic Reason, Mapping and the Geo-Coded World* (London: Routledge, 2004), 160.
31 For more information on post-representational cartography, see: Joost Grootens, 'Positioning', in: Grootens, *Blind Maps and Blue Dots*, op. cit. (note 7), 28–57.
32 René Koppert, email message to author, 20 July 2022.
33 Vincent J. Del Casino Jr and Stephen P. Hanna, 'Beyond the 'Binaries': A Methodological Intervention for Interrogating Maps as Representational Practices', *ACME: An International E-Journal for Critical Geographies* 4/1 (2006), 35.
34 Rob Kitchin, 'The Transformation of Cartographic Thought', lecture filmed September 2012 at 'Mapping Humans: From Body to Cosmos. An International Conference', Oxford, http://www.pulse-project.org/node/540, accessed 9 November 2018.
35 Sébastien Caquard, 'A Post-Representational Perspective on Cognitive Cartography', *Progress in Human Geography* 39/2 (2015), 225–235.
36 Kitchin, 'The Transformation of Cartographic Thought', op. cit. (note 33).
37 Jeremy W. Crampton and John Krygier, 'An Introduction to Critical Cartography', *ACME: An International E-Journal for Critical Geographies* 4/1 (2005), 11–33.
38 Rob Kitchin and Martin Dodge, 'Rethinking Maps', *Progress in Human Geography* 31/3 (2007), 331–344.
39 Shoshana Zuboff, 'The Surveillance Paradigm: Be the Friction—Our Response to the New Lords of the Ring', *Frankfurter Algemeine Zeitung*, 25 June 2013, http://www.faz.net/aktuell/feuilleton/the-surveillance-paradigm-be-the-friction-our-response-to-the-new-lords-of-the-ring-12241996.html?printPagedArticle=true#pageIndex_0, accessed 28 July 2022.
40 James Ball, 'Angry Birds and "Leaky" Phone Apps Targeted by NSA and GCHQ for User Data', *The Guardian*, 28 January 2014, https://www.theguardian.com/world/2014/jan/27/nsa-gchq-smartphone-app-angry-birds-personal-data, accessed 28 July 2022.
41 Simon Usborne, 'Disputed Territories: Where Google Maps Draws the Line', *The Guardian*, 10 August 2016, https://www.theguardian.com/technology/shortcuts/2016/aug/10/google-maps-disputed-territories-palestineishere, accessed 28 July 2022.

# THE DIGITA

## THE DIGITAL CLOUD: BETWEEN GEOLOGICAL EXTRACTION AND METEOROLOGICAL HYPER-RESPONSIVENESS

Natalie P. Koerner

### Introduction

Data doubles, space data, weather data, search histories, archived chats and reddit threads—it seems that the digital cloud is striving to become a representation and a record of the planet and its events, with no innate hierarchy of importance or meaning. The digital cloud is thus becoming ever vaster, like Borges's 1:1 map of the empire, but in four dimensions.[1] Its conceptualization, like the metaphor 'cloud', takes on planetary dimensions.[2] The Internet itself has become a complex, global ecosystem. And like the planet, it is not free. The digital cloud is largely owned, mined and administered by transnational corporations, who steer users' political opinions and taste in consumption via microtargeted content and consistent nudging. And yet, access to the digital cloud also represents an important privilege: the digital divide became especially clear during the coronavirus pandemic, when a study found that almost half of the world's population has no access to the Internet.[3] Lacking Internet access means missing important basic resources, such as medical information and financial opportunities. As the Internet continues to expand to encompass more users, it is worth paying attention to and potentially adjusting the structure of knowledge production, archival practices and mechanisms of expansion of today's Internet. Specifically the digital cloud, an elusive metaphor for the data centres that store much of the Internet's data, is an important site of investigation. For most users, data centres are physically inaccessible and shrouded in contrived secrecy, also through the obfuscating use of the term 'cloud'.

The proposed article explores the spatialities, materialities and temporalities embedded in the metaphor of the digital cloud and the data centres it denotes. I will investigate the spatiality of the digital cloud as a voracious and extremely responsive entity. Describing and unpacking the cloud through planetary analogies makes its mechanisms of data harvesting, transmission and archival more concrete and accessible. One important informant is the meteorological cloud, and how it visualizes or models a large array of environmental parameters. A second focus for this investigation is the materiality of digital technologies that form the physical backbone of the metaphorical digital cloud. How does the materiality of the cloud and its reliance on earth minerals, extraction and climate politics reveal innate colonialist injustices? This materiality contains archival qualities already inherent in the geological ground, prior to its extraction.

## An Alien Cloud and Non-Conscious Cognition

The digital cloud is so murky and inaccessible to most users that it could just as well be an alien entity that has descended from outer space to hover above planet Earth. Imagine a universe-travelling gaseous conglomerate that now scans the atmosphere for any radio signals emitted on our planet that it can intercept and modify to communicate in any language, visually and aurally. The cloud synthesizes, processes and stores immense amounts of information, unhindered by bonds to location or form. The described entity—as similar as it sounds to today's digital cloud—is the intelligent, communicative and interplanetary alien protagonist of the renowned astronomer Fred Hoyle's 1957 science fiction novel *The Black Cloud*.

This interstellar gas cloud, even though it is mostly characterized as wise and benevolent, is also a ruthlessly colonizing force that extracts energy and information from any entities it crosses. As the black cloud approaches Earth to harvest the sun's energy, an ice age descends upon the planet: Hoyle conjures imagery of frozen tropical forests in complete darkness, icy oceans of petrified wave crests, and a myriad of cadavers scattered under the looming and incomprehensible cloud body. Like the digital cloud and its high energy consumption, the black cloud's presence entails an environmental catastrophe for planet Earth. And like the digital cloud, the black cloud is data-hungry. Its first intelligible reply to a direct message containing the basics of Western science, mathematics and the English language is voracious and impatient: 'Message received. Information slight. Send more.'[4] And like the digital cloud, which promises endless storage and computing capacities, the black cloud's brain and memory can grow infinitely, as it plants intelligence in suitable gas clouds wherever it encounters them: 'And the logic of brain-building seems to have some relation to our programming of a computer,' observes one of the characters.[5] Like the condensed water of meteorological clouds, the black cloud's neurological structures are arranged on the surfaces of miniscule solid bodies—aerosols—of the gas, plasma and dust that compose interstellar clouds.

Pre-empting the apprehensions associated with the machine-learning algorithms of the digital cloud, data is absorbed by the black cloud in a way that proves incongruous with human cognition. And thus, the novel ends with the death of the human protagonist, scientist Dr Christopher Kingsley, who had insisted on attaching himself to a kind of telepathic communication device built according to the instructions of the cloud.[6] Kingsley's brain cannot assimilate the knowledge structures transmitted by the cloud and he dies from relentlessly increasing mental disruption.[7] Similarly, cloud users often perceive the world of Big Data as problematic because the complex workings and seeming autonomy of machine-learning algorithms are untransparent and incomprehensible. As she develops a 'cloud ethics' that treats algorithms as ethicopolitical entities reflecting the social and technical milieu in which they operate, political geographer Louise Amoore acknowledges that 'the most widespread concern in public and scholarly discourse on the operation of algorithms in society is that they are unaccountable or that they cannot meaningfully be held to account for harmful actions'.[8] Further, the sheer immensity of data accumulated and extracted exceeds the human imagination.[9]

However, the digital cloud's processes are not as alien from human cognition as they might seem. Machine learning refers to computer systems (software applications) that harness algorithms and statistical models to increase their accuracy at predicting outputs while updating these with new data. In literary critic and theorist N. Katherine Hayles's reasoning, cognitive processes in such technical systems resemble human cognition. She suggests that we have exteriorized—not the intelligence associated with our conscious thought processes (artificial *intelligence*)—but rather our non-conscious cognition. Thus, biological and technical cognitions are now 'deeply entwined'.[10] The non-conscious scans environmental signals, patterns input and filters what requires the attention of consciousness and uses the information to anticipate future events.[11] Non-conscious cognition is most immediately related to the world and a body's material activities as it directly processes raw data. Non-conscious cognitive activities are akin to those that govern cloud computing, archiving and processing. Hayles explores non-conscious cognition as an activity that humans, technical systems and all other organic beings alike engage in. The non-conscious proposes the more inclusive, post-anthropocentric categories of 'cognizer' and 'non-cognizer'. Separating cognition from consciousness opens up a rich field of information processing that is independent of a sense of self.[12] The non-conscious thus overcomes the Cartesian dichotomy between the adult human who *thinks* and all other non-thinking entities, including children and animals.

A comparable approach was taken by a contemporary of Descartes, who lived in 'an age when the geocentric and anthropocentric view of reality was breaking up'.[13] Yet unlike Descartes, Renaissance philosopher Tommaso Campanella (1568–1639) did not separate the adult human conscious 'I' from the surrounding world. Instead, in Campanella's inclusive understanding, the whole world is a perceptive entity pervaded entirely by just one common spirit. This *sensus rerum*—sense of things—'can be found in every thing'.[14] While perceiving a certain quality, the sensing entity is permeated by the perceived quality: if I touch fire with my finger, then my finger becomes hotter. Stones sense when they are magnetically attracted to one another, the air senses warmer temperature and rises in response, plants sense the sun and turn towards it. I will now show that clouds in all their manifestations, beginning with the meteorological, can be imagined as a large, 'extremely sensitive animal'[15] in the Campanellian sense, as they mutate continuously to visualize everything they *sense*: the geology they float above, the temperature, materiality and reflectivity of the ground, the winds, the cosmic rays, atmospheric pressure, and the myriad chemical interactions among the aerosols they transport. Their responsiveness resonates with Hayles's non-conscious cognition, which describes a direct and immediate engagement with the world. And similarly, it resembles the algorithm's continuous 'engagement with the world'.[16]

## Meteorological Clouds Are Analogue Computers

To some extent, all earthly meteorological clouds store data like the black cloud, retain the sun's energy, and harvest and transport resources from all over the world. Whether they are anthropogenic or naturally occurring, clouds mediate data and signal information. For example, smoke-signal clouds are an ancient long-distance communication method[17] and volcanic eruption clouds transmit news of geological events. Meteorological clouds are omnipresent as they continuously cover 67 per cent of the planet. And yet, clouds remain fundamentally elusive: they are unmappable, and difficult if not impossible to model digitally or predict accurately, because they respond to and model immense amounts of data.

Clouds occur when water droplets condensate on aerosols. Aerosols—fine anthropogenic or naturally occurring solids and droplets, also known as 'seeds' or cloud condensation nuclei—are found in the tropo- and stratosphere. They are miniscule, usually smaller in diameter than a human hair.[18] Aerosols travel on air currents for four to seven days and reach speeds of around 5 m per second. There are many different types of aerosols that are produced on the planet's surface, for example, oceans are veiled by a thin mist of salt and sulphates produced by whitecaps and microalgae, Africa's northern half is often covered in Saharan dust particles and black carbon particles enter the atmosphere where there are wildfires. These aerosols differ in their surface topography and chemical composition and interact with one another, which affects their ability to seed clouds, determines when the cloud precipitates as rain, and how much heat it retains or reflects. Aerosols can be understood as data points that infuse the atmosphere as well as the lungs of all breathing entities with physical information from all across the planet. Just as the black cloud stored data on aerosols, the data carried by meteorological clouds constitutes their materiality and affects their spatial formation. The random correlations and vicinities of aerosol data—such as Atlantic Sea salt crystals and Saharan dust next to factory exhaust soot that merged with mushroom spores—is representative of the digital cloud's information structures.[19] Like data in the digital cloud, the continuously transforming and moving aerosols in a meteorological cloud compose ever new adjacencies and juxtapositions. These chance correlations and unrelated thematic vicinities differ from the static spatial coherence between information and a corresponding chronological temporal order of pre-digital archival.[20]

Besides aerosol composition, the extremely responsive morphology of clouds continuously visualizes many changing parameters, such as the phase behaviour of water, atmospheric pressure, land formations below, wind speeds, temperatures, nearby ships and aeroplanes. Clouds were therefore difficult to classify, until pharmacist and amateur meteorologist Luke Howard (1772–1864) presented a nomenclature for clouds in 1802. Howard understood that clouds were the 'visible signs of vast atmospheric processes'.[21] His basic categories—cirrus, stratus, cumulus (and later nimbus)—therefore reflected atmospheric processes rather than the resulting cloud shapes. Today's long-term observations of cloud patterns, enabled by Nasa's Terra and Aqua satellites, show the clouds' responsiveness in ever more detail. For example, wave clouds—large V-shaped fields of alternating clear and clouded strips—form above tall icebergs or islands. These comparatively small elevations push circulating air masses upwards where they meet higher-travelling air. The bands visualize the co-presence of air masses of different temperatures and moisture content.

This example actualizes mathematician and inventor of early mechanical and programmable computers Charles Babbage's idea of the air as an archive. Babbage (1791–1871) declared that 'the air itself is one vast library'[22] of all the words that have been spoken, and all the winds and currents that have acted upon it. His view of the air as an archive was informed by a computational logic. According to Babbage, when we speak, we set airwaves into irreversible motion, affecting every atom of the atmosphere and changing their trajectories forever: 'The waves of air thus raised, perambulate the earth and ocean's surface, and in less than twenty hours every atom of its atmosphere takes up the altered movement.'[23] The air in Babbage's view is an archive as much as a predictive tool. Given enough computational power, knowledge of the air's behaviour and the causes acting on it, the atmosphere's past and future trajectories could be deduced.[24] The air, according to Babbage, is a visualization of the forces that act on it, just as the clouds are responsive visualizations of extensive information and processes.

Considering the modelling capacity of clouds and their continuous updates, meteorological clouds can be thought of as analogue computers. While digital computing relies on discrete units of ones and zeros, analogue computing, prevalent until the 1940s, evolved out of calculation devices and modelling technologies to provide 'continuous representation and physical analogy'.[25] For example, the Moniac, a famous example built in 1949 by Bill Phillips, modelled the UK's financial market. It was a large-scale, wall-sized model of tanks, tubes and pipes that represented different funds, monetary institutions and players. The hydraulic flow was mediated by valves and pumps representing the movement of money. Analogue computers were important tools in early meteorology. In 1918, influential meteorologist Lewis Fry Richardson (1881–1953) used a bowl of water spinning on a gramophone to model the rotational movement of the atmosphere (while serving as a pacificist ambulance driver at the Western Front).[26] These spinning bowls were fairly common. In 1929 at the United States Weather Bureau, meteorologist Carl-Gustaf Rossby used a 2-m-wide saltwater bowl with different dyes to model thermal currents. Ludwig Prandtl (1875–1953), remembered for his research on aerodynamics, created an entire rotating room, 3 m in diameter. Clouds as analogue computers model—or *compute*—the planet by continuously reconfiguring to manifest the conditions of the planetary assemblage of competing parameters. Modelling is the fundamental premise of climate and weather studies, which attempt to assimilate 'many interlocking systems, including the atmosphere, the oceans, the cryosphere (ice and snow), land surfaces (soil, reflectance), and the biosphere (ecosystems, agriculture, etc.)'.[27] As analogue computers, clouds respond to and model these entangled phenomena more accurately than any digital model could.

## The Digital Cloud, Electronic Colonialism and the Geology of Archives

The digital cloud shares characteristics both with the colonizing, expansive black cloud and the extremely responsive, data-visualizing meteorological cloud. The digital cloud amasses and transmits vast amounts of data—while some of this data has become indispensable for communication, navigation, knowledge making and governance, much of it is entirely incomprehensible and unused. The cloud is a metaphor for data centres, where the data is stored. These centralized administrative and archival units, operated by large corporations, are situated in large, nondescript and entirely inaccessible boxes that contain an environmentally controlled accumulation of servers. Only cleaning, security and maintenance teams—often 'imported' temporarily for cost and work efficiency—get to enter. From here, cables transmit the data either via cables or wireless connections out into the world.

Even though digital data is often organized according to a meteorological logic and transmitted via microwaves, the world of data infrastructure is decidedly geological. Media technologies consume a large amount of *the ground*—rare earths, minerals and metals, such as cobalt, gallium, indium, tantalum, antimony, platinum, palladium, niobium, neodymium and germanium.[28] The devices on which digital content in data centres is stored are composed of rearranged slithers of geological material. The first layer of a hard drive disk contains platinum group metals and functions as an optimized 'soft magnetic underlayer' for the perpendicular field on which the data is imprinted. Two CoNiFe (cobalt, nickel, iron) layers separated by a four-atom-thick layer of ruthenium constitute this stratum.[29] The next layer is for the magnetic storage. It is made of sublayers of CoCrPt (cobalt, chromium, platinum) alloys. Each material plays a precisely defined role in the reorganized archival strata of digitally deciphered disks: 'The cobalt provides the necessary orientation of the crystals; the chromium improves the signal-to-noise ratio, while the platinum provides thermal stability. Ruthenium [helps to] orientate the magnetic grains, as well as reducing interference between layers.'[30] The information is deciphered by a sensor head that scans the surfaces of the spinning disks layered with magnetic fields of zeros and ones.

The digital cloud is tightly anchored in the Anthropocene (or the Capitalocene), as it relies on data and material mining. If only geologist Richard Alley had been right when he oversimplified the geological ingredients of a phone to foster awareness of the geological materiality of digital technology: 'A cell phone is just a cup of oil, a handful of sand, a few of the right rocks.'[31] Unfortunately, any device that accesses the digital cloud also trails a convoy of gigantic mining trucks or alternatively an army of underpaid (child) labourers, exposed to radioactive collocated materials and collapsing tunnels. 'A few of the right rocks' engender tons of the wrong rocks, which were extracted in labour- and energy-intensive processes. A 'handful of sand' on the scale of the media technology industry entails collapses of riverbanks and of (human and animal) livelihoods. Mining geological material and rearranging it as slick digital devices and invisible infrastructures is an invasive, often cruel, and polluting undertaking. Its by-products are meteoroid sized craters, extensive $CO_2$ emissions and colonial structures of exploitation.

The reality of the mining industry that supports our digital technologies is as inaccessible as a data centre. Most of us cloud users do not know and cannot find out where the materials of our laptops, phones and the server discs of the data centres that back-up our emails and photos were mined. We do not know if child labour or deadly accidents were involved.[32] A certainty is, however, that some of these materials could be mined in the Western world but are not, because the work is much more expensive if it is done in a healthy, physically sustainable way. For example, rare earth elements are distributed relatively evenly throughout the world, but for many years the West preferred to buy them cheaply from abroad, rather than handle the extremely laborious procedures.[33]

Spinning the metaphor of 'cloud' for the world of big data and its geological materiality further, one could imagine that like the sun, which heats up the ground and causes evaporation of water later visible as clouds, the data economy causes a phase change of the ground beneath our feet to rearrange the extracted material as data carriers. And just as the meteorological cloud condensates on aerosols and thus reflects its material consistency, the digital cloud's morphology represents its corporate owners. 'It matters what thoughts think thoughts. It matters what knowledges know knowledges. It matters what relations relate relations. It matters what worlds world worlds. It matters what stories tell stories,' writes Donna Haraway.[34] Unfortunately, many of the thoughts, knowledges and relations that form the basis of the digital cloud and its geological precipitation are wrought with Western colonialist exploitation, Euro/US-centric values and the Capitalocene. This latter term, unlike the Anthropocene, focusses more on the human-initiated extractive practices of our current geological epoch than on humankind itself.

Under colonialism, networks of extraction were established that still inform the mining industry that supports digital technologies: 'The networks of sugar, precious metals, plantations, indigenous genocides, and slavery, with their labor innovations and relocations ... of peoples, plants, and animals; the leveling of vast forests; and the violent mining.'[35] In today's updated electronic colonialism, these abusive power networks are continued.[36] Critical voices, such as artist and healer Tabita Rezaire, have drawn attention to the spatial relationship between the Internet and colonialism, for example that the 'submarine fiber optic cables that carry and transfer our digital data' are 'layered onto former colonial shipping routes'.[37] In her 2017 video essay 'Deep Down Tidal', Rezaire explains that: 'Electronic Colonialism is the domination and control of digital technologies by the West to maintain and expanse their hegemonic power over the rest of the world.'[38] She elaborates: 'It also operates by sustaining the dependency of former colonized countries on the West, by the importation of hardware, software, engineers, know-how and information protocols.'[39] However, beyond this importation, electronic colonialism is also extractive—in terms of materials, but also in how the mechanisms of Big Data treat all data subjects as mineable for clicking labour and consumption.

The spatiality of the digital cloud is tentacular, like a colonial empire. In Rezaire's words, 'cables are spaces where labour, knowledge and capital sunk into the sea'.[40] And the Internet itself has been described as a series of tubes that sneak their way ever more densely across the planet.[41] I wonder if Donna Haraway's 'tentacular thinking' could serve as a spatial and theoretical antidote to the exploitative nature of the ever-growing infrastructure of cables and the ever-murkier digital cloud. To Haraway, strings and entangled string figures speak of a sympoietic togetherness, where 'human exceptionalism and bounded individualism' have become unthinkable against the reality of 'multispecies muddles'.[42] To Haraway: 'The tentacular are also nets and networks, it critters, in and out of clouds. Tentacularity is about life lived along lines—and such a wealth of lines—not at points, not in spheres.'[43] The ground—that territory of extraction and buried infrastructure—also offers temporal and spatial narratives that remind us of archive and media stories that by far exceed the human horizon.

Accessing information stored on devices of geological materiality conjures a rich history entangled with geostoriesof the geological ground as an archive and as media.[44] Media scholar Shannon Mattern explores the relationship between geological materiality and the archaeology of urban media: 'For millennia mud and its geologic analogs have bound together our media, urban, architectural, and environmental histories.'[45] Geologist Marcia Bjornerud describes rocks and landscapes as a palimpsest:

Everywhere on Earth, traces of earlier epochs persist in the contours of landforms and the rocks beneath, even as new chapters are being written. The discipline of geology is akin to an optical device for seeing the Earth text in all its dimensions.[46]

Similarly, artist and architecture historian John Ruskin (1819–1900) understood rocks both as paper to be written on by the sculptor-architect, and as archives of the material's own lore and history. In a passage that evokes today's philosophy of elemental media, Ruskin suggests that with increased exposure to the 'language' of marble, the material would soon reveal its archived data and 'even the unsculptured walls of our streets [would become] to us volumes as precious as those of our libraries'.[47] The founders of Western geology, James Hutton (1726–1797) and Charles Lyell (1797–1875), understood the geologist as the decipherer of the planet's archival structure of geological strata arranged across planetary deep time.[48] In his seminal book *The Principles of Geology* (1830), Lyell famously portrays the planet as its own archive-archivist—always changing and keeping track of its own changes. The tempo of geological events reveals archival temporalities, embedded in the geological material, that exceed human temporal horizons. The geological mode engenders spatialities of stratification and thematic vicinity, chronological temporalities and an archival drive to update. This mode is inherent in data centres that harness the archival capacities of geological matter, as data is stored in magnetic fields on data carriers that reassemble extracted geological strata and data transmission is run along cables embedded in the ground and below the sea.

### Elemental Media and Leaving the Cloud in the Ground

In conclusion, the global phenomenon 'cloud' is more than a fruitful metaphor for our online data practices supported by data centres and vast infrastructures. The meteorological cloud's morphological responsiveness corresponds to the machine-learning mechanisms of Big Data encompassed by the digital cloud. The geological materiality of data centres harnesses the qualities of the ground as an archival entity. Beyond the cloud metaphor, digital archives and infrastructures are physically embedded in globe-spanning networks of water, earth and air. These natural elements, along with fire, represent the origins of media, as media scholar John Durham Peters shows in his philosophy of elemental media.[49] Like more recent media institutions such as newspapers and the Internet, elemental media are extremely effective at storing and transmitting information. In a similar vein, architect and theorist Keller Easterling suggests an evocative media-infused understanding of the environment, superseding an object-focussed approach to space, where focus is 'given to the medium—to the field instead of the object, the ground instead of the figure'.[50] She sees this shift in perception regarding infrastructural media space as having the potential to lead to a more fluid, adaptable, resilient and real urbanism. Embracing the logic of media would align human endeavours with elemental forces, rather than positioning them in opposition and danger. Changing coastal lines could be a site of potential, not of bankruptcy; hurricanes would generate energy, not just destruction.[51] In a world where humanoid organisms read clouds as a 'wet information system more common than a digital cloud', they might appropriate the air, earth and water as organizational forms, rather than attempt in vain to contain them.[52]

Thinking spatially about the digital cloud with a philosophy of elemental media raises the speculation that it should be possible to simply leave the digital cloud in the ground. Extracting the digital cloud's lithic and metallic matter, along with the archival qualities inherent in the ground, is unsustainable and extremely complex (technologically, politically and scientifically). The finesse to which it is possible to extract, isolate and rearrange on an atomic level minerals and metals for today's digital technologies is astonishing. Would it then not be equally conceivable to develop technologies for injecting data into the ground instead?[53] Further, keeping in mind the shared processes of non-conscious environmental scanning and data processing of organic and technical entities alike, and that all living beings are overlapping 'communities of symbionts', it seems natural to tap into this innate potential, and to bring the data to the ground, rather than the ground into the data centre (after myriad reconfigurations).[54] I join Ruskin who wanted to learn the language of marble, Durham Peters and Easterling who read the clouds, in my hope for a more sympoietic coexistence with the ground and the sky and their rich and dense data scanning, transmission and archival.

Notes

1 Jorge Luis Borges, 'On Exactitude in Science', 1946.
2 The term 'cloud computing' was developed as an unsuccessful marketing name in 1996 by Compaq marketing executive George Favaloro and technologist Sean O'Sullivan, of the now defunct Netscape. The cloud had made visual appearances since the early 1970s as an icon on maps drawn by administrators of communication networks, to mark an unfixed conglomerate of devices and cable networks.
3 Douglas Broom, 'Coronavirus Has Exposed the Digital Divide like Never Before', World Economic Forum, 22 April 2020, https://www.weforum.org/agenda/2020/04/coronavirus-covid-19-pandemic-digital-divide-internet-data-broadband-mobbile/.
4 Fred Hoyle, *The Black Cloud* (Richmond: Valancourt Books, 2015), loc. 2875, Kindle.
5 Ibid., loc. 3170.
6 Ibid., loc. 3651–3652.
7 Ibid., loc. 3804–3806.
8 Louise Amoore, *Cloud Ethics: Algorithms and the Attributes of Ourselves and Others* (Durham, NC/London: Duke University Press, 2020), 18.
9 For similar occurrences of such immense scale, see: Timothy Morton, *Hyperobjects: Philosophy and Ecology after the End of the World* (Minneapolis/London: University of Minnesota Press, 2013).
10 N. Katherine Hayles, *Unthought: The Power of the Cognitive Nonconscious*, Kindle (Chicago/London: The University of Chicago Press, 2017), loc. 208.
11 Ibid., loc. 500.
12 See also the study of 'the sense of being sentient' in: Daniel Heller-Roazen, *The Inner Touch: Archaeology of a Sensation* (New York: Zone Books, 2009).
13 From the description of the journal *Bruniana & Campanelliana*.
14 Campanella, *Compendium physiologiae*, quoted in: Heller-Roazen, *The Inner Touch*, op. cit. (note 12), 171. There are similarities between vibrancy, agency—explored by Deleuze, Delanda and Bennett, among others—and Campanella's understanding of spirit. See, for example: Jane Bennett, *Vibrant Matter: A Political Ecology of Things* (Durham, NC/London: Duke University Press, 2010).
15 Heller-Roazen, *The Inner Touch*, op. cit. (note 12), 172.
16 Amoore, *Cloud Ethics*, op. cit. (note 8), 22.
17 For an exploration of sky media, see: John Durham Peters, *The Marvelous Clouds: Toward a Philosophy of Elemental Media* (Chicago/London: The University of Chicago Press, 2015), 165.
18 'Aerosols: Tiny Particles, Big Impact', Earth Observatory, 2 November 2010, https://earthobservatory.nasa.gov/Features/Aerosols/page1.php.
19 For the concept of correlation and digital data, see also: Wendy Hui Kyong Chun, 'Big Data as Drama', *ELH* 83/2 (2016), 363–382.
20 On archival structures, see also: Geoffrey C. Bowker, *Memory Practices in the Sciences*, Kindle Edition, Inside Technology (Cambridge, MA/London: MIT Press, 2005).
21 Richard Hamblyn, *The Invention of Clouds: How an Amateur Meteorologist Forged the Language of the Skies* (New York: Farrar, Straus and Giroux, 2001), loc. 1989.
22 Charles Babbage, *The Ninth Bridgewater Treatise* (London: John Murray, 1838), 113.
23 Ibid., 108–109.
24 Ibid., 111.
25 Charles Care, *Technology for Modelling: Electrical Analogies, Engineering Practice, and the Development of Analogue Computing*, History of Computing (London: Springer, 2010), 20, 39.
26 For Richardson's numerical work as a precursor of cloud computing, see: Seb Franklin, 'Cloud Control, or the Network as Medium', *Cultural Politics* 8 (2012), 452–454.
27 Paul N. Edwards, *A Vast Machine: Computer Models, Climate Data, and the Politics of Global Warming* (Cambridge, MA/London: MIT Press, 2010), xiv–xv.
28 On the relationship between geology and the materiality of digital technologies, see: Jussi Parikka, *A Geology of Media* (Minneapolis/London: University of Minnesota Press, 2015), Kindle.
29 Johnson Matthey, 'Properties of PGM', http://www.platinum.matthey.com/about-pgm/applications/properties-of-pgm, accessed 12 October 2018.
30 Ibid.
31 Richard B. Alley, *The Two-Mile Time Machine: Ice Cores, Abrupt Climate Change, and Our Future* (Princeton: Princeton University Press, 2000), ix.
32 'Congo, Child Labour and Your Electric Car', *Financial Times*, 7 July 2019.
33 See also: Guillaume Pitron, *The Rare Metals War: The Dark Side of Clean Energy and Digital Technologies* (Melbourne/London: Scribe US, 2020).
34 Donna J. Haraway, *Staying with the Trouble: Making Kin in the Chthulucene* (Durham, NC/London: Duke University Press, 2016), 35.
35 Ibid., 48.
36 See also: Michael Kwet, 'Digital Colonialism: US Empire and the New Imperialism in the Global South', *Race & Class* 60/4 (2019), 3–26.
37 Tabita Rezaire, 'Prologue—Decolonial Healing: In Defense of Spiritual Technologies', in: Kevin Smets et al. (eds.), *The SAGE Handbook of Media and Migration* (London: SAGE Publications, 2020), xxxv. See also: Nicole Starosielski, *The Undersea Network* (Durham, NC/London: Duke University Press, 2015).
38 Ibid., xxxv.
39 Ibid.
40 Ibid.
41 Andrew Blum, *Tubes: Behind the Scenes at the Internet*, Kindle Edition (London: Penguin Books, 2012).
42 Haraway, *Staying with the Trouble*, op. cit. (note 34), 30 ff. See also: Scott F. Gilbert, 'Holobiont by Birth: Multilineage Individuals as the Concretion of Cooperative Processes', in: Anna Lowenhaupt Tsing et al. (eds.), *Arts of Living on a Damaged Planet: Ghosts and Monsters of the Anthropocene* (Minneapolis: University of Minnesota Press, 2017), 73–90.
43 https://www.e-flux.com/journal/75/67125/tentacular-thinking-anthropocene-capitalocene-chthulucene/.
44 See also: Bruno Latour, *Facing Gaia: Eight Lectures on the New Climatic Regime* (Cambridge: Polity Press, 2017).
45 Shannon Mattern, *Code and Clay, Data and Dirt: Five Thousand Years of Urban Media*, Kindle (Minneapolis/London: University of Minnesota Press, 2017), loc. 2224–2229.
46 Marcia Bjornerud, *Timefulness: How Thinking Like a Geologist Can Help Save the World* (Princeton/Oxford: Princeton University Press, 2018), 22.
47 John Ruskin, *The Stones of Venice: The Fall*, iBooks ed., vol. III (London: Smith, Elder, and Co., 1851), 28. On elemental media, see: Durham Peters, *The Marvellous Clouds*, op. cit. (note 17).
48 See: John McPhee, *Basin and Range* (New York: Farrar, Straus and Giroux, 1981). Media theory has embraced this temporal depth. See, for example: Wolfgang Ernst, 'Radically De-Historicizing the Archive: Decolonizing Archival Memory from the Supremacy of Historical Discourse', in: Rado Ištok and Nataša Petrešin-Bachelez (eds.), *Decolonizing Archives* (L'Internationale Online, 2016), 12.
49 Durham Peters, *The Marvelous Clouds*, op. cit. (note 17).
50 Keller Easterling, 'The Year in Weather', Artforum International, https://www.artforum.com/print/201710/the-year-in-weather-72467, accessed 4 December 2018.
51 Ibid.
52 Ibid.
53 See also: Geoff Manaugh, 'Planet Harddrive', BLDGBLOG, 23 January 2009, http://www.bldgblog.com/2009/01/planet-harddrive/.
54 Gilbert, *Holobiont by Birth*, op. cit. (note 42), 73.

# ‘HERE BE D

## ‘HERE BE DRAGONS’: THE LIMINAL TOPOGRAPHIES OF STATISTICAL IMAGING

Gökçe Önal

A world does not predate its images. It is shaped by them.

Hashim Sarkis, Roi Salgueiro Barrio and Gabriel Kozlowski, *The World as an Architectural Project*[1]

Nineteenth-century cartographic imagination saw dragons, sea serpents, ‘men without heads with their faces in their chests’ and many other intimidating beasts of the lands unknown, *terre incognite*, abandon their corners of the Earth—permanently.[2] Popularized by the phrase ‘here be dragons’, the mapmakers’ exotic creatures occupied uncharted territories for centuries until the colonial expansion took hold, ousting the ambiguities and risks associated with venturing unknown lands. Today, the accuracy and detail of digital Earth Engines allow for little surface for such cartographic fiction to take hold, feeding a longstanding rhetoric of omniscience through the seamless ‘Google Earth’ interface. Yet, as this paper argues, the unknown, the unseen and the liminal continue to thrive within and across the infrastructures of environmental monitoring, without necessarily disturbing the cartographic surface. Resorting to Gayatri Chakravorty Spivak’s invocation of ‘planetarity’ alongside the ‘globe’, this paper asks: What remains unseen in the regime of ubiquitous surveillance? A media-material reading is engaged for responding to this question on the two operational levels of remote imaging: data gathering and data processing. The resulting framework of spectral and statistical (in)visibilities are further discussed in relation to the production of spatial knowledge.

### Introduction: Technoecologies of Images

In the midst of our serene new world of images, a descriptive revaluation of the conditions of imaging—its technical basis, and the gestures that divide it off from all previous forms of visuality—is a prerequisite for architecture to pose the question, to itself and to culture ... How can we learn to live differently? So differently that we might soon—very soon—become nonmodern?[3]

Exactly three decades ago, William J.T. Mitchell warns his reader of the tendency to (mis)take the electronic image as ‘simply a new, nonchemical form of photograph or as a single-frame video, just as the automobile was initially seen as a horseless carriage and radio as wireless telegraphy’.[4] Mitchell’s grip remains among the earliest problematizations of electronic surfaces with a ‘physical’ focus. To date, on the media studies front, scholarship on computational artefacts has gradually arrived at an elemental analysis of media infrastructures[5]—opening up an ecological field whereby questions of the *visual* extends from the screen into the 'deeper material processes' of imaging, or vice versa.[6] Elemental research as such conforms to a 'denaturalized, technologized ... and thereby generalized' concept of ecology—a technoecological present, to follow Erich Hörl—that no longer sustains an exhausted nature/culture divide or an alleged 'immateriality' of the digital.[7] It offers here relational ways for attending to our electronic surfaces as a growing-together of intensities, processes, and human and more-than-human agents—'settled temporarily into what passes for a stable state'.[8]

In the case of digital images, the emphasis on emergence and relationality brings into focus a motile *environment* of image objects—broadly the sites of production, dissemination and display through which 'computation is made [visible and] meaningful to us'.[9] In problematizing architecture's cartographic surfaces, I align with Jacob Gaboury's premise that elements of 'visibility, memory, simulation, relation, and history are each inscribed into the technical infrastructure of the medium of computer graphics itself'.[10] An environmentally-attentive reading here reveals digital graphic surfaces to be laden with extractive, reductive and distortive practices that today sustain the most prominent claim to the whole Earth and continually shape the visual economies of technoscientific narratives.

## Imaging Infrastructures and Architecture's Informational Surfaces

'We need to flip the figure and the background, and recognise infrastructures as one of the major planetary agencies we have at our disposal.'[11]



Today, one can't rigorously tackle questions of electronic media without engaging in a geographically distributed assemblage of resources, hardware, sensing and computing platforms, interfaces and human labour. Scholarly attempts to capture the growing complexity of computing infrastructures, of which imaging remains a part, have culminated in the many diagrams of 'sociotechnical' and 'planetary' stacks, and more recently, 'anatomical maps' of automated labour.[12] Varying in scale and focus, all investigations reveal interoperability to be a central requirement of technical systems, to the extent that 'prioritising any of these layers, at the expense of the rest of the stack, places a constraint on developing an accurate understanding of how a digital technology is conceived and works in practice'.[13] All of the layers of the imaging infrastructure—the remote sensing platform and the operating system employed for data acquisition, the type and the volume of gathered data, the algorithms and the statistical models used for processing data, the software application language, and at the outset, the interface and user interaction—perpetually co-constitute one another in the production of spatial knowledge.

The contemporary momentum of image production and screen labour have rendered computer-processable imagery a currency in design cultures. Architectural thought, reflexes and practice today are increasingly embedded in the graphic surfaces of this 'statistical-electronic' topography—a categorical term that will be adopted here, after John May, for referring to digital images.[14] To date, scholarship on the (in)visibilities of statistical-electronic display has extensively focused on pixilated censorship and blanked-out satellite imagery, no-fly-zones, aerial camouflage,

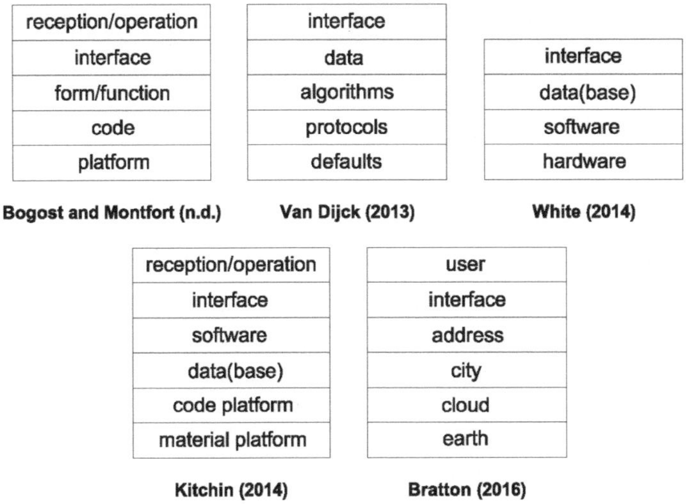


Fig. 1 Examples of 'digital technology stack' concepts from literature. Diagram adapted from Rob Kitchin, 'From a Single Line of Code to an Entire City: Reframing the Conceptual Terrain of Code/Space', in: Rob Kitchin and Sung-Yueh Perng (eds.), *Code and the City* (Abingdon, Oxon: Routledge, 2016), 20.



counter-surveillance practices and adversarial attacks—which, with the exception of the latter, conforms to a problematization of interface irregularities.[15] Of interest here, however, is not so much the censorship or curated misinformation, but rather the infrastructural layers into which the physical world is reduced, averaged, filtered and distorted before it is rendered as visual information. What gets lost—or cannot be transferred—in the act of sensing, processing and modelling Earthly energies into statistical-electronic objects? Setting out on this question, two lines of inquiry are defined: spectral (in)visibilities of sensing platforms, and statistical (in)visibilities of algorithmic modelling. The following response is structured around Spivak's invocation of 'planetarity' alongside the 'globe', which offers here a productive ground for attending to the gaps between the represented and the lived.

## The Globe and the Planetarity

'Globalization is the imposition of the same system of exchange everywhere ... The globe is on our computers. No one lives there.'[16]

Spivak's invocation of the planet doesn't posit an 'on the other hand' relation to the globe. The distinction, however, is growing increasingly integral to postcolonial studies, particularly for attending to the more-than-human timescales of climate histories. Against a scholarly habit of using the two words interchangeably, Dipesh Chakrabarty recounts the global as 'a singularly human story' and the planetary 'as a perspective to which humans are incidental'[17]—where the latter sustains the former. The globe, in Spivak's words, 'allows us to think that we can aim to control it'; it conforms to universals and standards, to master narratives, to an overarching geopolitical order. 'The planet,' she continues, 'is in the species of alterity'; it captures the globe as well as all things irregular, non-binary or ambiguous that escape its seamless surface, as did the mapmakers' beasts—'and yet we inhabit it'.[18] It is by paying attention to the alterity of the statistical-electronic globe that I problematize here the accelerating regime of electronic visual display—or screen—in architecture.

## The Excess of Globalization

The globe is an inherently human construct. The making of its figure is grounded in histories of colonial expansion and trade, as extensively discussed to date—systematically shaped by the instruments of (circum) navigation, mobility and cartography.[19] It is a statistical artefact sustained in 'the gridwork of electronic capital', as Spivak suggests, 'an abstract ball [of] latitudes and longitudes' representing the Earth as a homogenized surface of capital and material exchange. If we then take the globe as 'an integral pictorial metaphor of modernity', following Lukas Likavcan, and its contemporary image as the Google Earth Engine, the distinction between the global and planetary imaginations becomes an infrastructural question—one that brings into focus technics of Earth observation as 'one among many of modernity's optical apparatuses'.[20]

Digital Earth Engines today comprise the visible, screen end of this anthropocentric representation practice—sustaining the state-of-the-art claim to a longstanding European endeavour of imaging the whole Earth. Cultural histories of cosmic views argue that an airminded chronicle of seeing-as-knowing underlies, in the disclosed figure of the globe, this appeal to 'total information awareness'.[21] The epistemological 'authority' of globes and geographical maps rests on this model of absolute visibility.[22] Sloterdijk rightfully defines this condition as a state of 'shadowless[ness]' with no more timeouts.[23] Adorno and Horkheimer call it a 'fully enlightened earth'.[24] For Spivak, it is the logo of the World Bank where no one lives.[25] In either instance, the globe of globalization is a human artefact, illuminated to the degree of 'overexposure' in the process of its making—from the surface of which, as I'll argue, the trivial, non-profitable and non-extractable remain eliminated.[26] This paper locates the cartographic unseen in this overexposure act, attending to the realm of the technically eliminated through Spivak's invocation of planetarity vis-à-vis the globe.

'If globalization is characterized by excessive visibility,' Elizabeth DeLoughrey suggests, 'planetarity provides a means to think through—but not necessarily to represent—that which is rendered invisible.'[27] DeLuoghrey invokes Adorno's and Horkheimer's cautionary take on the 'fully enlightened earth' as an imperative for the visual economy of colonialism—the exploration, observation, mapping and ultimately distribution of all life 'peripheral to modernity', including its people, resources, fauna and flora.[28] Excessive illumination understood as such is deeply embedded in (a Western mode of) knowledge production—for surveying practices remain at the forefront of imperial growth and land exhaustion, as postcolonial scholarship has extensively shown. To reiterate a well-travelled but equally important point here, the image of the globe conforms to the mechanisms, technics and standards that have been historically instrumental in making the modern world, which today fundamentally informs the prominent technologies of visualizing the planet.

The problem resides not so much in the fact that the globe is an abstraction of the planet, but that it is a 'very poor' one. It conforms to a *uni*versal claim incapable of engaging a planetary profusion, for 'the planetary ... cannot be grasped by recourse to any ideal form'.[29] Spivak's genius in coupling planetarity with the globe lies in defining this gap, through which, as Auritro Majumder argues, she 'invokes a relationship between the human and the natural worlds that is in "excess" of capitalist globalization'.[30]

In what will follow, I'll engage this 'excess' by addressing the—chronic—limitations of the (statistical-electronic) globe in sustaining planetary diversities, and discuss the implications of this gap for the production of spatial knowledge.

## Spectral (In)visibilities: Reducing Life to Numerical Means

'A planetary thinking is primarily an imperative for diversities.'[31]

Contemporary planetary imagination is fundamentally shaped by a distributed and diverse operation of remote, or increasingly so, geo-sensing. Geo-sensing today has become a hypernym for the many practices of Earth observation. By design, (geo-)sensors always operate as part of a data acquisition system, either spaceborne, airborne, UAS or terrestrial. The sensing platforms involved in the task abide by different operational parameters for receiving and measuring environmental stimuli, comprising systems of 'detectors, signal conditioners, processors, memory devices, data recorders, and actuators'.[32] Regardless of the platform, all electronic sensors work by 'receiving a stimulus and responding with an electrical signal'—stimulus here implying (radiated) energy, be it optical, acoustic, mechanical or thermal, among others.[33] This practice of 'converting all of lived experience into discrete, measurable, calculable electrical charges (signals)', namely signalization, is examined here as an act of spectral elimination, which brings into focus a multi-layered operation of waves, signals and data conditioning.[34]

The processing sequence involved in remote imaging extends from the acquisition of image data to its calibration, analysis, modelling and integration, which exceeds the scope of any single article. The following discussion focuses on the initial stages of this sequence—namely on aspects of sensor design, data collection and pre-processing—as the site of a dynamic rift between the planetary and its digital abstractions. This rift, understood here as an adjustment and channelling of energies, grows together with a technoecology of signalization, where the incoming stimulus is filtered, divided, converted (into an electrical signal) and is ultimately sampled in digits. The energetic gaps that occur in the process, however, do not simply retreat from sight but simultaneously become part of a continuous subsurface equation—multiplying as approximations, illiteracies and errors along the many sequences of image processing.

### Selective Sensibility: Calibrations in Sensor Design and Data Acquisition

Sensor sensitivity (to stimuli) is necessarily a selective constant. By design, any sensor is a partial aperture that only gathers stimuli (that is, radiated energy) from a predefined—and narrow—spectral band, known as its dynamic range. All such calibration involves prioritizing one form of information by physically discarding the others' radiation at the outset, the range of which depends on customer demands. The majority of optical and thermal sensors used in Earth observation include components for further channelling and separating the received radiation into spectral subregions, namely by splitting the signal into narrower wavelength bands through prisms, filters, diffraction grating or spectroscopes before the signal is converted into electric current.[35] In the following stages of processing, the physical segmentation facilitates different channels to be treated individually, which broadly involves operations of correction, feature selection, standardization and dimensionality reduction.

Spectral calibrations are imperative to sensor efficiency, given the wide range of remote sensing applications from climate studies to smart urbanization. Of interest here is not so much the component of feasibility, but a latent practice of 'selective blindness' that has historically shaped the imperial project of resource mapping.[36] Geo-sensing implicitly accommodates technics of resource extraction through the processes of sensing, quantifying, classifying and mobilizing bodies in isolation, deeming certain information redundant in the process.[37] As a practice of elimination, this may broadly involve 'removing pixels unrelated to a particular study ... channelling ratios to highlight specific ground feature characteristics ... and reducing [data] dimensionality of the image by Principal Component Analysis'.[38] A politics of (in)visibility underlies this process of removal, or the erasure of 'all that doesn't belong to [modernity's] parameters of legibility and certainty', Rolando Vazquez suggests, as much as a disregard of all things unprofitable and in-exchangeable.[39]

Datafication can't occur without making some forms of information more valuable than others; collecting all of the data all of the time doesn't automatically open out new potentialities for knowledge creation; rather, it solidifies the interests of the actors who are most able to take advantage of it.[40]



The increasing refinement of the sensors' stimulus range today not only multiplies this selective blindness, but simultaneously renders it more precise and processable by systematically eliminating the 'undesired' planetary energy with increasing efficiency. It echoes, following Louise Amoore and Alexandra Hall, an Enlightenment compulsion with a 'stripping away of excess by decomposition and fragmentation for the purposes of control'.[41] Today, technics of spectral and statistical separation, elimination and processing continually intensify a regime of information control by intermediaries.[42] This marks our entry point for attending to the gap between the global (here, the cartographic visible) and the planetarity (here, the unseen).

### Unrecoverable Energies: Gaps and Binaries

At the outset of any remote sensing operation, once the Earthly radiation is received by the sensor, the process of electrical conversion allows it to be quantified as a numerical value, relying on the two possible states of electricity moving through a (computer's) circuit: flow or no flow, that is, 'on' or 'off', hence the binary system of 'ones' and 'zeros'. Fed into the base-2 numeral system of computing devices, the received ones and zeros are recorded as patterns of binary digits, more commonly known as 'bits'. A binary code is thus the smallest unit of the—complex—computational language. 'This language is electrical in its nature,' as Jacob Fraden suggests, and 'sensors intended for the artificial

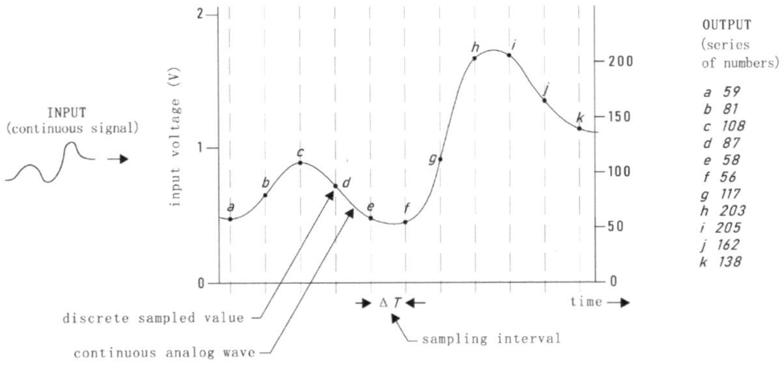


Fig. 2 Analogue-to-digital conversion process. Image drawn by the author based on Thomas M. Lillesand, Ralph W. Kiefer and Jonathan W. Chipman, Remote Sensing and Image Interpretation (New Jersey: Wiley, 2015), 26.



systems must speak the same language as the systems "speak" ... responding with the output signals where information is carried by displacement of electrons.'[43] The figure below represents the process by which the (continuous) electronic signal received from the sensor is sampled in evenly fixed time intervals and recorded at each sample point as the corresponding (discrete) number.

The conversion of radiated energy here is twofold: first, into an electric current (input voltage), and secondly, into a numerical value (output). The statistical-electrical transfer depicted above marks a critical threshold between what Lucía Jalón Oyarzun describes as 'cartographic overexposure' and its 'remainders'.[44] The diagram renders tangible a physical process of leaving out (the ΔT), as well as the technique of numerical sampling involved in the act. As this conversion always conforms to an 'energy transfer between the object of measurement [and] the sensor', the overexposed—or eliminated—here is understood as planetary life itself, and only secondarily, its visibility.[45] This energetic gap is irreducible and multiplies further into the stages of image processing. May defines the technics of this signalization act as a particular form of memory and storage that is always already quantified, inscribed in the 'gestural-mental routines' of our (cultural) being.[46] However small, the gap—technically—remains intrinsic to the statistical-electronic surfaces of contemporary architecture cultures, to follow May, and ultimately to our ways of thinking and doing (through images).

In her discussion of civic activism within the 'big data optimized [smart] city,' Alison Powell points to this limitation by illustrating how the 'world-making potential of techno-systemic thinking winds its way even into the definitions and potential for civic action'—despite the sensors' technical agency in bringing to light new and multiple ways of knowing. Even in their most dissident, DIY and bottom-up form, sensing practices ultimately support the urban platform model, Powell observes, 'because [they] contribute to the same extractive and calculative dynamics of the big data optimized city'.[47] Processes of electrical conversion, sampling and datafication are technically limited in capturing the breadth of planetary life, energies and spectra. To acknowledge this partiality, I argue, is to renounce the globe-as-a-claim-to-the-whole.

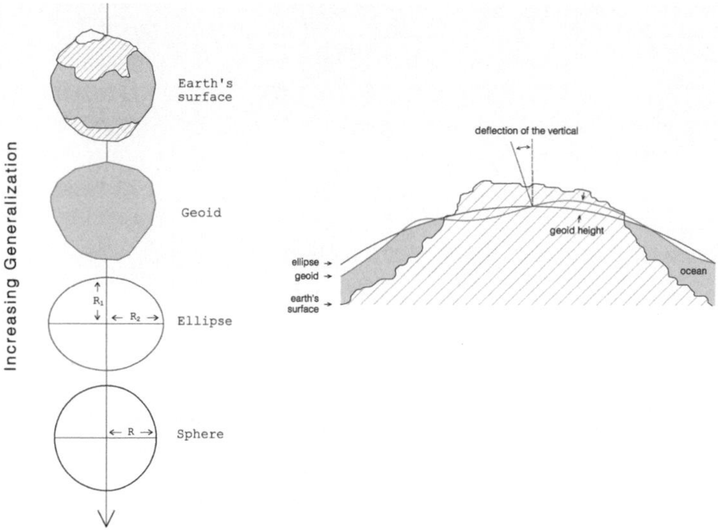


Fig. 3 [left] Modelling the shape of the Earth. Image drawn by the author based on Harrison et. al., *Earth Observation.*, Vol. 1A, 56. [right] Deflection of the vertical from the geoid to the spheroid. Image drawn by the author based on *Encyclopædia Britannica, Inc.*, s.v. 'geoid', 17 March 2021, https://www.britannica.com/science/geoid, accessed 16 June 2022.

## Statistical (In)visibilities: Fitting Life to Standards

As much as the electrical conversion constitutes a gateway to platform interoperability, the integration of the electrical signal into the computer infrastructure hardly remains smooth. By design, a sensor may produce output signals in the form of a voltage, a current or a charge, and these may be further registered as amplitude, polarity, frequency, phase or digital code.[48] The incompatibility of different types of sensors, measurements and platforms grows exponentially into image processing—as 'variability in formats, spatiotemporal granularity, access methods, and differing semantic definitions hinder attempts to ... compare, and combine data sets'.[49] The task is further complicated by a range of (spectral, spatial, temporal or radiometric) errors that are innately present in all remotely-sensed raw data—which may be either platform, sensor and/or acquisition-related, or caused by the Earth's curvature and rotation. Addressing operational interruptions as such requires the observation data to be calibrated at all stages of remote imaging.

In the base-2 numeral system of computing devices, the recorded patterns of binary digits (bits) allow the registered phenomena to be further transmitted, stored and processed in numerical sets. Algorithms and statistical models are designed to convert these datasets into quantifiable outputs with respect to the object space in question. Calibration here involves the application of several correction and standardization models for establishing platform consistency and minimum levels of differentiation between datasets. This is also referred to as homogenization. The task of homogenization—as an attempt at interoperability—brings into focus the infrastructural logic that today, on a planetary scale, 'standardise tendencies and regimes of engagement between bodies in space: postal address systems, languages and scripts, railways, transoceanic cables, time zones, international business standards, cloud platforms, and distributed ledgers'.[50] Remotely sensed datasets are similarly calibrated to pre-defined standards and measurement models for further processing.[51] Regardless of their intended use—agriculture, resource mining, climate monitoring, urban planning, etcetera—all processed image data are registered in present-day infrastructures of data analysis, interpretation and display. The globe is an outgrowth of this standardized, corrected and homogenized infrastructure space (of exchange)—resting on mathematical models of what signals, bands and surfaces, as well as errors and deviations should ideally look like.

### Earth to Globe: Approximating Life

Standardization of the Earth's surface as a statistical-electronic ground (for all cartographic and positioning tasks) requires the construction of a global coordinate system—the precondition of which is the accurate measurement of the Earth's size and shape. The operations involved in modelling the shape of the Earth, however, points to a mathematical process of (surface) homogenization, namely a technique of standardization-by-approximation that yields the 'idealized' figure of the globe. Unlike the widespread, spherical imagery of the Earth Engines, Earth's shape is most accurately represented as a *geoid*. The geoid model corresponds to the Earth's gravitational forces, which vary in magnitude and direction due to the planet's irregularly distributed mass—hence the geoid's irregular shape.[52] The following section recounts the algorithmic gestures of modelling Earth by addressing the reductive-distortive instruments of surface fitting (over and under), approximating (by interpolating or extrapolating) and filling (in the blanks).[53]

The geodetic exercise of Earth modelling should be recalled here in the context of post-war Earth sciences and the research programmes initiated during and after the Cold War. Geodesy in the past century came into focus as part of an interdisciplinary endeavour to build planetary models, fuelled by 'the [post-war] availability of instruments and programs that could provide a truly global data set'.[54] This project of 'global model-building' was an imperative for the mid-twentieth century rhetoric of the whole Earth, as Robert Poole suggest, for it not only contributed to the establishment of the planet as a set of integrated systems, reconciling the growing fields of geosciences, but also yielded an interdisciplinary ground for establishing 'the interaction of science and technology with politics, society, economy, and the environment, and ... an integrated mode of thinking about global developments'.[55] Decades-long efforts of accurately reconstructing the shape of the planet thus contributed to 'a new vision of the globe as an integrated political, technological and environmental space'.[56] Here, the military-scientific nexus of the geoid model brings into focus the underlying rhetoric of globalization.

The geoid, as stated above, is considered to be the most accurate representation of the planet's shape and indicates its gravitational potential at any given point. Integral to gravity field calculations is the method of interpolation, namely the statistical estimation of 'unknown values between known values of irregularly spaced control points'.[57] The geoid is hence an interpolated model, bearing an approximated yet highly 'complex undulating surface [that is] impossible to define in simple mathematical terms', proving too irregular for map projections.[58] Earth sciences employ (versions of) the *ellipsoid* model for computational simplicity, which is 'fitted' onto the geoid to substitute the Earth's actual size and shape either locally or globally, and to construct a precise coordinate system to support all cartographic and geodetic calculations.[59] The (approximated) reference surface thus constitutes an elemental precondition of platform interoperability in remote (Earth) imaging. What is ultimately established in the ellipsoid is a standardized, mathematically definable surface that only comes close to the shape of the Earth, presenting a smoother yet more distorted approximation of the planet than the geoid.[60]

The above diagram of 'increasing generalization' begins to render tangible the figure of the globe as problematized throughout this essay: a homogenized surface of 'overexposure', an 'abstract ball', a computerized artefact. Here, the process of refinement is coupled with a distortive dimension—that of approximation-by-interpolation and surface fitting. What gradually appears is not merely the globe-as-an-idealized-model-of-the-planet, but more fundamentally a cultural technique that abstracts the Earth by separating, converting, reducing, approximating and distorting its (gravitational) forces. The alterity moves through these energetic gaps, residues and models as they multiply into the algorithmic construction of planetary images

## Civic Life to Statistical Model: Latent Complexities

'The problem of bias has mostly originated from the fact that machine learning algorithms are among the most efficient for information compression, which engenders issues of information resolution, diffraction and loss.'[61]

As monitoring tasks become more specific and multi-layered—and datasets exponentially larger—algorithmic calculations involved in processing and modelling observation data are attributed further dimensions. Broadly, algorithms are finite sets of instructions to be followed in calculations. They are also generative agents that function in tandem, sustaining the building blocks of Artificial Intelligence (AI).

AI is most commonly defined as a pattern recognition mechanism comprising neural and machine learning networks, deep learning, robotics and computer vision. The product of algorithmic computation is a statistical model, namely a cluster of probability distributions. Statistical models today remain at the forefront of big data optimization, including urban innovation platforms. Yet, as existing scholarship has extensively shown, AI is technically limited by (historical, dataset-related or algorithmic) biases, assumptions and weights, which affect city planning and management at various levels of decision making.[62]

'The cybernetic systems that underpin the framework of the data-optimized smart city are intended to reduce complexity in various aspects of city life,' Powell suggests.[63] Optimization in data-driven systems implies a 'capacity to generate an actionable output from a set of attributes', reducing probabilities to one single—optimal—conclusion.[64] And while the 'condensed output' is put forward from a set of 'rejected alternatives', as Amoore argues, 'the branching pathways continue to run beneath the surface'. However singular, the algorithmic output is a fragile and contingent value: shift the weights in the 'hidden layers' and it is modified.[65] At the city scale, this may involve rather mundane instances such as the 'shortest' routes proposed by your navigation software, or dining suggestions made by search-and-discovery apps that align with your previously-visited restaurants or consumer habits—suppressing the non-optimal or less relevant in the process. It may, however, entail sharper consequences such as reproducing the racialized practices of municipal redlining in data-driven planning, as Sara Safransky illustrates. Her work on proprietary market value assessment reveals bias in urban governance to be largely shaped by the categorical choices of algorithm designers—more specifically, by their understanding of social problems, risks, census track analysis, value estimations and investment opportunities, among other things.[66] On various levels of decision making, the statistical reductions, approximations and assumptions multiply into the everyday information infrastructures, comprising today the visual economy of Earth Engines and digital twins.

## Conclusion: Emergent Relations

Sensors are capable of making environments 'present and interpretable' across spatial, spectral and temporal scales, to recall Jennifer Gabrys, bringing human and more-than-human subjects and processes into (emergent) material relations.[67] This bringing together, however, is a numerical operation growing on 'unrecoverable gaps'.[68] As this paper illustrated, outside of the planet's statistical-electronic topography lies an excess that is unextractable, non-binary and unprofitable.

The statistical-electronic technics of engaging life bring certain objects into focus and eliminate others by moving earthly intensities such as waves, signals, digits, algorithms and statistical models: comprising a technoecology of images. An elemental reading of (the extents and limits to) electronic sensing here begins to reveal a growing-integral of the seen and the unseen in quantifiable terms. The statistical-electronic unseen—namely the ecology of gaps, residues, approximations and rejected paths—does not imply an absence of surface visibility alone, but itself becomes part of a continuous subsurface equation. The invisible today grows alongside the visible in increasing precision, multiplying into the sequences of image processing and ultimately into our knowledge infrastructures.

Cartographic imagination infiltrates the surface via infrastructural rifts, I argue here, as did the dragons and the beasts. What is left out (of the image) then becomes the function of an accelerating quantification of subsurface invisibilities. The unseen invades the margins of planetary speculation with deepening precision, leaving little room for alternative registers of knowledge production to prosper. The statistical-electronic visible here remains a 'threshold of inscription', rendering the surface a function of numerical weights.[69] It is on these thresholds of the (in)visible that I locate Spivak's planetarity as a means to attend to the 'excess'—understood here as all forms of engagement that have been marginalized from the statistical-electronic regimes of architecture's surfaces, much like the beasts, sea serpents and dragons of other registers.

Notes

1 Hashim Sarkis, Roi Salgueiro Barrio and Gabriel Kozlowski, *The World as an Architectural Project* (Cambridge, MA: MIT Press, 2019), 1.
2 Chet Van Duzer, 'Bring on the Monsters and Marvels: Non-Ptolemaic Legends on Manuscript Maps of Ptolemy's Geography', *Viator* 45/2 (2014), 312.
3 John May, *Signal. Image. Architecture. (Everything Is Already an Image)* (New York: Columbia University Press, 2019), 28.
4 William J. Mitchell, *The Reconfigured Eye: Visual Truth in the Post-Photographic Era* (Cambridge, MA: MIT Press, 1992), 2.
5 Nicole Starosielski, 'The Elements of Media Studies', *Media+Environment* 1/1 (2019).
6 Jacob Gaboury, *Image Objects: An Archaeology of Computer Graphics* (Cambridge, MA: MIT Press, 2021), 11.
7 Erich Hörl, 'Introduction to General Ecology: The Ecologization of Thinking', translated by Nils F. Schott, in: Erich Hörl and James Burton (eds.), *General Ecology: The New Ecological Paradigm* (London: Bloomsbury Academic, 2017), 3.
8 Sy Taffel, *Digital Media Ecologies: Entanglements of Content, Code and Hardware* (New York: Bloomsbury Academic, 2021), 1.
9 Gaboury, *Image Objects*, op. cit. (note 6), 10.
10 Ibid., 5.
11 Lukáš Likavčan, *Introduction to Comparative Planetology* (Moscow: Strelka Press, 2019), chapter 1, Kindle.
12 For digital stacks, see: Rob Kitchin and Rob and Sung-Yueh Perng (eds.), *Code and the City* (Abingdon, Oxon: Routledge, 2016). For planetary stacks, see: Benjamin H. Bratton, *The Stack: On Software and Sovereignty* (London: MIT press, 2016). For 'anatomical maps', see: Kate Crawford and Vladan Joler, 'Anatomy of an AI System: The Amazon Echo As An Anatomical Map of Human Labor, Data and Planetary Resources', AI Now Institute and Share Lab (September 2018); https://anatomyof.ai; Matteo Pasquinelli and Vladan Joler, 'The Nooscope Manifested: Artificial Intelligence as Instrument of Knowledge Extractivism', KIM HfG Karlsruhe and Share Lab (May 2020), http://nooscope.ai.
13 Rob Kitchin, 'From a Single Line of Code to an Entire City: Reframing the Conceptual Terrain of Code/Space', in: Rob Kitchin and Sung-Yueh Perng (eds.), *Code and the City* (Abingdon, Oxon: Routledge, 2016), 8.
14 According to May's technical categorization of architecture's surfaces, an image is a statistical-electrical form of recording, storing and recovering information, commonly exemplified as (but not limited to) computer-processable digital images. Photography, as the technique of inscribing light on film, conforms to a chemical-mechanical gesture, and lastly, drawing to a hand-mechanical one.
15 For pixilated censorship and blanked-out satellite imagery, see: Trevor Paglen, *Blank Spots on the Map: The Dark Geography of the Pentagon's Secret World* (New York: Penguin, 2009); Laura Kurgan, *Close Up at a Distance: Mapping, Technology and Politics* (New York: Zone Books, 2013); Lisa Parks and Caren Kaplan (eds.), *Life in the Age of Drone Warfare* (London: Duke University Press, 2017). For no-fly-zones, see: Hagit Keysar, 'Who Owns the Sky? Aerial Resistance and the State/Corporate No-Fly Zone', *Visual Studies* 35/5 (2020), 465–477. For counter-surveillance practices and adversarial attacks, see: Adam Harvey, 'HyperFace project', 2016, https://ahprojects.com/hyperface; Eyal Weizman, *Forensic Architecture: Violence at the Threshold of Detectability* (Oxford: Princeton University Press, 2017); Kate Crawford, *The Atlas of AI* (London: Yale University Press, 2021).
16 Gayatri Chakravorty Spivak, *Death of a Discipline* (New York: Columbia University Press, 2003), 72.
17 Dipesh Chakrabarty, 'Climate and Capital: On Conjoined Histories', *Critical Inquiry* 41/1 (2014), 23.
18 Spivak, *Death of a Discipline*, op. cit. (note 16), 72.
19 Elizabeth M. DeLoughrey, 'Satellite Planetarity and the Ends of the Earth', *Public Culture* 26/2 (2014), 265; Dipesh Chakrabarty, 'The Planet: An Emergent Humanist Category', *Critical Inquiry* 46/1 (2019), 6.
20 Likavčan, *Comparative Planetology*, op. cit. (note 11), chapter 3; Lucía Jalón Oyarzun, 'Nightfaring & Invisible Maps: Of Maps Perceived, but Not Drawn', *The Funambulist* 18 (2018), 40.
21 Peter Adey, 'Aeromobilities: Geographies, Subjects and Vision', Geography Compass 2/5 (2008), 1322.
22 Denis E. Cosgrove, *Apollo's Eye: A Cartographic Genealogy of the Earth in the Western Imagination* (Baltimore: Johns Hopkins University Press, 2003), 206.
23 Peter Sloterdijk, *In the World Interior of Capital: For a Philosophical Theory of Globalization*, translated by Wieland Hoban (Cambridge: Polity Press, 2013), 141.
24 Theodor W. Adorno and Max Horkheimer, *Dialectic of Enlightenment* (London: Verso Books, 2016).
25 Gayatri Chakravorty Spivak, *An Aesthetic Education in the Era of Globalization* (Cambridge, MA: Harvard University Press, 2013), 338.
26 'Overexposure' used in Oyarzun, 'Nightfaring', op. cit. (note 20).
27 Elizabeth M. DeLoughrey, *Allegories of the Anthropocene* (Durham, NC: Duke University Press, 2019), 65–66.
28 Ibid., 67.
29 Likavčan, *Comparative Planetology*, op. cit. (note 11), chapter 3; Chakrabarty, 'The Planet', op. cit. (note 19), 25.
30 Auritro Majumder, 'Gayatri Spivak, Planetarity and the Labor of Imagining Internationalism', *Mediations* 30/2 (2017), 20.
31 Yuk Hui, 'For a Planetary Thinking', *e-flux Journal* 114 (December 2020), accessed 2 March 2022, https://www.e-flux.com/journal/114/366703/for-a-planetary-thinking/.
32 Jacob Fraden, *Handbook of Modern Sensors: Physics, Designs, and Applications* (New York: Springer, 2016), 2.
33 Ibid.
34 May, *Signal. Image. Architecture.*, op. cit. (note 3), 80.
35 Barb A. Harrison et al. (eds.), *Earth Observation: Data, Processing and Applications. Volume 1A: Data—Basics and Acquisition* (Melbourne: CRCSI, 2017), 180–181.
36 Daniela Bleichmar, 'Visible Empire: Scientific Expeditions and Visual Culture in the Hispanic Enlightenment', *Postcolonial Studies* 12/4 (2009), 441.
37 Gökçe Önal, 'Media Ecologies of the "Extractive View": Image Operations of Material Exchange', *Footprint* 14/2 (2020).
38 Barb A. Harrison et al. (eds.), *Earth Observation: Data, Processing and Applications. Volume 2A: Processing—Basic Image Operations* (Melbourne: CRCSI, 2017), 147.
39 Rolando Vázquez, 'Translation as Erasure: Thoughts on Modernity's Epistemic Violence', *Journal of Historical Sociology* 24/1 (2011), 31.
40 Alison B. Powell, *Undoing Optimization: Civic Action in Smart Cities* (New Haven: Yale University Press, 2021), 154.
41 Louise Amoore and Alexandra Hall, 'Taking People Apart: Digitised Dissection and the Body at the Border', *Environment and Planning D: Society and Space* 27/3 (2009), 450.
42 Powell, *Undoing Optimization*, op. cit. (note 40), 110.
43 Fraden, *Modern Sensors*, op. cit. (note 32), 2.
44 Oyarzun, 'Nightfaring', op. cit. (note 20), 41.
45 Fraden, *Modern Sensors*, op. cit. (note 32), 3.
46 May, *Signal. Image. Architecture.*, op. cit. (note 3), 39, 47–48.
47 Powell, *Undoing Optimization*, op. cit. (note 40), 8, 109, 141.
48 Fraden, *Modern Sensors*, op. cit. (note 32), 2–3.
49 Taha Selim Ustun et al., 'Data Standardization For Smart Infrastructure in First-Access Electricity Systems', *Proceedings of the IEEE* 107/9 (2019), 1793.
50 Likavčan, *Comparative Planetology*, op. cit. (note 11), Introduction.
51 Harrison et al., *Earth Observation. Vol. 1A*, op. cit. (note 35), 21.
52 George D. Garland and Urho A. Uotila, 'Geoid', *Encyclopedia Britannica*, 17 March 2021, https://www.britannica.com/science/geoid.
53 Pasquinelli and Joler, 'The Nooscope', op. cit. (note 12).
54 Soraya Boudia, 'Observing the Environmental Turn through the Global Environment Monitoring System', in: Simone Turchetti and Peder Roberts (eds.), *The Surveillance Imperative: Geosciences During the Cold War and Beyond* (New York: Palgrave Macmillan, 2014), 195.
55 Robert Poole, 'What Was Whole about the Whole Earth? Cold War and Scientific Revolution', in: Simone Turchetti and Peder Roberts (eds.), *The Surveillance Imperative: Geosciences During the Cold War and Beyond* (New York: Palgrave Macmillan, 2014), 224, 220.
56 Joseph Masco quoted in: ibid., 215.
57 Terry A. Slocum et al., *Thematic Cartography and Geovisualization* (Harlow: Pearson Education Limited, 2014), 307.
58 Ibid., 141.
59 Harrison et al., *Earth Observation. Vol. 1A*, op. cit. (note 35), 56.
60 Slocum et al., *Thematic Cartography*, op. cit. (note 57), 142.
61 Pasquinelli and Joler, 'The Nooscope', op. cit. (note 12).
62 Louise Amoore, *Cloud Ethics: Algorithms and the Attributes of Ourselves and Others* (Durham, NC: Duke University Press, 2020); see also: Pasquinelli and Joler, 'The Nooscope', op. cit. (note 12).
63 Powell, *Undoing Optimization*, op. cit. (note 40), 62.
64 Amoore, *Cloud Ethics*, op. cit. (note 62), 160.
65 Ibid., 162.
66 Sara Safransky, 'Geographies of Algorithmic Violence: Redlining the Smart City', *International Journal of Urban and Regional Research* 44/2 (2020), 213–215.
67 Jennifer Gabrys, *Program Earth: Environmental Sensing Technology and the Making of a Computational Planet* (London: Minnesota Press, 2016), 29–32.
68 Amoore and Hall, 'Taking People Apart', op. cit. (note 41), 459.
69 Brighenti Andrea Mubi, 'The Visible: Element of the Social', *Frontiers in Sociology* 2 (2017).

# BLOCKCHA

## BLOCKCHAIN AND THE REAL-ESTATE-MEDIA COMPLEX

Marija Marić

From the introduction of fractional ownership to the emergence of entire markets for developing and trading digital land, the real estate domain has become the latest frontier for start-ups and companies organized around the uses and promises of blockchain technologies. Since its first appearance in 2008, when a pseudonimously published paper introduced Bitcoin—a digital currency generated and circulated through a peer-to-peer network that promised to revolutionize financial transfers by excluding third parties such as central banks—blockchain has received an almost mythical status of a system that could allow for decentralized and transparent transactions of information in all of its imaginable forms: from money to data, digital cats or real estate shares.[1] And while Bitcoin, and more importantly, the blockchain protocol that underlined it, did involve a technological innovation, still, what brought this technology to the public's attention was a solution to the social coordination problem it proposed. In other words, blockchain 'introduced a practical way of coordinating computational activity *without* explicit and sustained participation, and with unknown and untrustworthy actors'.[2] As a social technology, blockchain thus came in at the right historical moment—the 2008 economic crash that brought with it a final kick of disillusionment with the existing financial and political institutions—and successfully offered an alternative: an unchallenged trust into the objectivity of the machines and a big return of the technological solutionism.

Claiming the capacity to execute data transfers without friction, or more precisely, without people, blockchain was quickly embraced by the real estate industry, still burdened by a strong presence of intermediaries, as well as slow and expensive transaction processes. Coupled with burgeoning land and real estate prices and a growing problem of housing affordability, but never challenging the paradigm of home ownership and private property—the PropTech industry framed blockchain as the latest, go-to disruptive technology that would revolutionize our relationship with the built environment as we know it. Soon, the new technology found its place on the covers of a number of future-of-real-estate reports and articles, advertising many of its potential applications in the property sector.[3]

Thus, for instance, the management of land registers—national systems that store information on the ownership, control and other recorded rights over land—appeared as one of the first testing grounds for blockchain technologies. As a peer-to-peer protocol that makes all transactions of information traceable, blockchain has been hailed as an effective tool both for the easier management of land ownership transactions, as well as for making them more accessible to the wider public, providing a better overview of practices of land accumulation and patterns of speculation. With countries such as Georgia or Sweden, who were among the first to test the possibilities of blockchain on their national land registry systems, it became clear that the promises of full transparency and automation behind the new technology are still a form of a techno- (and legal) fiction. In reality, the shift mainly included digitization of existing land registers and their anchoring onto the blockchain—a process that is still dependant on the power and trust in a central authority, in this case, that of the state.[4]

Offering 'fractional and frictionless real estate investing' and claiming 'ownership reinvented', the Prop Tech market has also seen an explosion of start-ups that specialize in real estate tokenization, referring to the usage of blockchain technology for buying and selling digital real estate shares.[5] Drawing from the history of Real Estate Investment Trusts in the 1960s United States, when REITs appeared as 'financial intermediaries' that allowed 'individuals to invest in real estate and mortgages ... in the same way they can buy common stocks through mutual funds', real estate tokenization effectively utilized blockchain technology to allow the very objects of real estate to be traded in a new way: as a set of fractions that could be bought, sold or rented out, independently of each other, by different stakeholders.[6] In other words, traditional property units, such as an apartment, a house or a plot of land, could now be abstracted into assemblages of digital tokens, each essentially constituting a share, or rather a *fraction* of a given property, detached from any of its spatial or functional aspects. Designed to lower investment thresholds and therefore allow a much broader pool of prospective real estate owners to purchase only fragments instead of entire property units, this shift towards fractional ownership of real estate has been presented by its native industry as a 'democratization' of the non-accessible housing market—a 'paradigm shift' enabled by a blockchain technology.[7]

Next to the management of tangible properties that real estate tokenization and digitization of land registries provided for, blockchain's presence in the real estate domain has perhaps been most commonly associated with an emergence of less physical but equally material forms of property: the non-fungible tokens, or NFTs, referring to the non-interchangeable digital assets, such as images, videos, music and even virtual real estate. One such NFT is the now already famous Mars House, designed in 2020 by a Toronto-based artist Krista Kim as the first digital-only architectural NFT project. Sold for 288 Ethereums (worth approximately half a million US$ at the time), Mars House is now being rented out as a space for virtual events, such as wedding ceremonies or family reunions, serving also as a trigger for a booming architectural and urban NFT market.[8]

Beyond digital houses and singular objects, different companies are also competing in constructing entire virtual worlds operating as new marketplaces for the trade of digital land, real estate and other commodities. Thus, for instance, projects like Decentraland—a virtual reality platform that features 90,000 digital, 16 × 16-m plots of land that each initially sold for 0.1 Ethereums—provide new imaginaries of land speculation, whose value is now grounded solely in the attention economy that powers these initiatives.[9] No matter how intangible the digital land of Decentraland is—traded as LAND tokens and defined only by the unique x, y coordinates of its Cartesian world, as well as a reference to a content the owner wishes to provide on their virtual land—its economies are more than material. In 2021, the Metaverse Group purchased a parcel in Decentraland for 2.4 million US$, just more than half the price of the most expensive piece of virtual land sold in Sandbox (another platform that offers trade of digital properties) for 4.3 million US$ to Republic Realm—one of the largest real estate developers in the Metaverse.[10] The profitable formulas of projects such as Decentraland or Sandbox, however, do not end with their property portfolios, but also feature their own, custom-made cryptocurrencies. Thus, for instance, the digital parcels in Decentraland are traded through MANA, whose market capitalization surpassed 5 billion US$ only four years after the platform's launch.[11]

Similarly, the trade and speculation of virtual land and real estate is not just a matter of digital environments. Projects such as Earth 2.0, designed as a digital twin of the 'analogue' Earth, is split into 5.1 trillion land tiles, 'in which real-world geolocations on a sectioned map correspond to user-generated digital virtual environments', and that, most importantly, can be bought, owned and traded.[12] By mirroring the geography of the physical world, Earth 2.0 and other similar platforms also mirror its on-the-ground geopolitics and systems of value, thus making the projection of power hierarchies and inequalities embedded into these geographies possible. As such, the virtual domain of projects like these appears not as a final destination for the trade of the digital land, but rather as a tool or an intermediary that could shape the value, based precisely on the relationship between the digital and the physical.

## The (Spatial) Politics of Blockchain

Drawing its origins from the cypherpunk movement of the 1990s, organized around a belief that technology could serve as an alternative form of governance to that of the state, blockchain has been deeply shaped by the cyberlibertarian political ideologies. As David Golumbia observed in his book *The Politics of Bitcoin: Software as Right-Wing Extremism*, cryptocurrencies have incorporated some of the most basic elements of a right-wing worldview, even if their broader narrative does not immediately relate to the rhetorics of the right.[13] From understanding of the central bank as a 'deliberate plot to "steal value" from the people to whom it actually belongs', to its role in keeping the world on the brink of collapse, and belief that '"hard" currencies such as gold provide meaningful protection against that collapse'—as Golumbia points out, the cyberlibertarian ideology has a lot in common with the right-wing movements, especially in Europe and the United States.[14] By positioning technology as a means of enabling and protecting freedom, and thus framing IT products, services and markets as necessary for maintaining democracy, cyberlibertarianism has successfully aligned itself with the logic of capitalist production. Similarly, and despite its mainstream narrative, the blockchain itself has been eagerly adopted and embraced by capitalist institutions, ranging from various industries to banks themselves, serving as an infrastructure for a more efficient flow of money, trade and speculation.[15]

Blockchain is, therefore, a social technology.[16] In (re)producing power hiearchies and new social relationships, blockchain is an 'inherently political technology', and as such it has helped shape our imaginaries of property.[17] In its most bare form (such as Bitcoin and other cryptocurrencies), it has successfully managed to propagate the social model of a 'techno-Leviathan', where cryptocurrency users 'establish novel forms of social contracts'—the ones in which they 'agree to "contract" with cryptographic algorithms that guarantee their (financial) peace and safety'.[18]

Following from there, blockchain technologies and their discontents, such as cryptocurrencies and NFTs, as Felix Stalder pointed out, managed to produce 'a trivial version of Adam Smith's society of independent shop-keepers as "creators"', creating an ideal model of an 'ownership society', in which all social relations are embedded in the logic of private property and distrust in social institutions—a position that could also be seen as central to the current neoliberal economic and political paradigm.[19]

Its extractivist logic, articulated by its language of mining, has also found its place in the real estate domain, itself characterized by extractive practices of growth and profit making.[20] Starting from here, one could ask: How will the wider acceptance of blockchain technologies that facilitate real estate trade and speculation, impact the already unstable values and unequal distribution of land and housing on-the-ground? What is the role of technology in hyper-commodification of the built environment—a state in which, as David Madden and Peter Marcuse observed, 'all of the material and legal structures of housing—buildings, land, labor, property rights—are turned into commodities?'[21] And finally, which new forms of living and belonging could emerge out of these new relationships between humans and space, and technical and financial infrastructures that communicate between them?

Going beyond current real estate market's rendering of lived space as square metres, and towards translating real estate into tokens, designed to be easily traded and speculated upon—real estate tokenization has led to a form of abstraction of housing, thus allowing for a new way of seeing properties: as metrical units of money flows, instead of spatial units of habitation. Parasitizing on the existing inequalities embedded in our dysfunctional housing system, PropTech narratives of property democratization through real estate tokenization not only help to maintain the current economic status quo, but also actively contribute to the radicalization of speculative real estate practices. This leads to the question: Does the translation from use-value to trade-value of housing have the capacity to produce another kind of translation: from inhabitants to the new kind of space occupants—a family of real estate token owners?

Similarly, standing on the opposite spectrum of what artist and media theorist Hito Steyerl described as 'poor images', referring to the compressed and pixelated free and anonymous digital objects on the Internet, whose condition of circulation and far-reaching dispersion has the potential to redefine the value of an image—NFTs like Mars House are essentially based on the speculative logic of private property and starchitecture.[22] Appearing as ultimate commodities whose value is constructed solely on the grounds of programmed scarcity, the proliferation of architectural NFTs marked a beginning of a new kind of real estate market—the one in which buildings are finally liberated from the burden of the use-value and are instead replaced by images that can be freely traded. With blockchain as 'the medium in which an artwork is inscribed, valorized, traded, archived, and even displayed' itself becoming a 'financial technology', it becomes clear that 'digital art practices, then, are themselves currently in the process of becoming financial technologies'.[23] As such, NFTs could essentially be seen as a radicalized form of private ownership that now takes the digital sphere as its new frontier.

At the same time, platforms like Decentraland could be seen tools that are providing us with ideal real estate visions: a grid reduced to a bare spatial organization of economic potential; an urban machine for land and real estate speculation, unconstrained by any regulatory or ethical obstacles. An advanced form of commodification of urban imaginaries, these real estate fictions are also actively shaping material land. The launch of digital twins of Earth, through platforms such is Earth 2.0, not only led to an acquisition rush for digital parcels that geographically correspond to the location of 'value' on the ground, but could also easily be seen to work the other way around too—strategically acquiring and developing land in the digital space in order to increase its value in the physical one.

## The Possibility of Urban Commons beyond the Real-Estate-Media Complex

The proliferation of 'real estate products', such as real estate tokenization, architectural and urban NFTs or the virtual lands of Decentraland and Earth 2.0, depends heavily on the construction of attention, and the cultivation of its economies. Language, in this process, plays a key role. The discoursive language of start-ups and companies whose profit relies on dreams of growth, rather than existing material grounds, operates on a discursive technology, described by anthropologist Anna Tsing as 'the economy of appearances', referring to the 'self-conscious making of a spectacle ... necessary for gathering the investment funds'.[24] In other words, and as Tsing suggests, profit must be imagined in order to be obtained. The same goes for the expansive visions of property. Drawing from the language of the housing crisis on the one hand and real estate growth on the other, the PropTech industries have positioned blockchain as a tool for 'democratization' of property markets by inventing ever new and more radical forms of private property ownership. In a real-estate-media complex in which architecture, cities and entire urban imaginaries are being produced today, language appears as an alternative construction site, where projects are being built by means of narrative, storytellng and above all, fiction.[25]

In their essay 'Smart Cities as Corporate Storytelling', Ola Söderström, Till Paasche and Francisco Klauser positioned smart cities as 'part of contemporary language games around urban management and development', devised to position tech and real estate companies who offer smart city products and services as 'obligatory passage points' in the 'transformation of cities into "smart" ones'.[26] Tech-industry's vision of smartness could thus be seen as an economically performative narrative, and above all a financial model, that 'allows real estate firms to charge a "smartness premium", thus driving up already prohibitive costs of real estate'.[27] What started as a 'smartness mandate' after John Palmisano's famous IBM talk 'A Smarter Planet', could also be seen as a base for all the iterations of technological solutionism that followed until today, including blockchain technologies as they claim to revolutionize real estate domain.[28]

Asking why, throughout history, philosophers and economists who shaped our common understandings of property so often chose a narrative instead of a scientific approach to making their argument, legal theorist Carol M. Rose suggested:

> We may only be able to understand property arrangements through narrative discourses like literature and history, discourses that construct a story of how things got to be that way—a story in which there were genuine choices along the way and in which things were not really predictable in advance and did not have to wind up the way they did.[29]

Understanding property as a story, it becomes clear that, in order to change the narrative and allow for unexpected twists and turns, we need to rethink who the narrators are. In other words, if the Decentralands of the world are conceived as real estate fairy tales written by companies and developers, could we also imagine the opposite—blockchain technology as a testing ground for imaginaries of another relationship with property, or its undoing altogether? Moving beyond the dichotomy of good and bad technologies, as Felix Stalder pointed out, the question of 'whether the technological potential of the blockchain can be used to advance the social vision of more-than-human commons', remains open.[30]

This reminds us of the fluidity of capital, its modes and sites of reproduction, that once intersected with novel technologies lead to ever-new forms of commodification of the built environment, providing us, at the same time, with the tools and possibilities to rethink the notion of the commons, resistance and critical spatial practices in (and through) the digital domain. In asking, 'what can urban commons learn from the free software hackers?', Dubravka Sekulić observed how 'the digital, more than any other field, discloses property as being inappropriate for contemporary relationships between production and reproduction, and additionally, proves how it is possible to fundamentally rethink it'.[31] Learning from NFTs, LANDs and 'fractional ownership' as new paradigms of private ownership created in the intersection with novel technologies, it also becomes necessary to think of new modes of digital commons these technologies could help outline, seeking, at the same time, their productive relationships with the struggles for urban commons on-the-ground.

In outlining an 'expanded activist repertoire in infrastructure space', architecture theorist Keller Easterling suggested that, in order to effectively counter the complex economic, political and technological forces that stand at the core of production of social and spatial injustice today, we need 'techniques that are less heroic, less automatically oppositional, more effective, and sneakier—techniques like gossip, rumor, gift-giving, compliance, mimicry, comedy, remote control, meaninglessness, misdirection, distraction, hacking, or entrepreneurialism'.[32] In other words, as she further observes, we need an approach 'to both form-making and activism that is more performative than prescriptive'.[33] With media strategists, tech industries and cryptocurrency brokers moving to the forefront of production of both our built environment and collective urban imaginaries, we need to radically rethink and extend the scope of design and research tools in order to reclaim architecture as a critical and political practice, by asking: What is the relationship between the built and the fictional, the Earth and its digital twins, the physical and virtual land, and how are architects to mediate these materialities critically?

Notes

1 Satoshi Nakamoto, 'Bitcoin: A Peer-to-Peer Electronic Cash System' (2008), https://bitcoin.org/bitcoin.pdf.
2 Blockchain protocol is based on a chain-like pattern of data transaction grouped into blocks of information, in which each block contains a cryptographic hash of the previous one, thus making the blockchain resistant to subsequent modifications. From Quinn DuPont, *Cryptocurrencies and Blockchains* (Cambridge: Polity, 2019).
3 Deloitte, 'Blockchain in Commercial Real Estate: The Future Is Here!' (Deloitte Center for Financial Services, 2017), among others.
4 See more in: Georg Von Wangenheim, 'Blockchain-Based Land Registers: A Law-and-Edonomics Perspective', in: Amnon Lehavi and Ronit Levine-Schnur (eds.), *Disruptive Technology, Legal Innovation, and the Future of Real Estate* (New York: Springer, 2020), 103–122.
5 From RealT, 'Fractional and Frictionless Real Estate Investing', https://realt.co/, accessed 16 February 2022.
6 For the history of REITs in the United States, see: Peter A. Schulkin, 'Real Estate Investment Trusts', *Financial Analysts Journal* 27/3 (1971), 33–40: 33.
7 On real estate tokenization, see: Marija Marić, 'Bricks, Blocks, and (Block)Chains', *Log* 55 (2022), 85–89.
8 Sophia Herring, 'Now on Sale: The World's First Digital NFT Home', *Architectural Digest*, 14 March 2021, https://www.architecturaldigest.com/story/worlds-first-digital-home.
9 Decentraland, 'Decentraland's LAND Parcel Size', 25 January 2019, https://decentraland.org/blog/platform/land-parcel-size/. See also: Catalina Goanta, 'Selling LAND in Decentraland: The Regime of Non-Fungible Tokens on the Ethereum Blockchain Under the Digital Content Directive', in: Lehavi and Levine-Schnur, *Disruptive Technology*, op. cit. (note 4), 139–154.
10 Elizabeth Howcroft, 'Virtual Real Estate Plot Sells for Record $2.4 Million', Reuters, 24 November 2021, https://www.reuters.com/markets/currencies/virtual-real-estate-plot-sells-record-24-million-2021-11-23/; and Nat Rubio-Licht, 'The Metaverse Real-Estate Boom Is On', Protocol, 30 November 2021, https://www.protocol.com/bulletins/republic-realm-sandbox-real-estate.
11 Market capitalization of MANA in November 2021 reached its peak at 5.5 billion US$. From 'Decentraland', CoinMarketCap, https://coinmarketcap.com/currencies/decentraland/, accessed 25 November 2022.
12 Earth 2, 'Earth2®', https://earth2.io/, accessed 21 October 2021.
13 David Golumbia, *The Politics of Bitcoin: Software as Right-Wing Extremism* (Minneapolis: University of Minnesota Press, 2016).
14 Ibid.
15 DuPont, *Cryptocurrencies and Blockchains*, op. cit. (note 2).
16 Ibid.
17 On the political reading of technology, see: Langdon Winner, 'Do Artifacts Have Politics?', *Daedalus* 109/1 (1980), 121–136: 123.
18 See: Brett Scott, 'Visions of a Techno-Leviathan: The Politics of the Bitcoin Blockchain', *E-International Relations* (blog), 2014, http://www.e-ir.info/2014/06/01/visions-of-a-techno-leviathan-the-politics-of-the-bitcoin-blockchain/; and DuPont, *Cryptocurrencies and Blockchains*, op. cit. (note 2)..
19 Felix Stalder, 'From Commons to NFTs: Digital Objects and Radical Imagination', *Notes & Nodes* (blog), 31 January 2022, http://felix.openflows.com/node/642.
20 In describing the extractivist logic behind cryptocurrencies, Nigel Dodd writes about 'digital metallism' of Bitcoin. See: Nigel Dodd, 'The Social Life of Bitcoin', *Theory, Culture & Society* 35/3 (2017), 35–56: 42.
21 David Madden and Peter Marcuse, *In Defense of Housing: The Politics of Crisis* (London/New York: Verso, 2016).
22 Hito Steyerl, 'In Defense of the Poor Image', *E-Flux Journal* 10 (November 2009), http://www.e-flux.com/journal/10/61362/in-defense-of-the-poor-image/.
23 Martin Zeilinger, 'Digital Art as "Monetised Graphics": Enforcing Intellectual Property on the Blockchain', *Philosophy and Technology* 31 (2018), 15–41: 21.
24 Anna Tsing, *Friction: An Ethnography of Global Connection* (Princeton/Oxford: Princeton University Press, 2005), 57.
25 Marija Marić, *Real Estate Fiction: Branding Industries and the Construction of Global Urban Imaginaries* (Doctoral Dissertation, Zurich, ETH Zurich, 2020); and Marija Marić, 'Real Estate Fiction: An Outline for a New Literary Genre', *Thresholds* 46 (2018), 320–327.
26 Ola Söderström, Till Paasche and Francisco Klauser, 'Smart Cities as Corporate Storytelling', *City* 18/3 (2014), 307–320: 307–308.
27 Evgeny Morozov and Francesca Bria, *Rethinking the Smart City: Democratizing Urban Technology*, City Series (New York: Rosa Luxemburg Stiftung, 2018), 14.
28 Orit Halpern, Robert Mitchell and Bernard Dionysius Geoghegan, 'The Smartness Mandate: Notes toward a Critique,' *Grey Room* 68 (2017), 106–129.
29 Carol M. Rose, 'Property as Storytelling: Perspectives from Game Theory, Narrative Theory, Feminist Theory', in: Carol M. Rose (ed.), *Property and Persuasion: Essays on the History, Theory, and Rhetoric of Ownership* (Colorado: Westview Press, 1994), 25–45: 30.
30 Stalder, 'From Commons to NFTs', op. cit. (note 18).
31 Dubravka Sekulić, 'Legal Hacking and Space: What Can Urban Commons Learn from the Free Software Hackers?', *Eurozine*, 11 April 2015, http://www.eurozine.com/legal-hacking-and-space/.
32 Keller Easterling, *Extrastatecraft: The Power of Infrastructure Space* (London/New York: Verso, 2014).
33 Ibid.

# LIFE NEEDS

LIFE NEEDS INTERNET

Handwritten Letters on Digital Culture

Jeroen van Loon

*Life Needs Internet* is an ongoing project that archives digital culture through handwritten letters. Each letter answers the question: How does the Internet influence your daily life and how does it make you feel? I started this project in 2010 and since then have collected more than 1,500 letters with personal and anecdotal stories from 40 different countries covering a period of more than a decade.

The goal is to preserve the global influences of the Internet through local and personal testimonials. The letters, all handwritten in the author's native languages, have mostly been collected during lectures, exhibitions, workshops and some through the Internet. The larger the archive becomes, the more chance these letters will create an archaeological insight into digital culture, showing how digital culture differs per country, culture and generation.

Every five years (2012, 2017, 2022), I use a number of letters from the archive in a video installation showing the original letters in combination with video portraits featuring the authors. These video installations often focus on newly collected letters. The selection presented here in Datapolis offers a larger overview of the whole archive. It shows three letters from 2010 to 2021 and is a first attempt to dive into the entire archive and construct a larger narrative by means of the individual letters. The selection process was primarily based on the told stories, anecdotes or experiences, while simultaneously trying to ensuring a fair balance between the different years and locations. Lastly, the letters were rearranged in a non-chronological order to avoid any historical or linear narrative, instead composing a series of combinations through the course of time. That way, the three letters react to each other, offering a unique insight into the archive of *Life Needs Internet*.

NAME: PRITHVIK SEN CHOUDHARY
CITY: UTRECHT, THE NETHERLANDS
APRIL 2022

| Male (23) | Digital media proffesional | 2022 |
| --- | --- | --- |
| Utrecht | The Netherlands/India | |

For me, it would not be an exaggeration to suggest that my life is incomplete without the internet, as is probably the case with a lot of people. However, for me there exists a compulsion. As someone who moved all the way from India to the Netherlands to pursue my higher education and subsequently begin my professional career, living without the internet is completely unimaginable. Ironically, I came to the Netherlands to pursue an education in New Media and Digital Culture, a field of study that enables one to think more critically about the relationship between society and digital media objects. With my educational background, it is often a conflict for me to use and interact with a lot of digital media platforms like Facebook, WhatsApp, Zoom, and Instagram that are otherwise known to be notorious about aspects like data protection, privacy encroachment, and algorithmic biases among other things. When I read about how these platforms operate, I can't help but think that by using them, I am in some ways enabling their unethical behavior. On the other hand, the services they provide, most crucially that of connectivity and closeness allows me to remain connected to not just my family back home but also maintain communication with the relationships I have been able to make across the globe during my time as a student at an international university. Therefore, the use of such platforms remains an ideological conflict for me but at the same time something that I cannot imagine living without.

Recently, I changed phones and the number associated with my WhatsApp account on my new phone also changed. However, I decided to keep my previous phone alive just so that my grandfather would not have go through the painful ordeal of changing my number on his phone and getting used to contacting me on a new WhatsApp number. Remaining connected with him is extremely important for me. Because he faces a lot of difficulty in operating his smartphone and only uses WhatsApp to call me, I decided I will keep my previous phone active just so that I can maintain regular contact with him. As a result now, I have two different WhatsApp accounts. While this increases my interaction with WhatsApp, I don't think I would have it any other way.

Ultimately, when I think more closely about this, boycotting these platforms or restricting our own usage don't seem like long-term solutions because that deprives us as a society to improve our lives and experience the benefits of technological advancement. It would certainly be a lot more useful to move our attention towards digital policymaking to overcome such conflicts that arise from using these platforms. In an ideal world, one should not really feel any conflict in using digital media platforms but should gladly embrace them. I am, however, optimistic that with all the attention that internet culture now receives from various factions in society, we might be heading towards a more inclusive culture.

いまの実態
　私がワメナに初めてやってきたのは１９８１年で、１９９９年以降、２００５年までのほとんどの期間をこのワメナ、およびその周辺の山々ですごしました。そうして２００６年以降、現在までこのワメナでインターネットカフェを営んで暮らしています。
　そんな私から見て、パプア人（ここではインドネシア人を対象から除く）とコンピューター、とりわけノートパソコンとの出会いは、彼らの生活に大きな変化を齎しました。これまでのテレビや音楽のない世界から一変し、自分たちの知らなかった世界を簡単に目や耳にすることができるようになったのです。
　それではそんな高価なもの（ワメナでのノートパソコンの購入価格は約５００ドル以上。

| Male (58) | Internet cafe owner | 2011 |
| --- | --- | --- |
| West-Papua | Indonesia | |

Current situation

It was 1981 when I came to Wamena for the first time. For most of time between 1999 to 2005, I lived there and also in the mountains nearby. In 2006, I settled in Wamena and started running an internet-café.

In my point of view, the encounter with computers, especially laptops, has brought a big change to Papuan lives (except for Indonesian). It enabled them to see and hear the unknown world out of the one without TV or music. Why do they manage to get such an expensive thing? (The price of computers is more than 500 dollars in Wamena. Monthly income of public employees is about 150 dollars.) My internet-café in Wamena offers fixing service. More than 2,000 computers are brought in a year. This situation can tell that there supposes to be about 10,000 computers around Wamena, half of which are provided by the government.

Computers have become one of the essential parts of life, but there are some problems.

First, because computers are provided for free, Papuan people don't take care of it very properly. Especially, the life with electric appliances is still new for them, its operation and treatment is a big problem. Second, although laptops give profit to villages in inconvenient areas, most of their electricity supplies (except for Wamena) are dependent on their own generator. Its unreliable electricity supply sometimes damages the inside equipments of computer. Third, people start understanding the basic software such as WORD or EXELE, which are the main purpose to use computer, still there are very few people use or learn Internet and other soft wares.

Vision for the future

It would take much longer time until Papuan people really understand computers and maximize its benefit. Because of a sudden change from the life without books or electric appliances, they look just confused to have the latest technology and eventually sometimes break it. That's the situation for now.

The biggest problem is that people lack the will to be better. I've found out through 4 years experience of running an internet-café that they are easy to forget even very simple operation.

New technology can be easily brought by money, but learning it takes time. Papuan people, who have been away from taking efforts to learn new things, would have to face difficulties from now on.

# الحياة تحتاج الى الانترنت - توثيق للثقافة الرقمية من خلال رسائل مكتوبة بخط اليد

المدينة: امستردام — الجنس: ذكر

المدينة / البلد: هولندا — السن: ٢٥

---

كيف تشعر حيال تزايد أهمية وجود الكمبيوتر والانترنت في حياتنا اليوم وكيف يؤثر ذلك على حياتك اليومية؟

انا بتذكر صوت طلب الاتصال الخطي بالانترنت من خط الهاتف.. كنا ننتظر احيانا ساعة او

| Male (25) | Art student | 2017 |
| --- | --- | --- |
| – | Syria/The Netherlands | |

I remember quite well how the connection to the internet was made by using the regular telephone line... Sometimes we had to wait for an hour or more, before receiving a response... 15 years ago; it took a very long time. Around 10 kilobyte per second, incredibly slow, but at least there was actual contact through the internet. So how is it possible to see everything in a distant part of the world by simply clicking a button? The Internet was actually a symbol of wealth at the time. The Internet was not for everyone. Whether or not you used the Internet depended on your finances and technological possibilities. After a relatively short time, more technology became available throughout the country. The mobile phone and computer were rapidly introduced into every home. It made communication through the Internet much easier. The Internet demanded more and more time both at work and at school, as well as when meeting up with friends. It actually became the way to follow the news. It is also a method of finding out something. Each question was asked on Google. For the majority, the questions were about our lives as Syrians, about what was happening in our country and how we could escape from that high-risk place. Nowadays, we are aware of the importance of the Internet. Even now, it is sometimes difficult to communicate through the internet due to the lack of electrical power and/or the bad connection. The Internet has helped us to understand what we need to do, how we should do it and where we are heading. The Internet is one of the most powerful media that has helped many people to survive. I now remember my first introduction to the Internet some 15 years ago and how it has become an essential part of our daily life. I do wonder what the situation will be like in 15 years' time; will I be living without the Internet?

Datum: 2018

**LIFE NEEDS INTERNET** afstammeling: Indonesië

Man: ☒ Leeftijd: 11 Woonplaats: Baleoren Naam:

Vrouw: ☐ Beroep: School Land: Nederland Mail *:

**Wat voor effect heeft het internet op je leven en hoe voel je je hier bij?**

ik speel vaak games en ~~Youtube~~ Luister Youtube

En

vaak

ik ved

te dik

gedoe

te spe

vind he

ben u

school

dan ga

vaak b

de nac

Bedankt vo
Je brief za
brieven coll

*niet verplic

Male (11) | Elementary school pupil | 2018
Utrecht | The Netherlands

I often play games and listen to Youtube. And that's super cool, but it often excites and stresses me and then I eat a lot and then exercise again to not get too fat. But that is always a hassle. It's hard to play another game or not use Youtube. I like it better when I'm no longer addicted because then I don't release my stress at school and at home and I don't want that because then things go wrong again……… It often goes wrong when I secretly play games in the night or when my parents are away too.

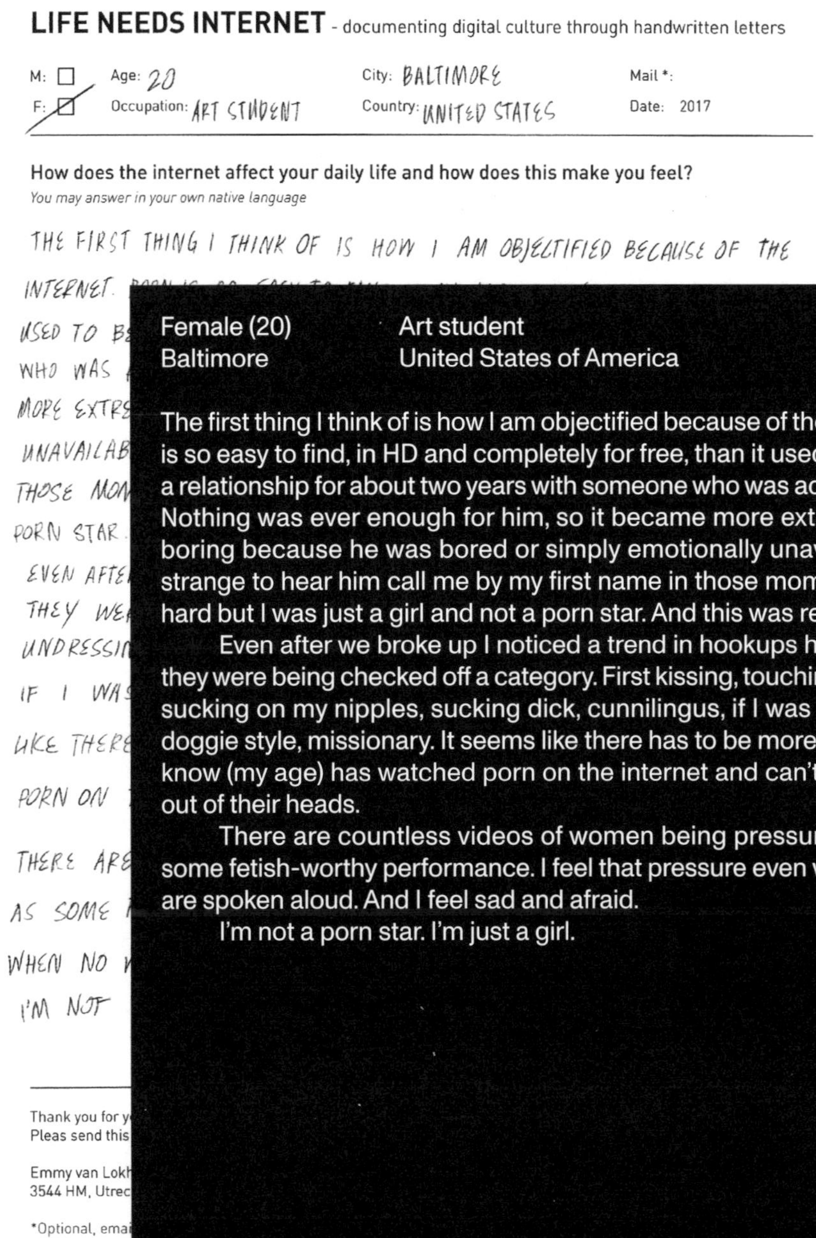
**LIFE NEEDS INTERNET** - documenting digital culture through handwritten letters

M: ☐ Age: 20 City: BALTIMORE Mail *:
F: ☒ Occupation: ART STUDENT Country: UNITED STATES Date: 2017

How does the internet affect your daily life and how does this make you feel?
*You may answer in your own native language*

THE FIRST THING I THINK OF IS HOW I AM OBJECTIFIED BECAUSE OF THE

| Female (20) | Art student | 2017 |
|---|---|---|
| Baltimore | United States of America | |

The first thing I think of is how I am objectified because of the internet. Porn is so easy to find, in HD and completely for free, than it used to be. I was in a relationship for about two years with someone who was addicted to porn. Nothing was ever enough for him, so it became more extreme. Sex was boring because he was bored or simply emotionally unavailable. It was strange to hear him call me by my first name in those moments. I tried so hard but I was just a girl and not a porn star. And this was real life.

Even after we broke up I noticed a trend in hookups happening. Like they were being checked off a category. First kissing, touching, undressing, sucking on my nipples, sucking dick, cunnilingus, if I was lucky, cowboy, doggie style, missionary. It seems like there has to be more but everyone I know (my age) has watched porn on the internet and can't seem to get it out of their heads.

There are countless videos of women being pressured into sex as some fetish-worthy performance. I feel that pressure even when no words are spoken aloud. And I feel sad and afraid.

I'm not a porn star. I'm just a girl.

# VISIONS: SP

## VISIONS: SPATIAL SPECULATIONS ON HOW TO LIVE WITH THE LEGACY OF DATAPOLIS

Datapolis Studio

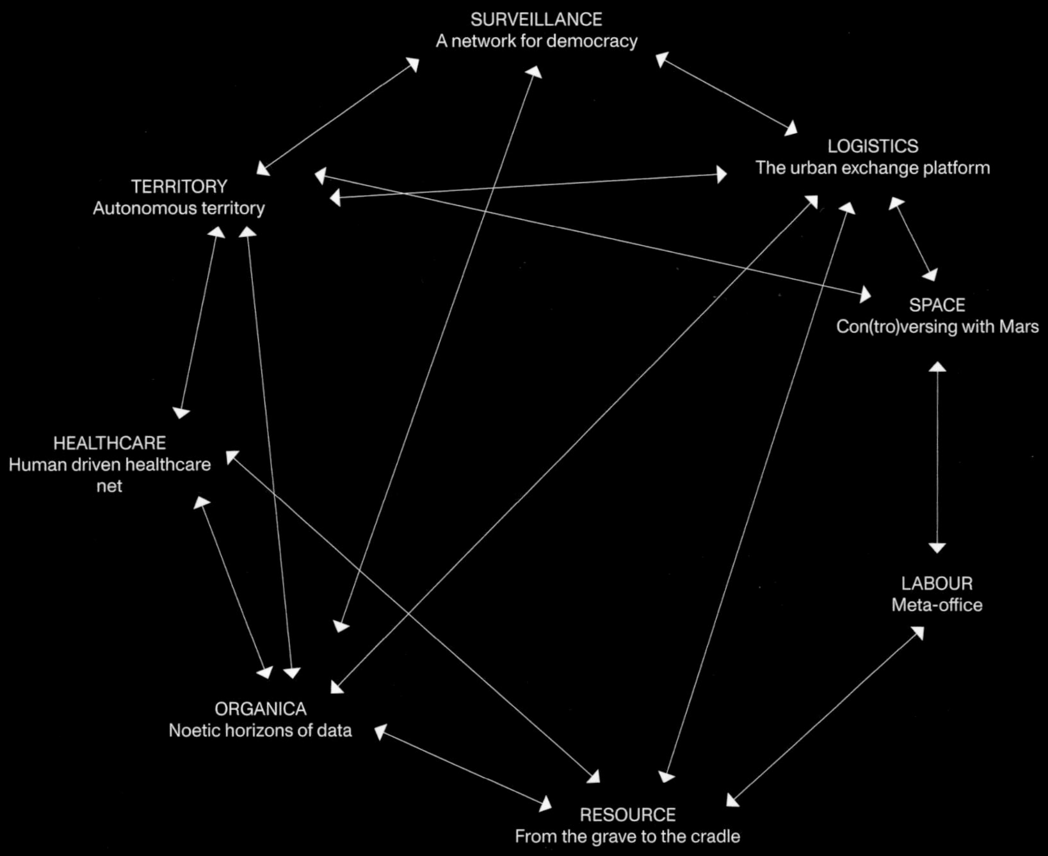

# AUTONOMOUS TERRITORY

Arda Ertan Yildiz, Wu Yongyi

By the year 2020, about 1.7 megabytes of new information will be created every second for every human being on the planet. As the digital transformation of our society expands, data centers play the critical role of providing the technology backbone that supports our digital lifestyle. In fact, data centers are accommodating these needs by occupying almost two billion square feet of facility floor space across the world.[1]

Due to the huge growth in data demand worldwide, the traditional data centre model has shown its limitations. But from the trends in the data centre typology evolution and geographical distribution, we can see a possibility of a completely independent system. Benefits from advanced computer technology, high-speed networking and artificial intelligence, the next generation of data centres can be close to the most advantageous external factors and develop to the most effective form.

In the research, three main positions have been put forward, which are:

1 In the foreseeable future, global data demand will grow exponentially, just like it has in the past few years.

2 A more intensive (hyperscale) and automated data centre typology will become a necessity for reasons of efficiency.

3 Energy and environment issues that massive data centres bring will affect our world and the data centres themselves.

As a relatively new human invention, the data centre seems to be incompatible with the natural environment and living environment in many ways. A typical example of this is e-waste: the non-degradable or even toxic parts that are the unavoidable consequence of the earth hidden behind the Internet we use.

These trends provoke new possibilities for data centres. Is it possible to better integrate data centres into universal or specific natural systems, or even industrial systems? How can data centres benefit from these natural or artificial environments, for instance by increasing efficiency or reducing adverse effects, and even achieve a mutually beneficial relationship with them?

The project proposes three specific locations for data centres:

A Under the Atlantic Ocean.

B On the outskirts of a city in southern China.

C In the depths of the Svalbard mountains.

The primary consideration in choosing these three locations is that they all have an existing connectivity to the Internet submarine cables: the Atlantic Ocean contains the existing connection between United States and Europe; South China is an important hub for the Pacific Rim; Svalbard has a submarine cable directly connected to Norway in the remote Arctic.

Except for Internet connectivity, each data centre presents distinct characteristics, such as relationship to natural environment, population density, industrial composition, political attributes and so on. These bring about different discussion topics at each location, which can be used to explore the different possibilities and potential adaptability of the data centre.

1 Steven Carlini, 'Data Center', Schneider-Electric 2017, https://blog.se.com/datacenter/2017/11/17/data-center-iot-architecture/.

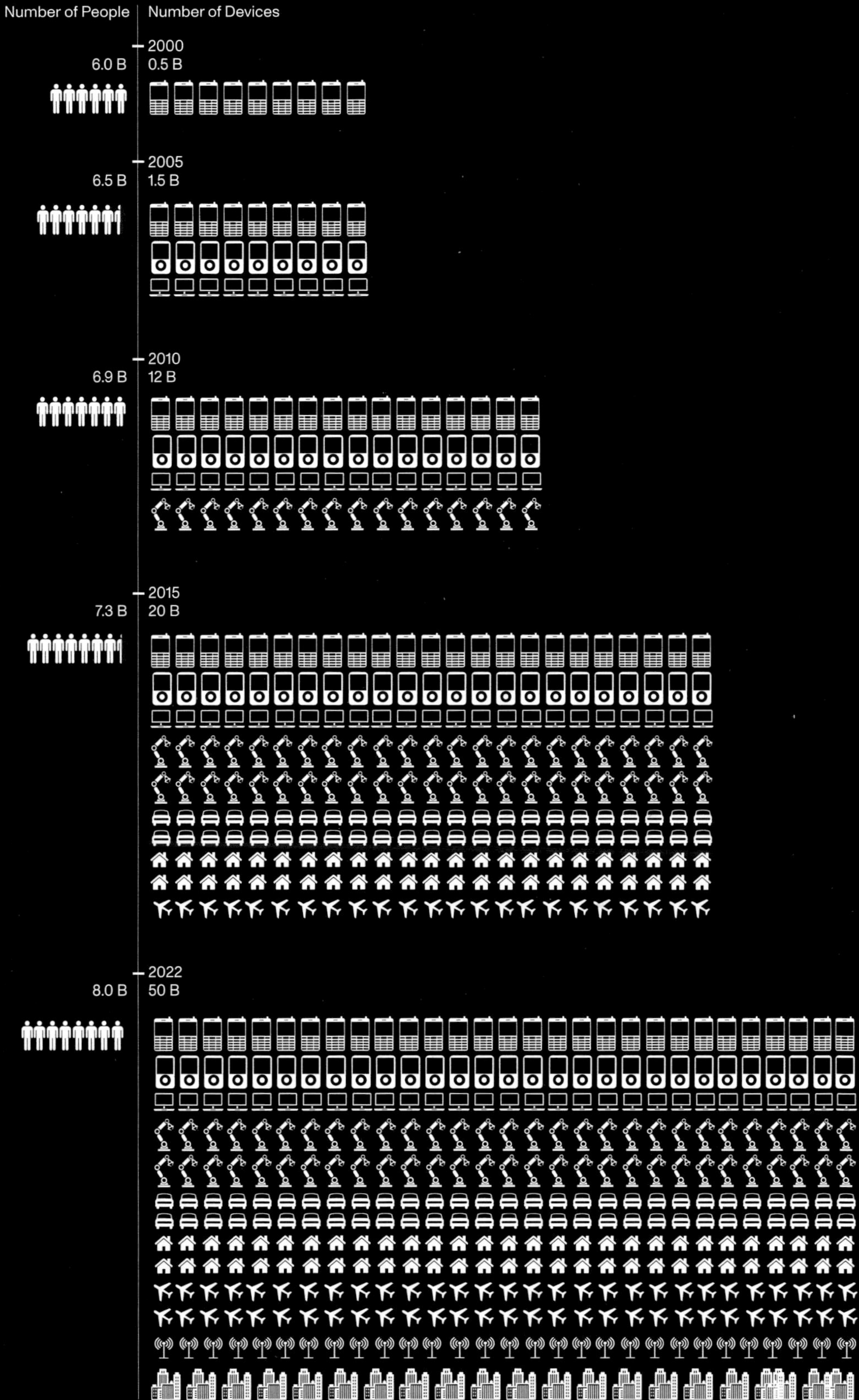


Fig. 1 Population growth versus the increase in number of connected devices between 2000 and 2020. Drawing by Federica Longoni and Filip Promaniuk.

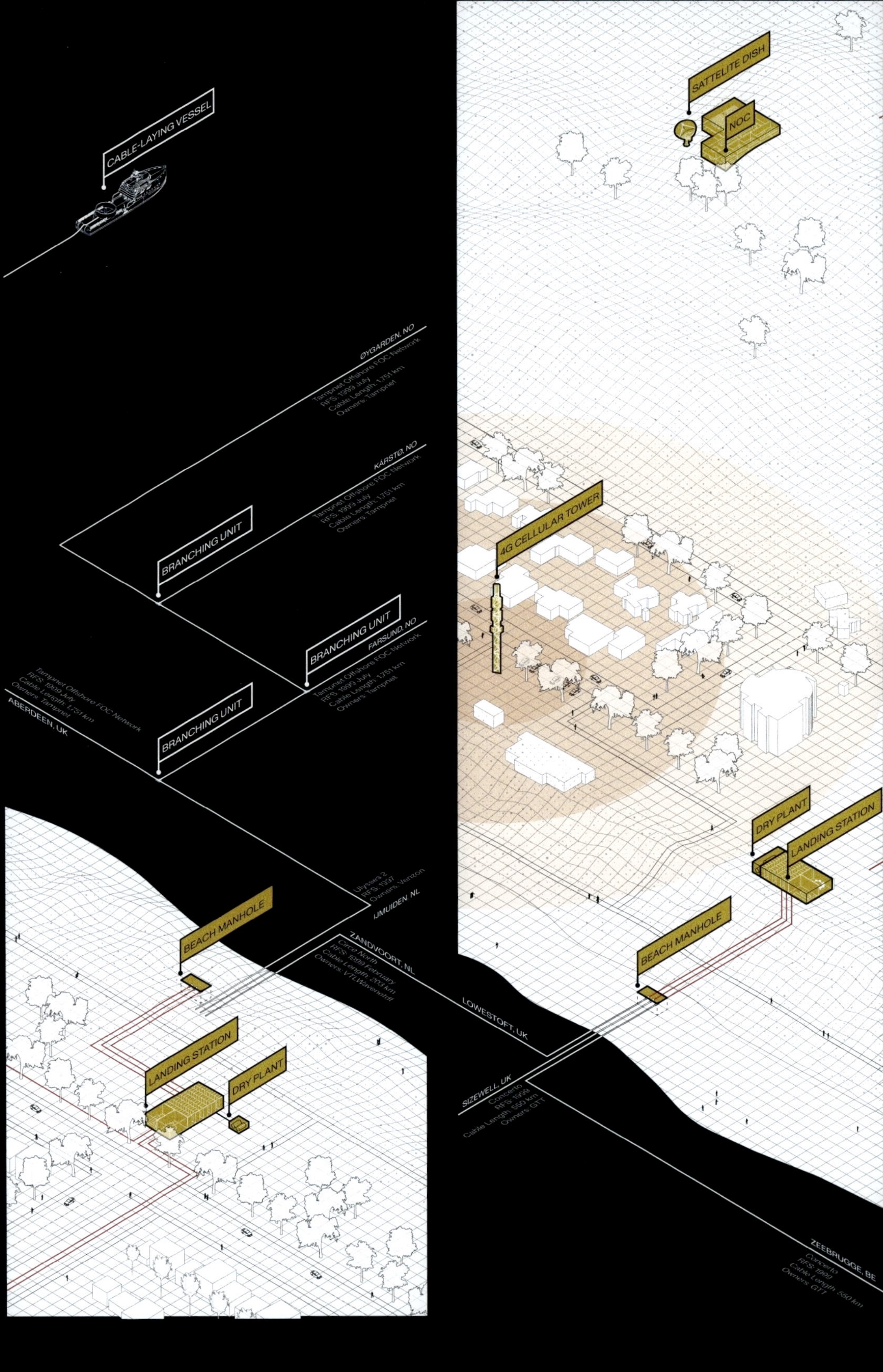
CABLE-LAYING VESSEL
SATTELITE DISH
NOC
ØYGARDEN, NO
Tampnet Offshore FOC Network
RFS: 1999 July
Cable Length: 1,751 km
Owners: Tampnet
KÅRSTØ, NO
Tampnet Offshore FOC Network
RFS: 1999 July
Cable Length: 1,751 km
Owners: Tampnet
BRANCHING UNIT
4G CELLULAR TOWER
BRANCHING UNIT
FARSUND, NO
Tampnet Offshore FOC Network
RFS: 1999 July
Cable Length: 1,751 km
Owners: Tampnet
Tampnet Offshore FOC Network
RFS: 1999 July
Cable Length: 1,751 km
Owners: Tampnet
ABERDEEN, UK
BRANCHING UNIT
DRY PLANT
LANDING STATION
Ulysses 2
RFS: 1997
Owners: Verizon
IJMUIDEN, NL
BEACH MANHOLE
ZANDVOORT, NL
Circe North
RFS: 1999 February
Cable Length: 203 km
Owners: VTLWavenet
BEACH MANHOLE
LOWESTOFT, UK
LANDING STATION
DRY PLANT
SIZEWELL, UK
Concerto
RFS: 1999
Cable Length: 550 km
Owners: GTT
ZEEBRUGGE, BE
Concerto
RFS: 1999
Cable Length: 550 km
Owners: GTT

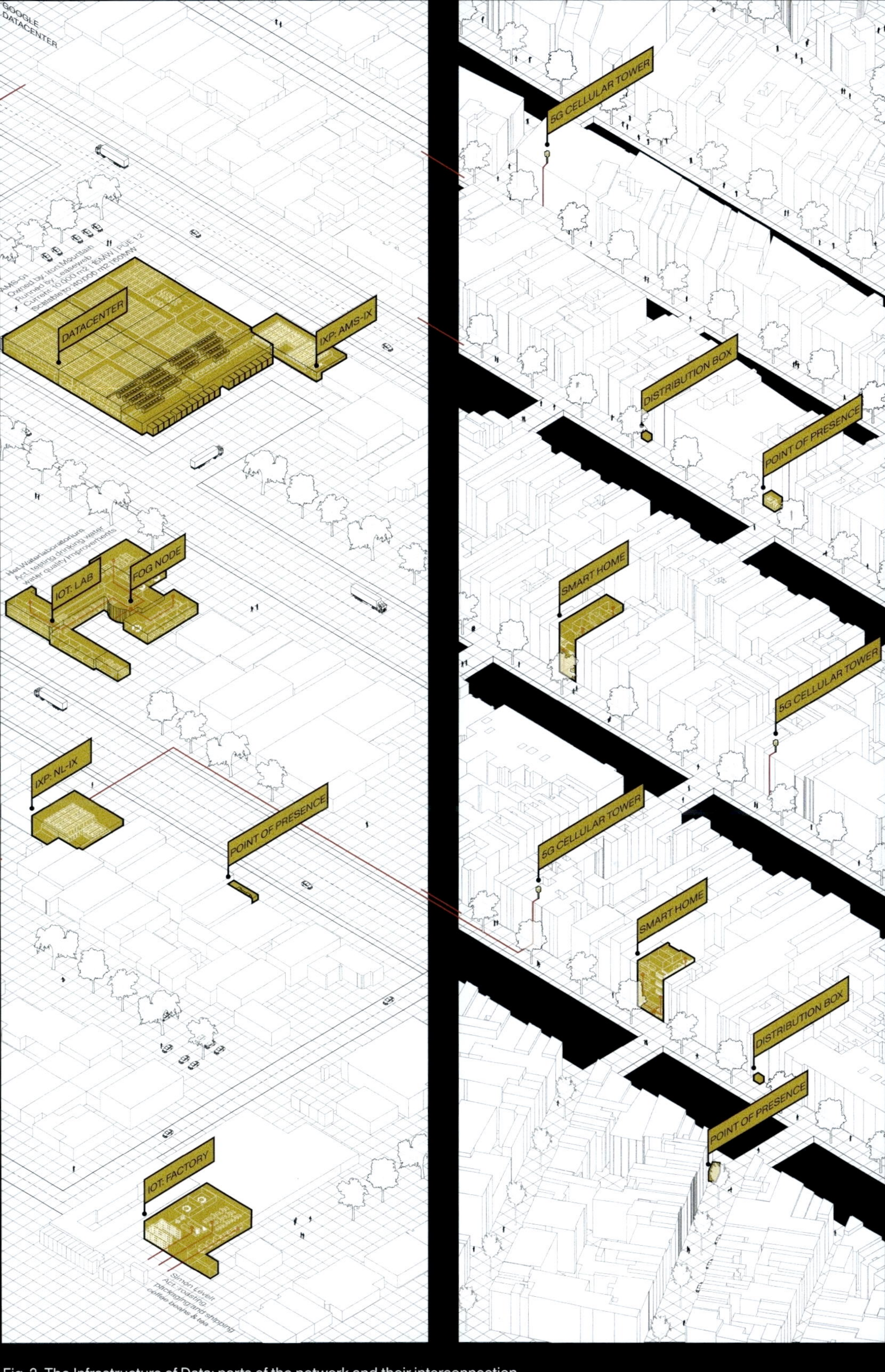


Fig. 2 The Infrastructure of Data: parts of the network and their interconnection. Drawing by Jasmijn van der Harst & Coen de Vries.

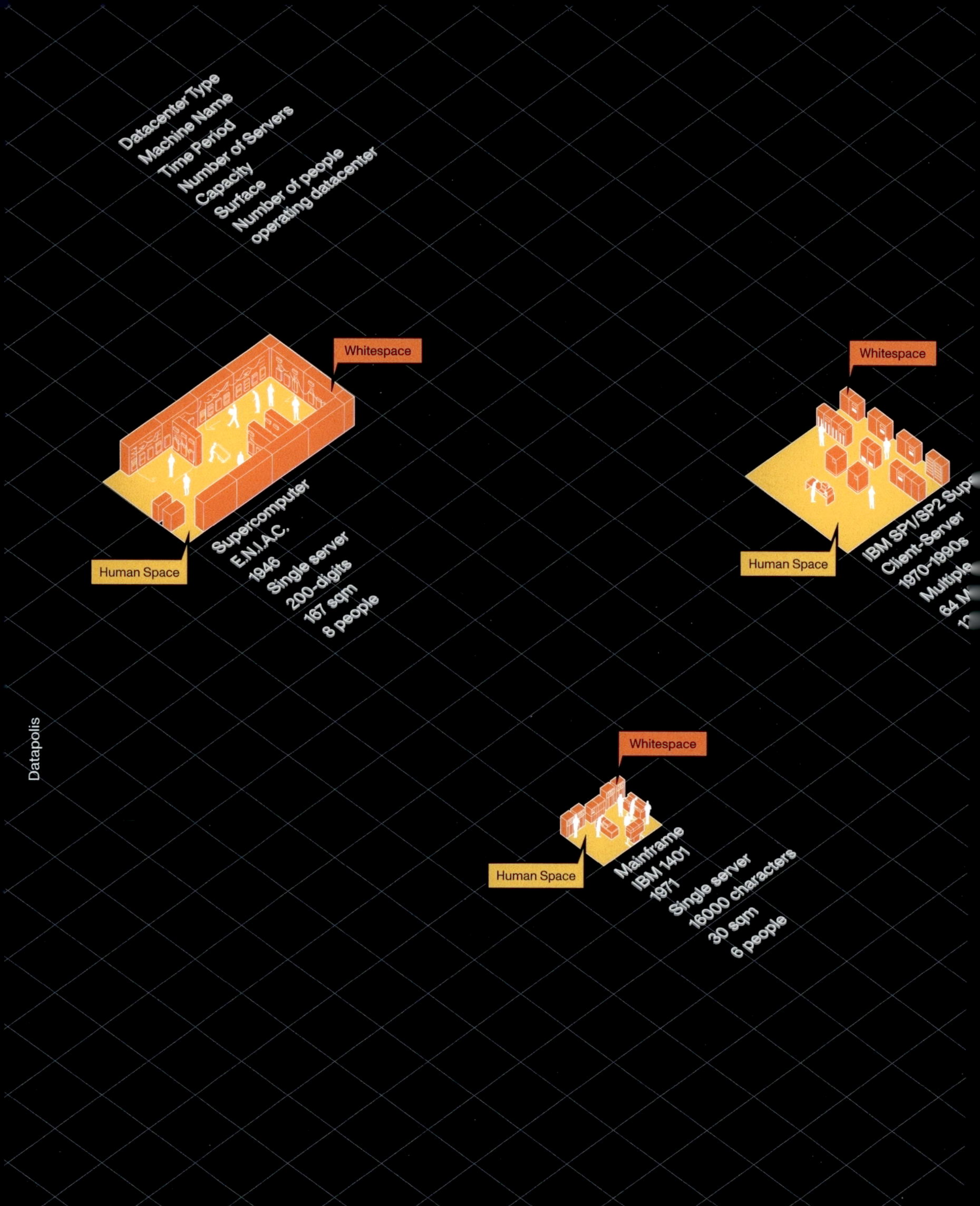


Fig. 3↑ The evolution of Data centres in 5 steps, changing relationship between people and computer.
Fig. 4↓ Energy Forecast suggests that the total electricity demand of information and communications technology (ICT) will accelerate in the 2020s, and that data centres will take a larger portion of this consumption.

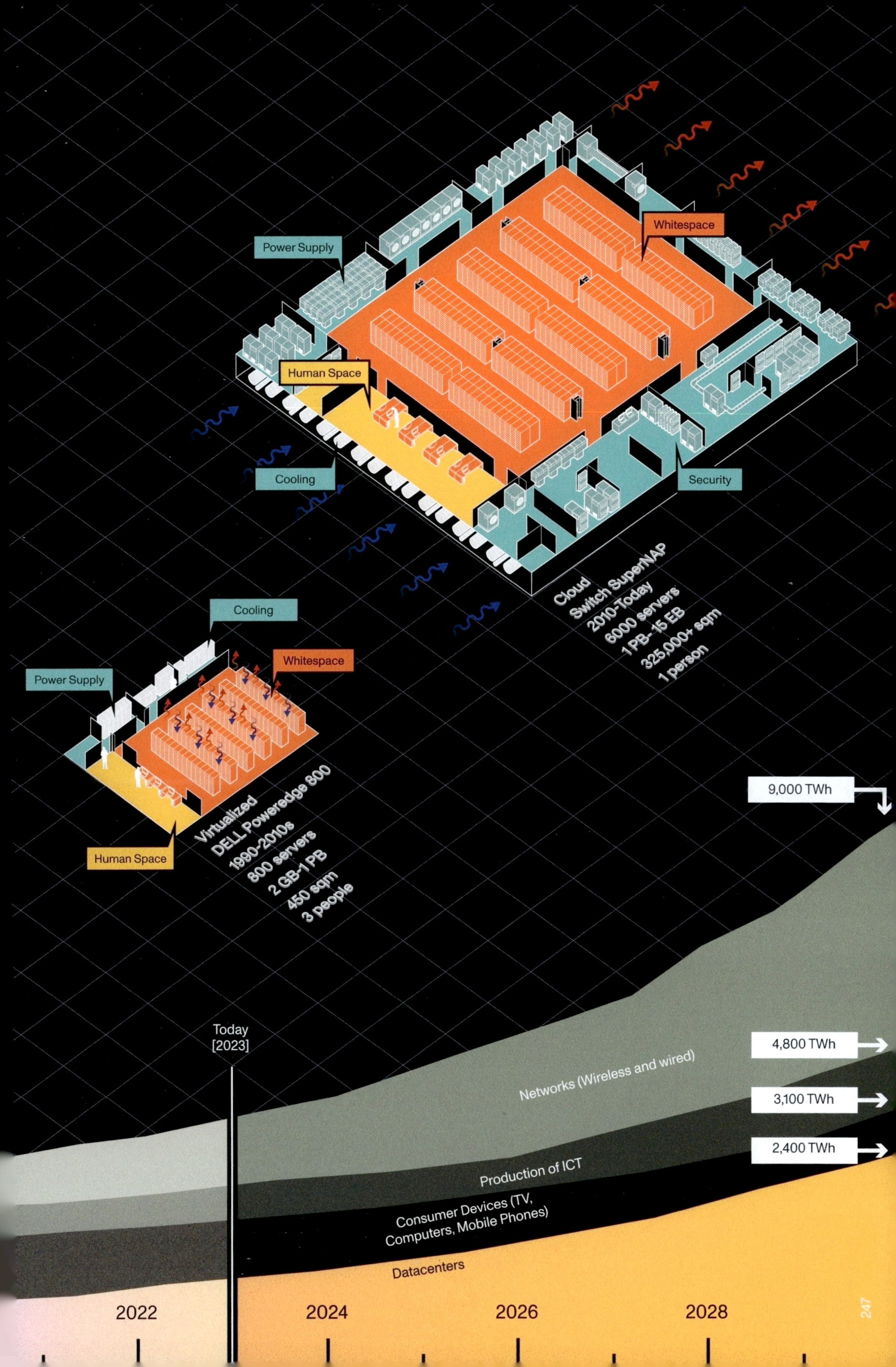
Power Supply
Whitespace
Human Space
Cooling
Security
Cloud
Switch SuperNAP
2010-Today
6000 servers
1 PB-15 EB
325,000+ sqm
1 person
Cooling
Power Supply
Whitespace
Human Space
Virtualized
DELL Poweredge 800
1990-2010s
800 servers
2 GB-1 PB
450 sqm
3 people
9,000 TWh
Today
[2023]
4,800 TWh
Networks (Wireless and wired)
3,100 TWh
2,400 TWh
Production of ICT
Consumer Devices (TV,
Computers, Mobile Phones)
Datacenters
2022
2024
2026
2028

Fig. 5  Three visions for future data centres, integrated into global, natural and industrial systems.

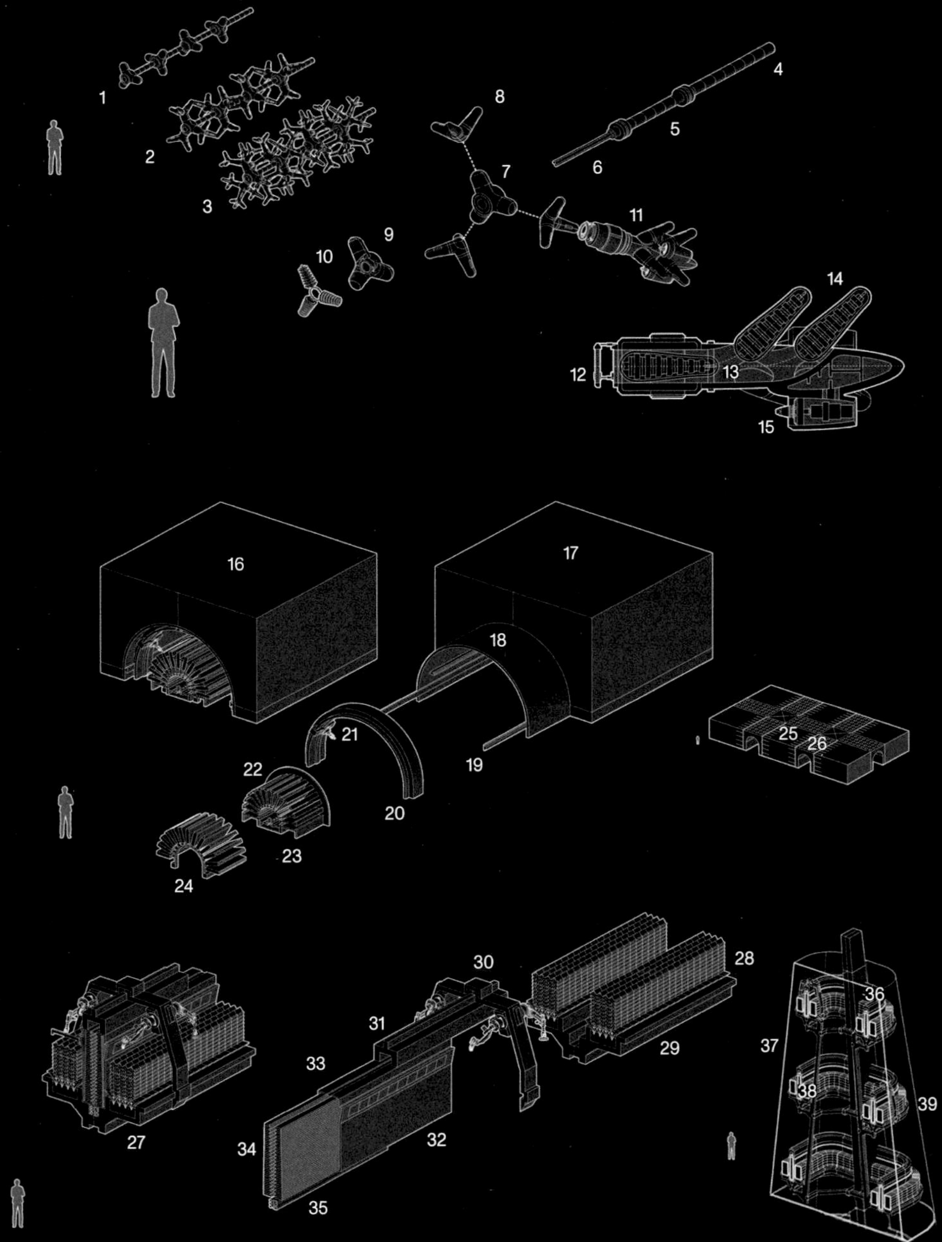


SERVER DESIGN I (Top)
Modular servers are connected in series by cables on different branches which are connected to submarine cables.

1–3 Artificial Reef Server Arrays,
4 Insulating tube
5 Connecting ring
6 Internet cable assembly
7 Sub module, Tier-I
8 Sub module, Tier-II
9 Shell
10 Triple-cone server
11 Water drone
12 Capture port
13 Transport route
14 Server dock
15 Battery and turbine

SERVER DESIGN II (Middle)
A confidential data centre is set up in the reinforced structure of an underground coal mine.

16 Archive's arch
17 Reinforced concrete arch
18 Artificial reef server array, Aggregation-III
19 Slide rail
20 Gantry robot
21 Robot arm
22 Arch rack
23 Cables and hot corridor
24 Servers
25 "Pillar": Remained coal
26 "Room": server

SERVER DESIGN III (Bottom)
The multi-storey greenhouse with servers and crops, taken care by robots which are moving on the slide rails.

27 Greenhouse data array segment
28 Crops
29 Supporting structure
30 Gantry robot
31 Slide rail
32 Server shell
33 Heat outlet
34 Server with cold corridor
35 Electrical and Internet cable
36 Facility tower
37 Crops-server ring
38 Supporting structure
39 Glass curtain

# THE URBAN EXCHANGE PLATFORM

Farian Anwar, Callum Birchall

The complex operations of logistics demonstrate the flows and exchanges of capital, resources and information over vast territorial landscapes. Clare Lyster describes the space of logistics as a 'large multi-scalar assemblage, whose territory extends from the planetary (satellites) to the local (a doorstep)'.[1] Logistics; encompassing the boundaries of trade negotiations to the operation of transportation networks, not only allows for the material connectivity of the physical world, but extends to the realm of the digital and immaterial. In this sense, the abstract 'spatiotemporal phenomenon' of logistics, its impact on the Earth and its associated footprints, can coincide with Lyster's understanding of the 'Hyperobject' defined by Timothy Morton. Footprints in the form of barcodes and packaging are indicative fragments of a wider and more complex procedure from which consumers experience a sense of detachment. Moreover, the parallel and closely connected transfers of data and information that correspond to physical objects have allowed large corporations to attain more efficient transfers that reduce friction within the supply chain process. However, the capitalist desires of large corporations have resulted in real world consequences, through the exploitation of labour and natural resources.

The desire for faster and more efficient logistics to accommodate 'same-day' delivery turnarounds, as well as the changing demands of products during peak months and in the case of a global crisis, have highlighted the flaws in the last-mile delivery model. Companies are increasingly spending on transport costs to bridge the distance between distribution centres and their consumers. Moreover, the unsustainable amounts of transport emissions and packaging waste that are produced as an outcome of increasing demands have resulted in the need for more integrated logistics systems that highlight the consequences of our consumption.

Informed by research, The Urban Exchange Platform (UXP) proposes a future for a decentralized logistics model. It envisions the integration of automated distribution modules within a network of shared, reused footprints in the city, demonstrating the flows of data, waste and products. The UXP strives to create a relationship between people, technology and their wider context, capturing the scales of logistical exchanges and encounters. Through the rising interest in autonomous vehicles and flexible work culture, existing buildings such as car parking structures and offices offer potential vacancies in the future. For exchanges to operate sustainably at a local city scale, the UXP aims to utilize this opportunity to reuse footprints for distribution in primely located retail and business districts. Furthermore, the project imagines a system that is woven into the city and the lives of its inhabitants, engaging with public space and the history of the existing urban landscape. This includes a waste management and recycling scheme whereby a circular economy is promoted through the physical exchange of packaging waste for consumer goods. The modular system aims to flexibly manage distribution while balancing the demands of both the environment and society. Through the final visual, the UXP depicts a future of coexistence and productivity, the extremes of which are open to interpretation.

1 Clare Louise Lyster, 'Territories of Equivalence: Objects of the Logistical Apparatus', Footprint 23 (Autumn/Winter 2018), 25, https://doi.org/10.7480/footprint.12.2.2100.

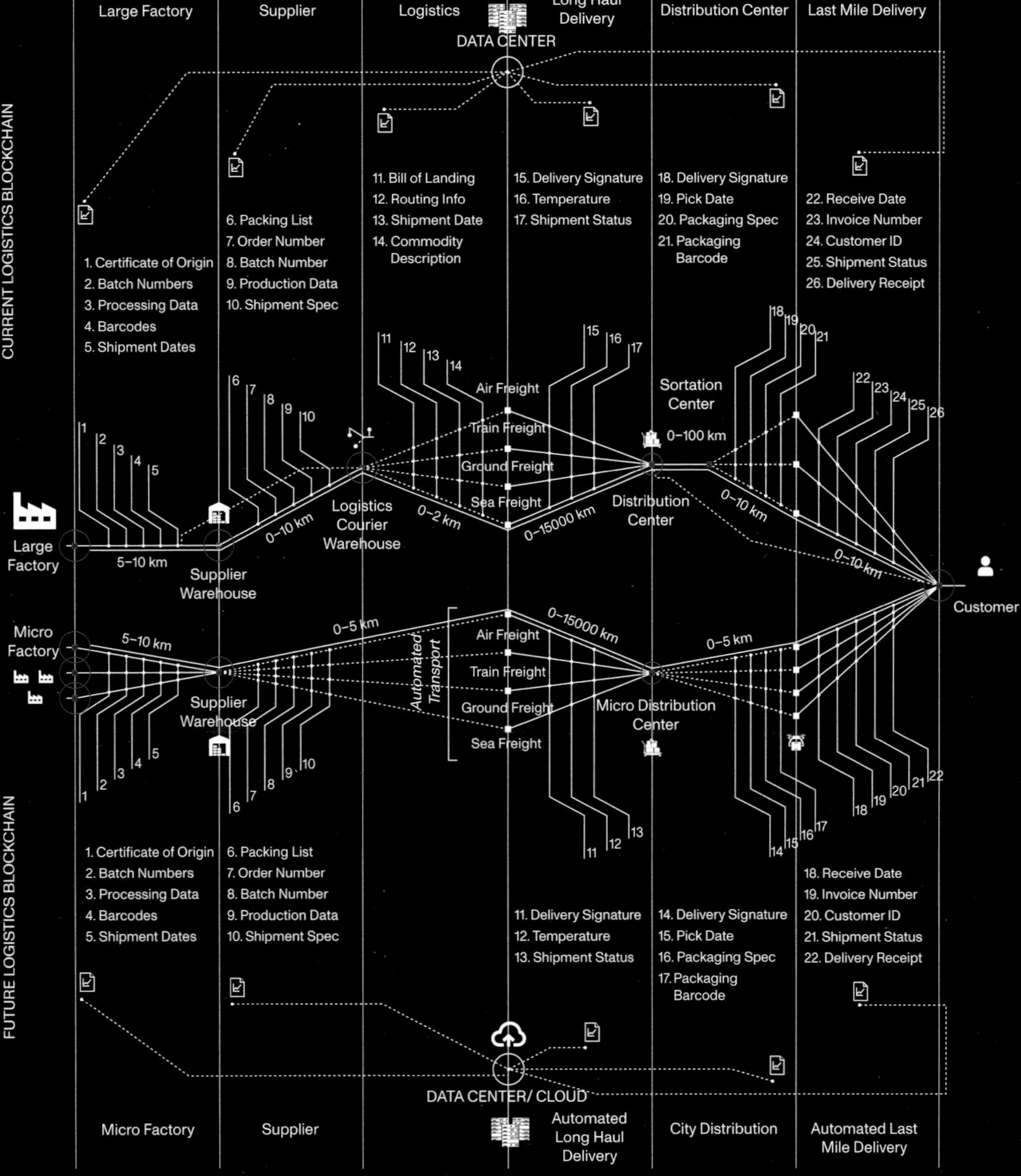


Fig. 1 The systematic relationship between data and the physical infrastructure of logistics and data recording throughout the supply chain process, current and projected future. Source: IRC Group.

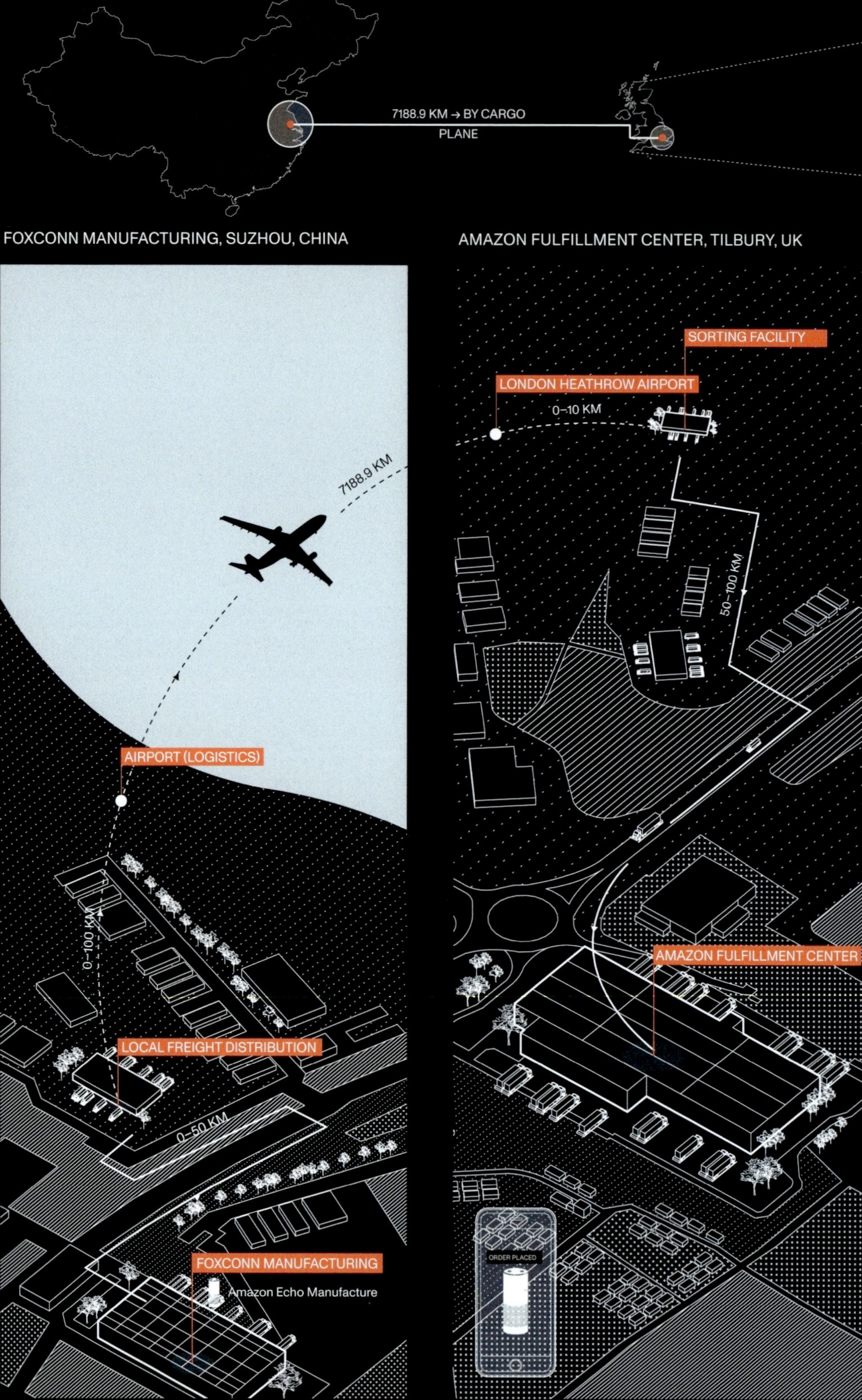
7188.9 KM → BY CARGO PLANE
FOXCONN MANUFACTURING, SUZHOU, CHINA
AMAZON FULFILLMENT CENTER, TILBURY, UK
7188.9 KM
AIRPORT (LOGISTICS)
0–100 KM
LOCAL FREIGHT DISTRIBUTION
0–50 KM
FOXCONN MANUFACTURING
Amazon Echo Manufacture
SORTING FACILITY
LONDON HEATHROW AIRPORT
0–10 KM
50–100 KM
AMAZON FULFILLMENT CENTER
ORDER PLACED

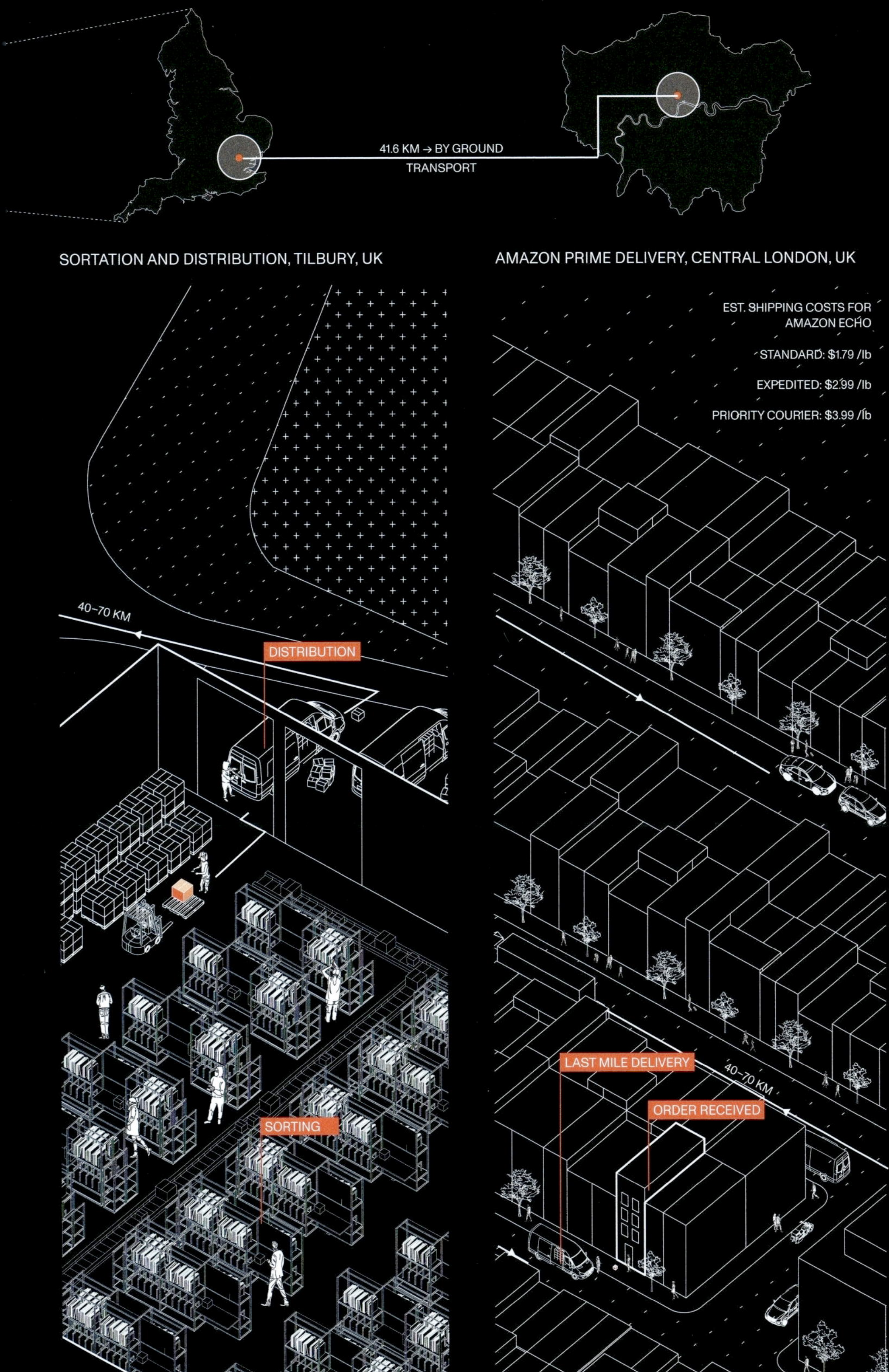


Fig. 2 Amazon Product Journey to UK. Supply Chain overview of Amazon's product journey—cost and distance travelled for the delivery of an Amazon Echo.

KIVA SYSTEMS DRIVE UNIT

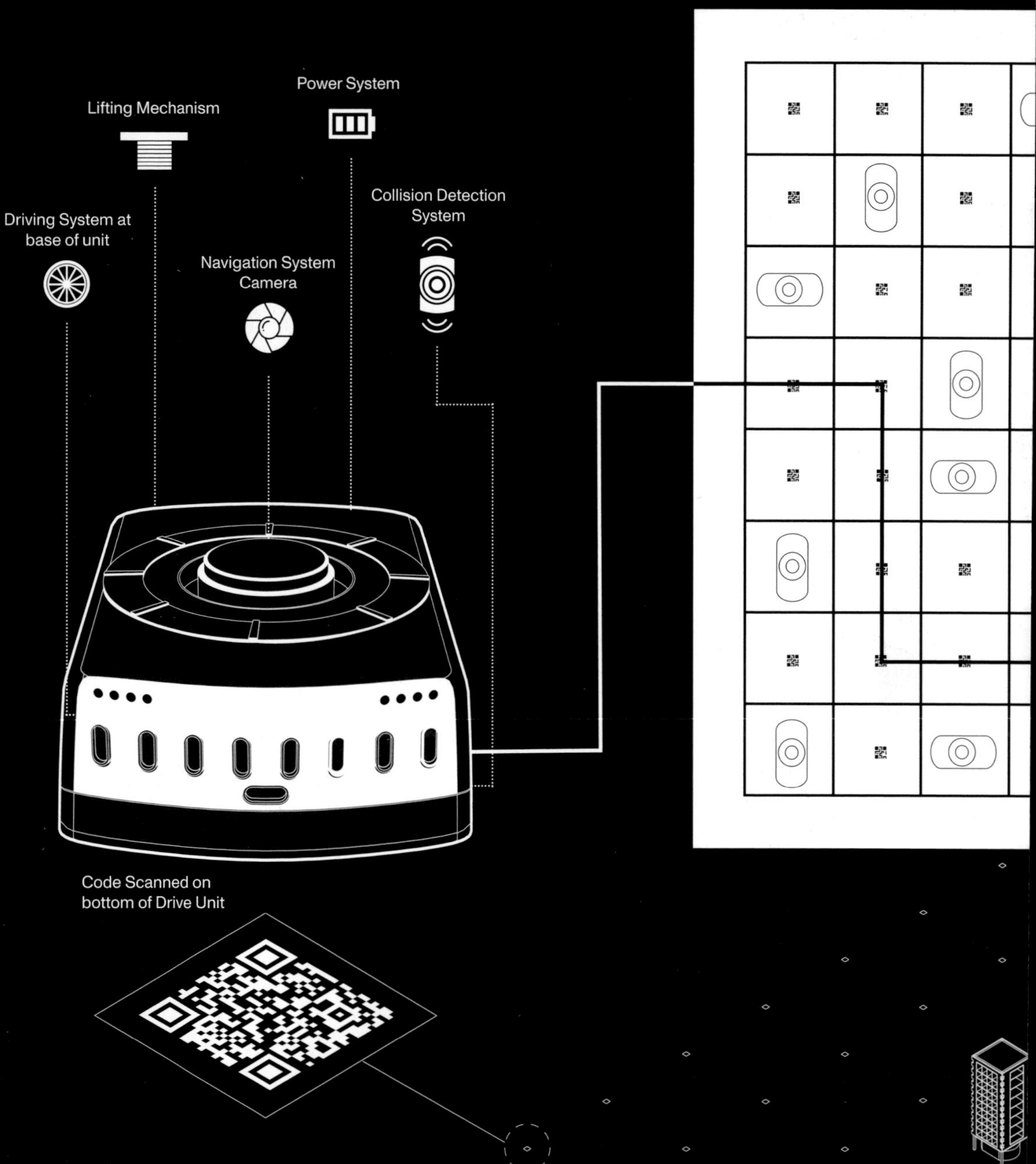

Fig. 3 Distribution Centre Order Picking: Amazon Kiva Robots.

Camera sensors at the base of the programmed KIVA robots allow them to follow QR codes that are arranged on the floor grid. This enables them to move goods horizontally to the human pickers, who collect and pack the products for distribution.

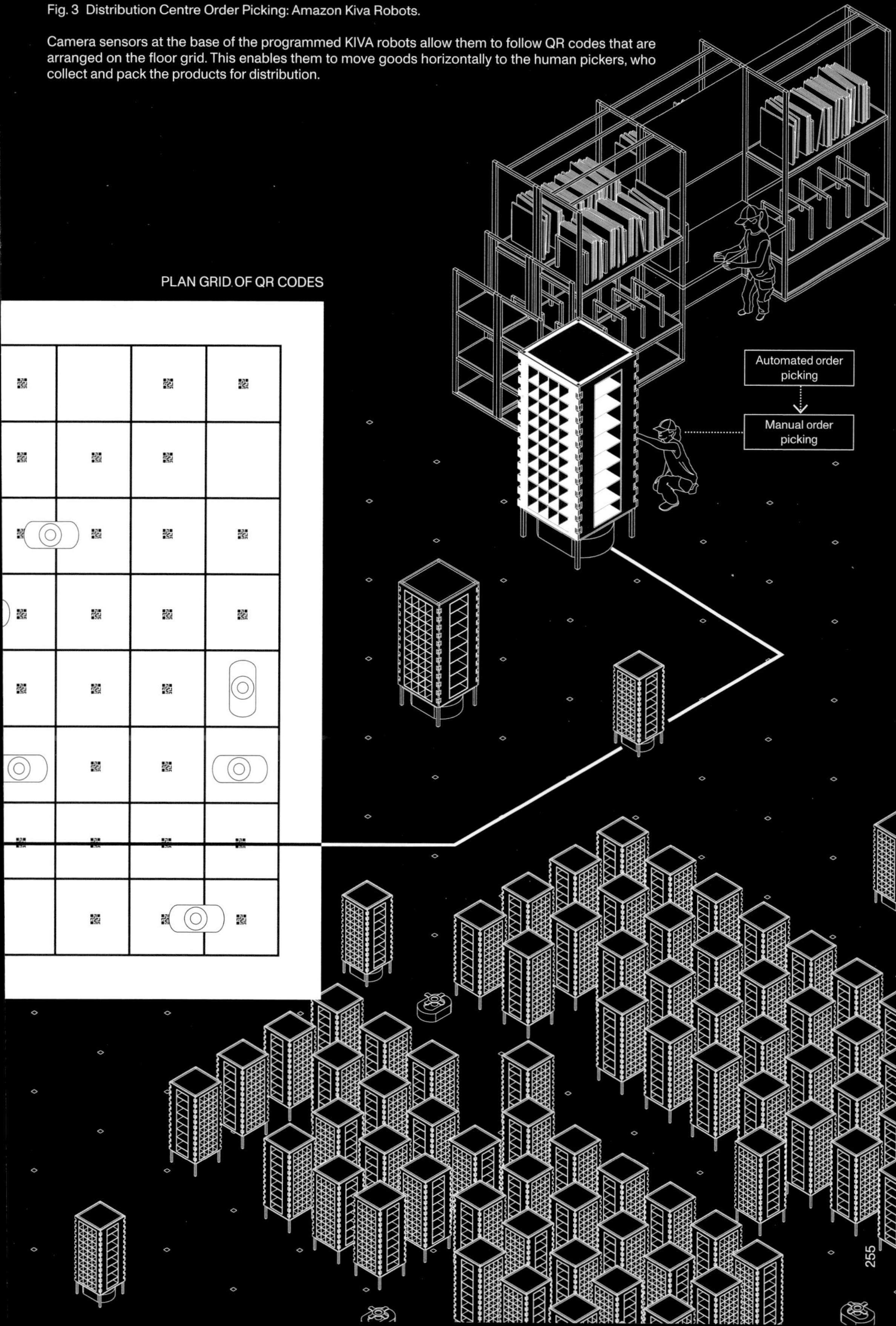

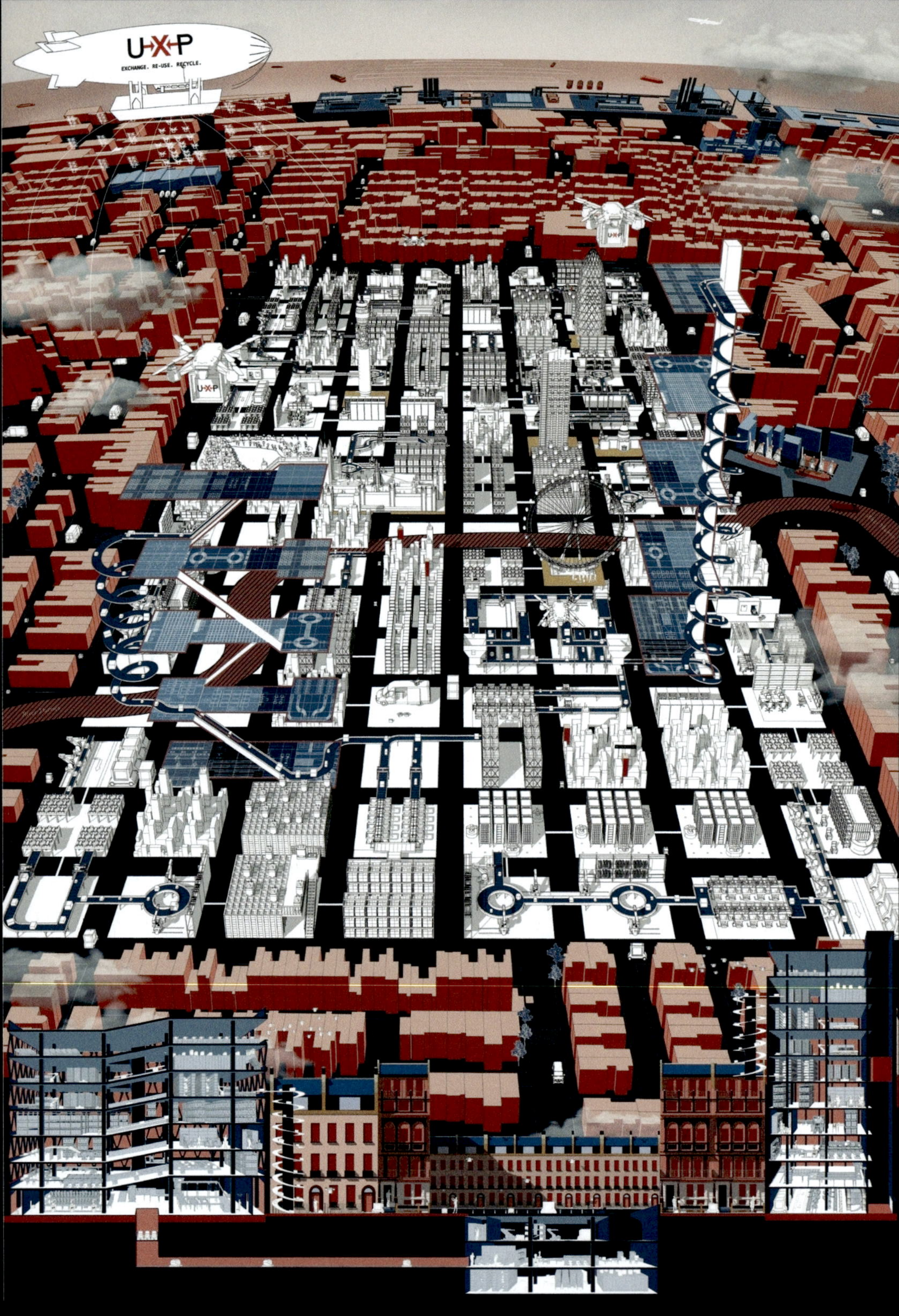





Fig. 4 ↑ The Urban Exchange Platform.
Fig. 5 → Programme Modules.

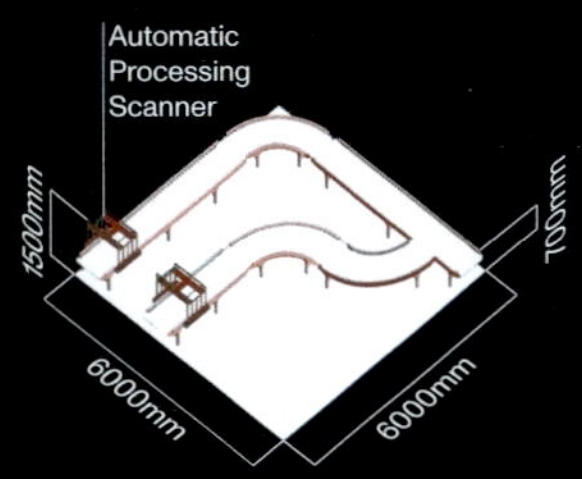

PROCESSING

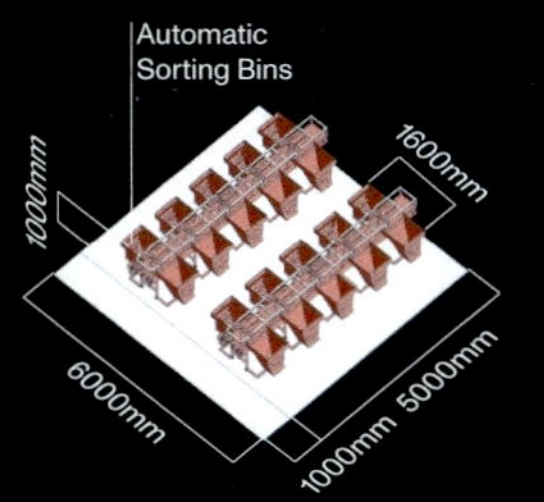

SORTING

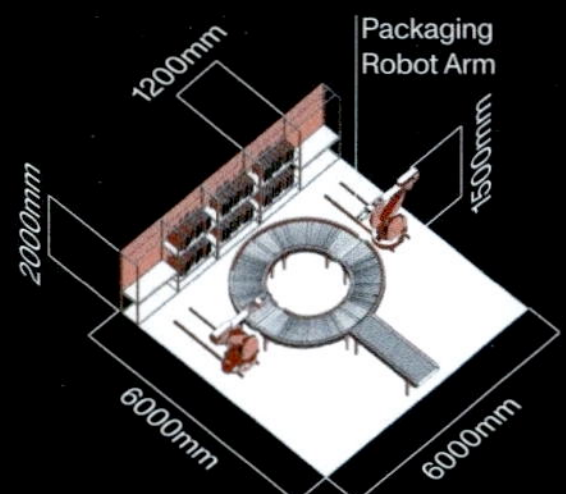

PACKAGING

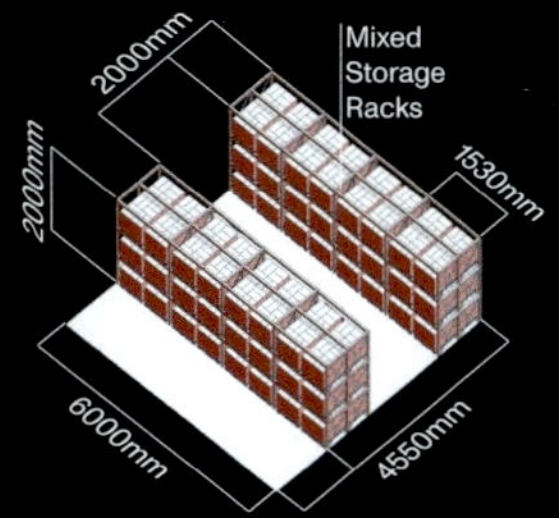

STORAGE

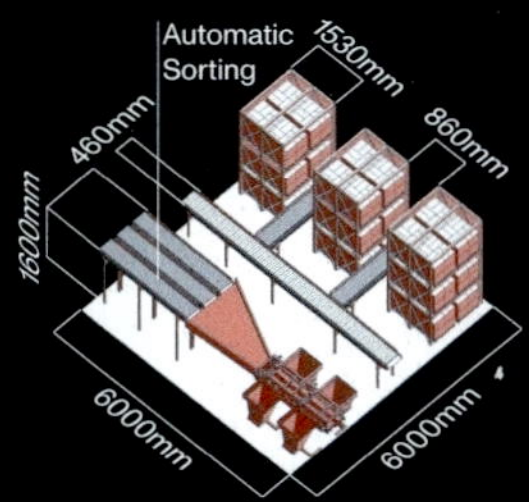

RETURNS

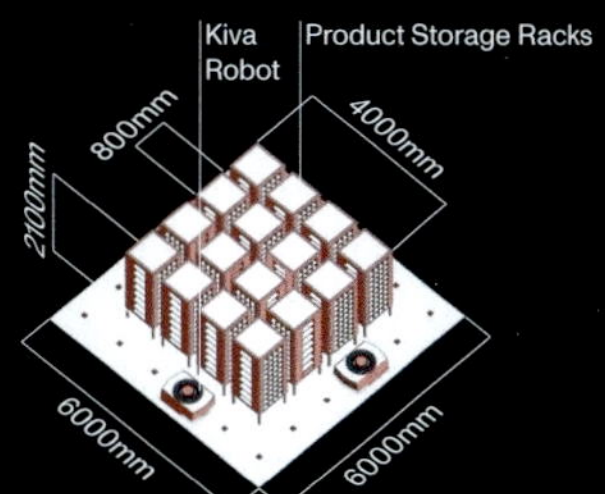

KIVA ROBOTS

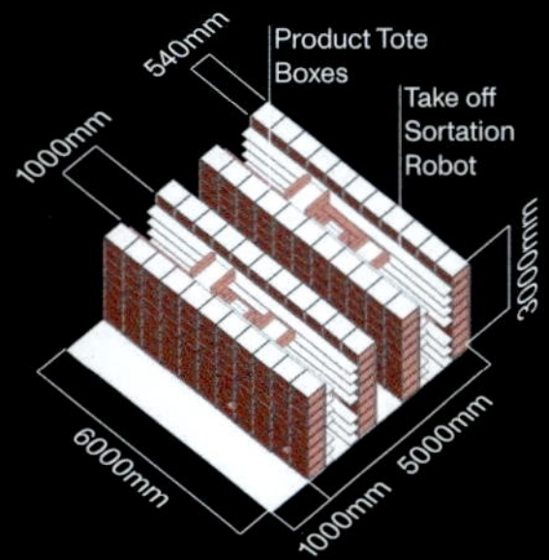

TAKEOFF
TECHNOLOGY

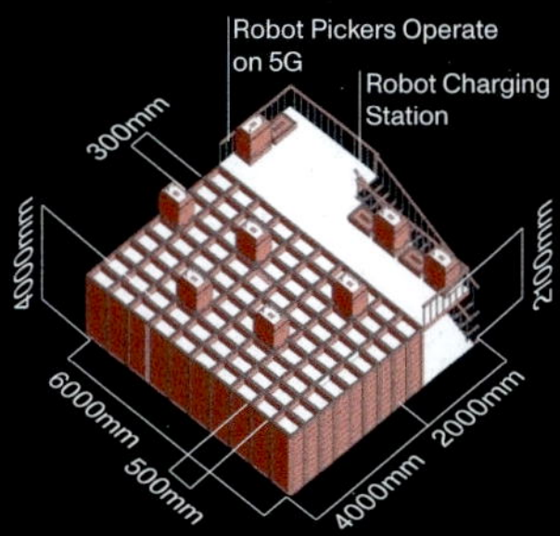

OCADO SMART
PLATFORM

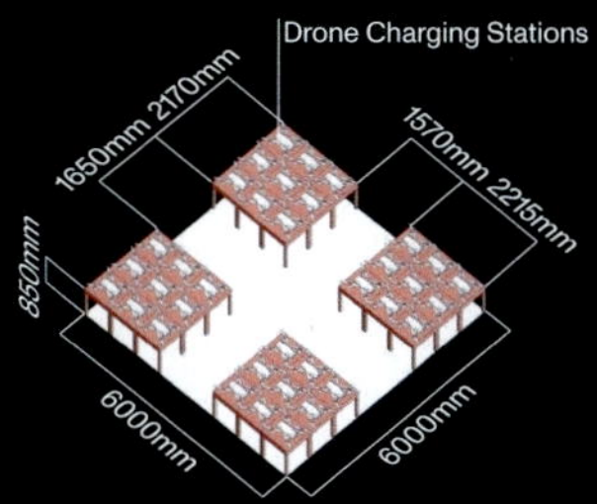

DRONES

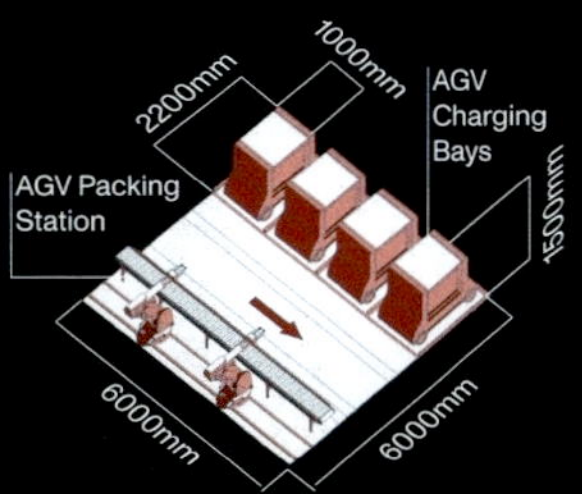

AUTOMATED GROUND
VEHICLES

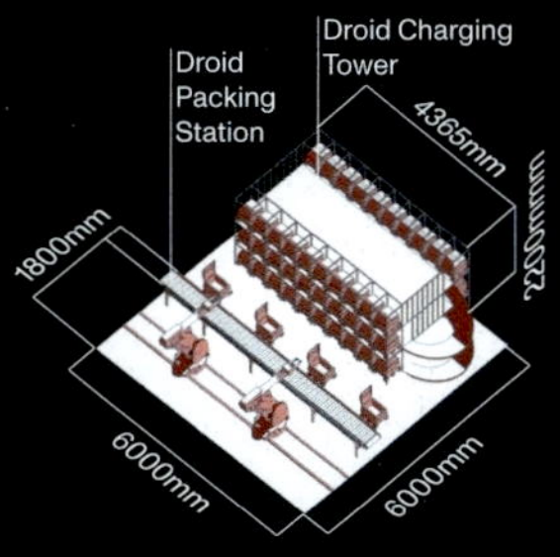

DROIDS

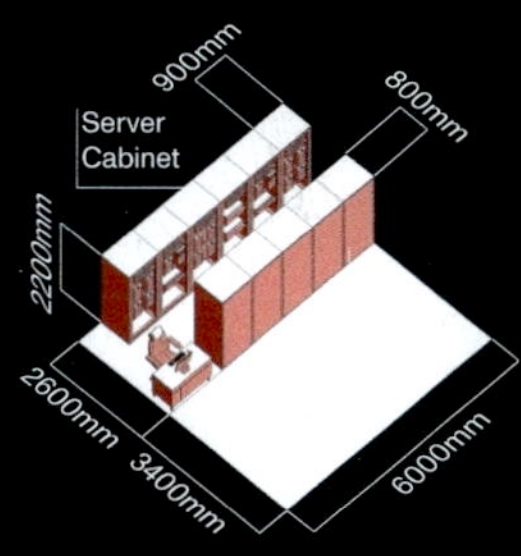

SERVER
ROOM

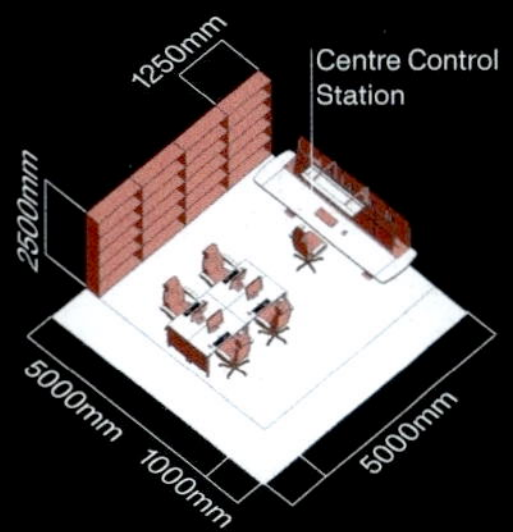

CONTROL
ROOM

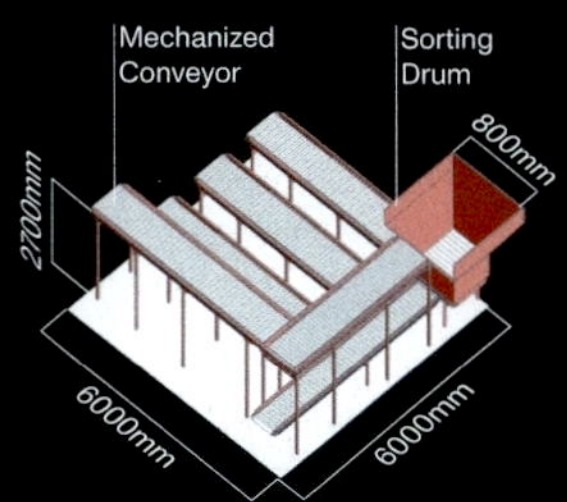

RECYCLING – SORTING/
PROCESSING

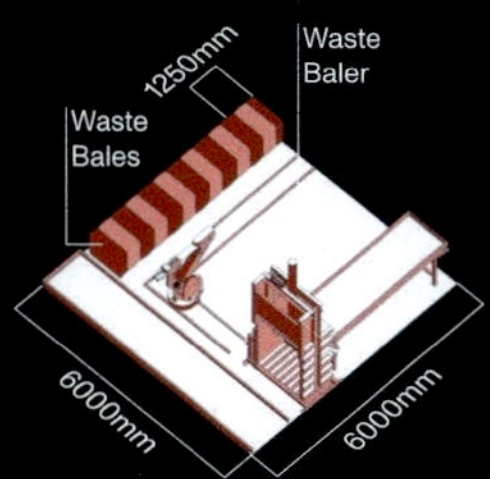

RECYCLING –
PACKAGING

## FROM THE GRAVE TO THE CRADLE
Federica Longoni, Filip Romanuik

What kind of data do we get from space, and what is the infrastructure behind it?

Fifty years ago we had images of outer space, but what we did not have were images of Earth seen from space. NASA released the first such picture, taken by a satellite, in 1966, but it only captured part of the planet. A photograph of the whole Earth was taken in 1972: later named The Blue Marble, it constitutes one of the most valuable successes in space exploration. One astronaut who was on the mission said: 'When we originally went to the moon, our total focus was on the moon: we weren't thinking about looking back at the earth. But now that we've done it, that may well have been the most important reason we went.' [1]

The Whole Earth image started an important reflection regarding our impact on the planet, raising awareness of its fragility: from outer space, Earth could be seen for the first time as a sphere, an enclosed environment with limited resources in comparison with the infinite availability we were used to thinking of. Since the launch of the first satellites, Earth suddenly became a 'planet', a 'satellite in itself'. [2]

Nowadays, satellites are responsible for collecting data for monitoring the environment, giving us an objective and distant picture of our impact on Earth.

Until recently, however, we did not realize that the process of satellite launching was causing a significant amount of debris in space and CO2 emissions in the atmosphere: when satellites reach the end of their lifespan they are either sent into a graveyard orbit, burned in the atmosphere or sent back to Earth to a dedicated point in the Pacific Ocean: the Nemo point, the farthest point on Earth from the human race.

Since the space age started, we have enlarged our footprint to the whole cosmos, moving from the condition of the Anthropocene to the one of Anthropocosmos. While aware of our impact on Mother Earth, we did not realize we were just redistributing our mode of consumption to space. Can we make the whole process of obtaining data from space sustainable?

The project aims to envision a post-Anthropocosmos condition of satellite launching interested by a zero-waste total approach from energy to materials, from the ground to the space.

A solar panel plant positioned in outer space is the main source of energy, which is transmitted to the rocket through microwaves. The same energy from space is also used to divide water into its primary components through electrolysis, creating the fuel used by the rocket: liquid hydrogen.

Satellite debris at the bottom of the oceans is recovered and divided into its main materials, which constitute valuable resources at risk of depletion on Earth. Likewise, satellite debris in space is reassembled and made functional again in recycling stations.

The whole project is positioned in the Nemo point, transforming what was a spacecraft graveyard into the cradle of a new sustainable system.

1 David Beaver, OVERVIEW from Planetary Collective on Vimeo.
2 Eric McLuhan and Frank Zingrone, Essential McLuhan (London: Routledge, 1997).

**"One's garbage doesn't go 'away' it just goes somewhere else"**

**TIMOTHY MORTON**

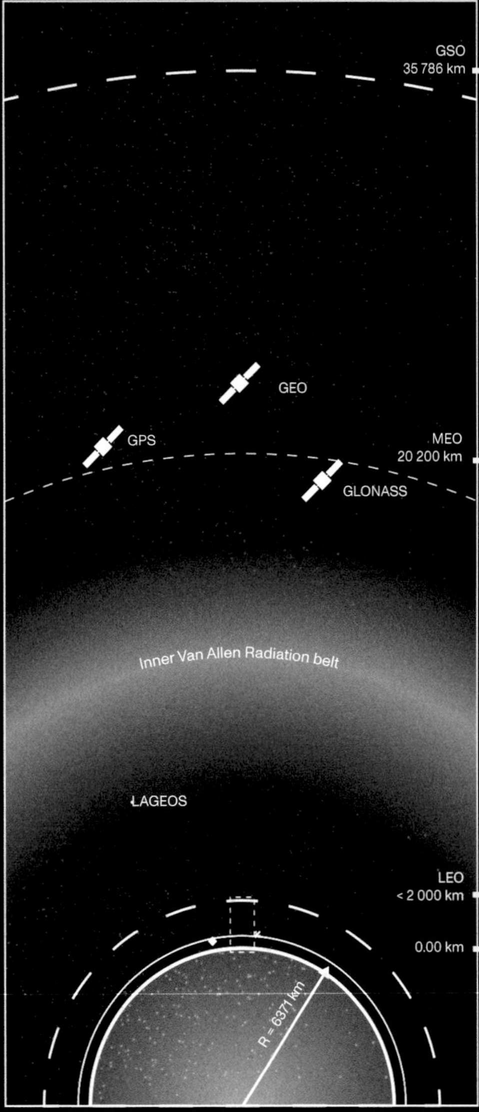


Fig. 2a The planet Earth in relation to its biggest natural satellite: the Moon.
Fig. 2b Section through planet Earth with its three primary types of orbits, Geostationary, Medium, and Low Earth Orbit.

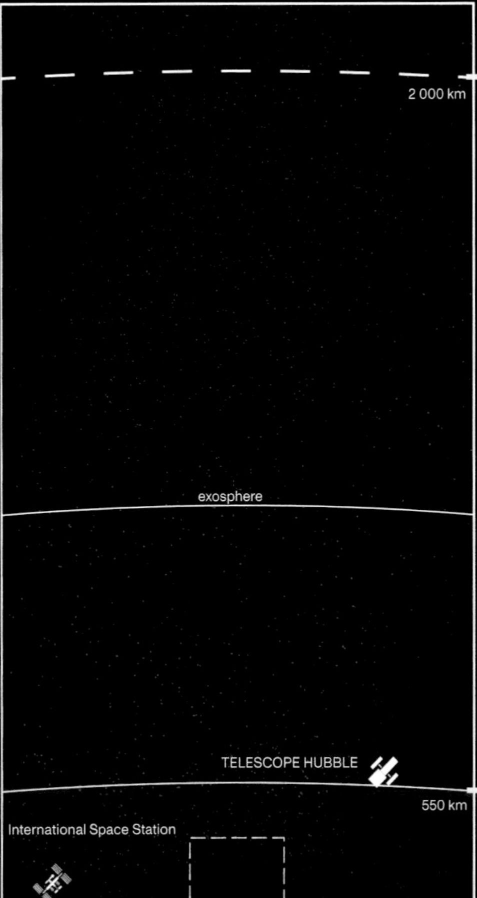


SCALE 1: 15 000 000

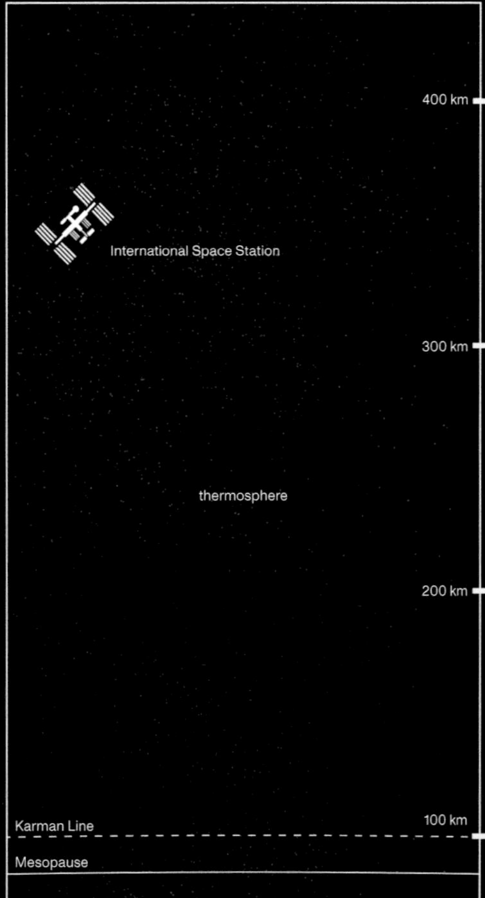


SCALE 1: 3 000 000

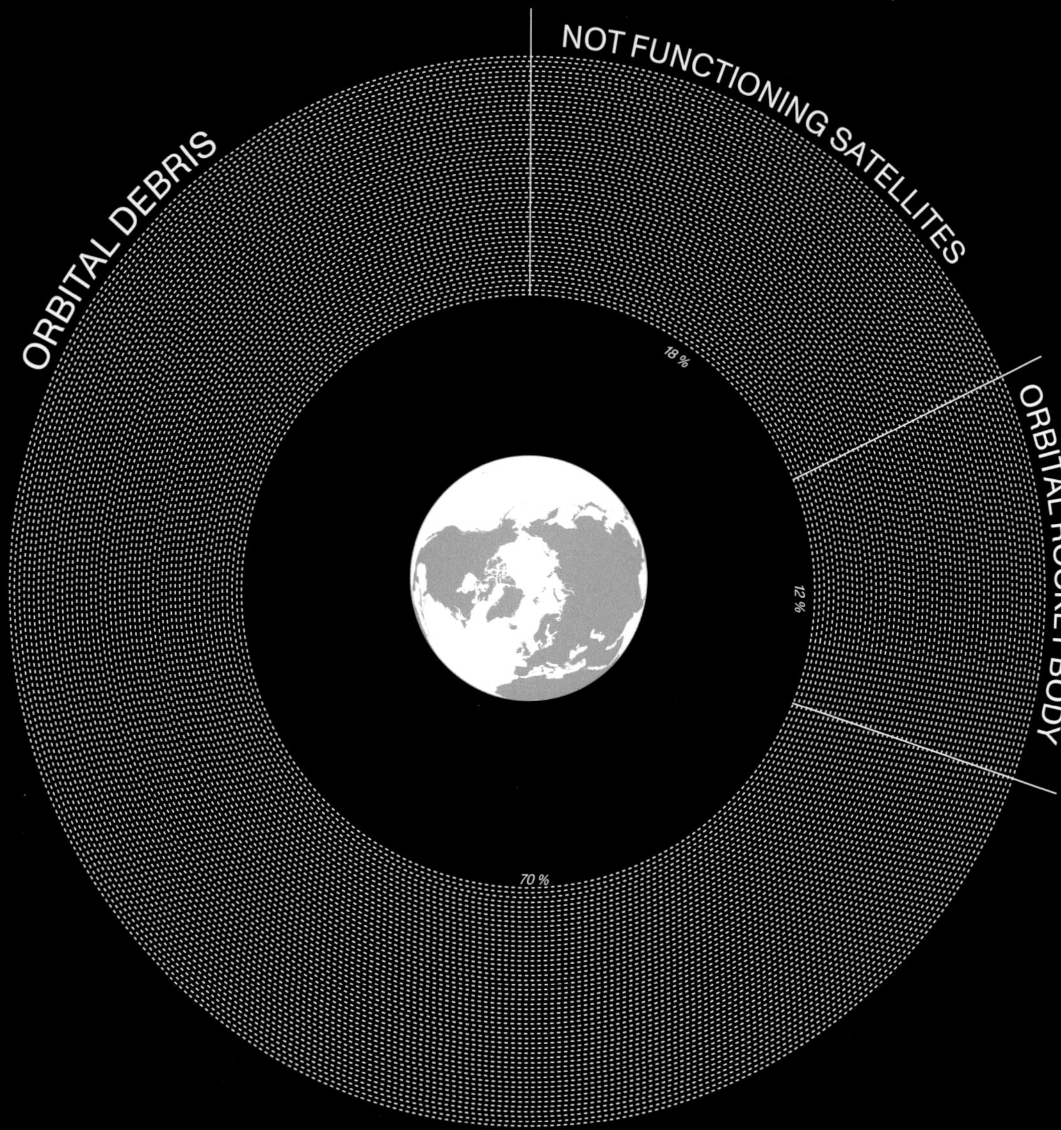


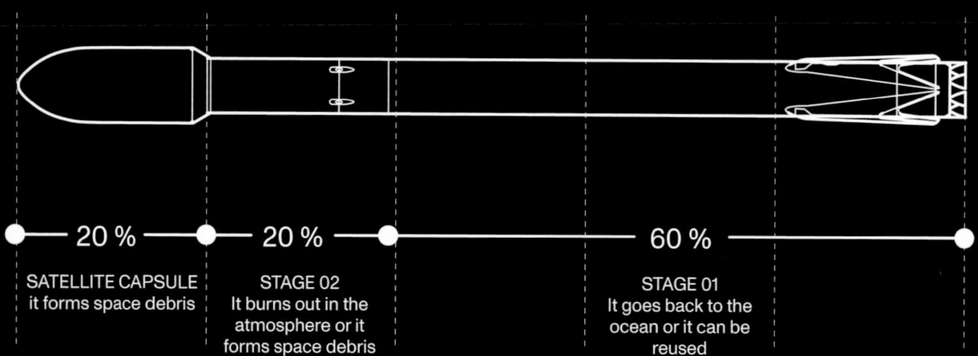


Fig. 3 ↑↑ Percentages of active and inactive satellites. Data Source: ESA's Space Debris Office—Space debris by the numbers.

Fig. 4 ↑ Rocket overview: the satellites constitute only a small piece of a much bigger entity: the rocket. The majority of its components are displaced in either space or down to Earth forming a relevant quantity of debris.

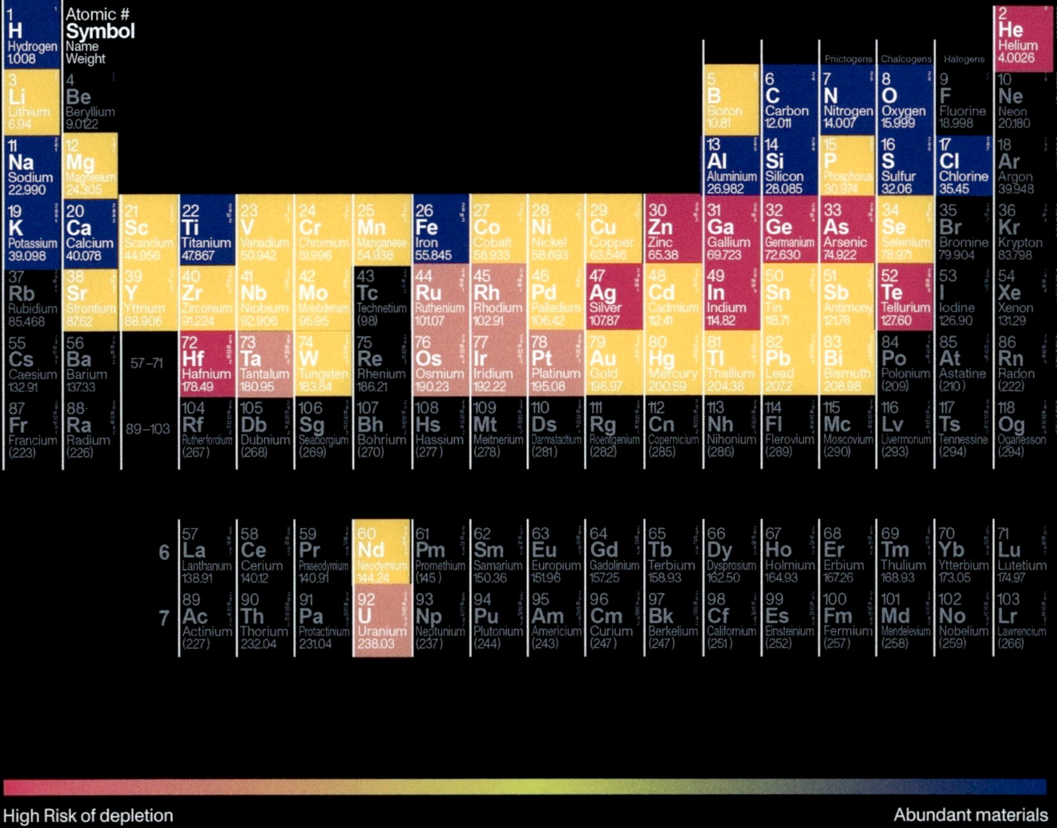


Fig. 5 Endangered elements periodic Table—Graphic Reinterpretetion by the authors. Source: Endangered elements periodic table by Mike Pitts, Director of Sustainability at Chemistry Innovation, UK. This version of the Periodic Table provides an assessment of the elements that will be in short supply and endangered over the next 100 years. The majority of the materials used in the construction of satellites are falling within the range of limited resources.

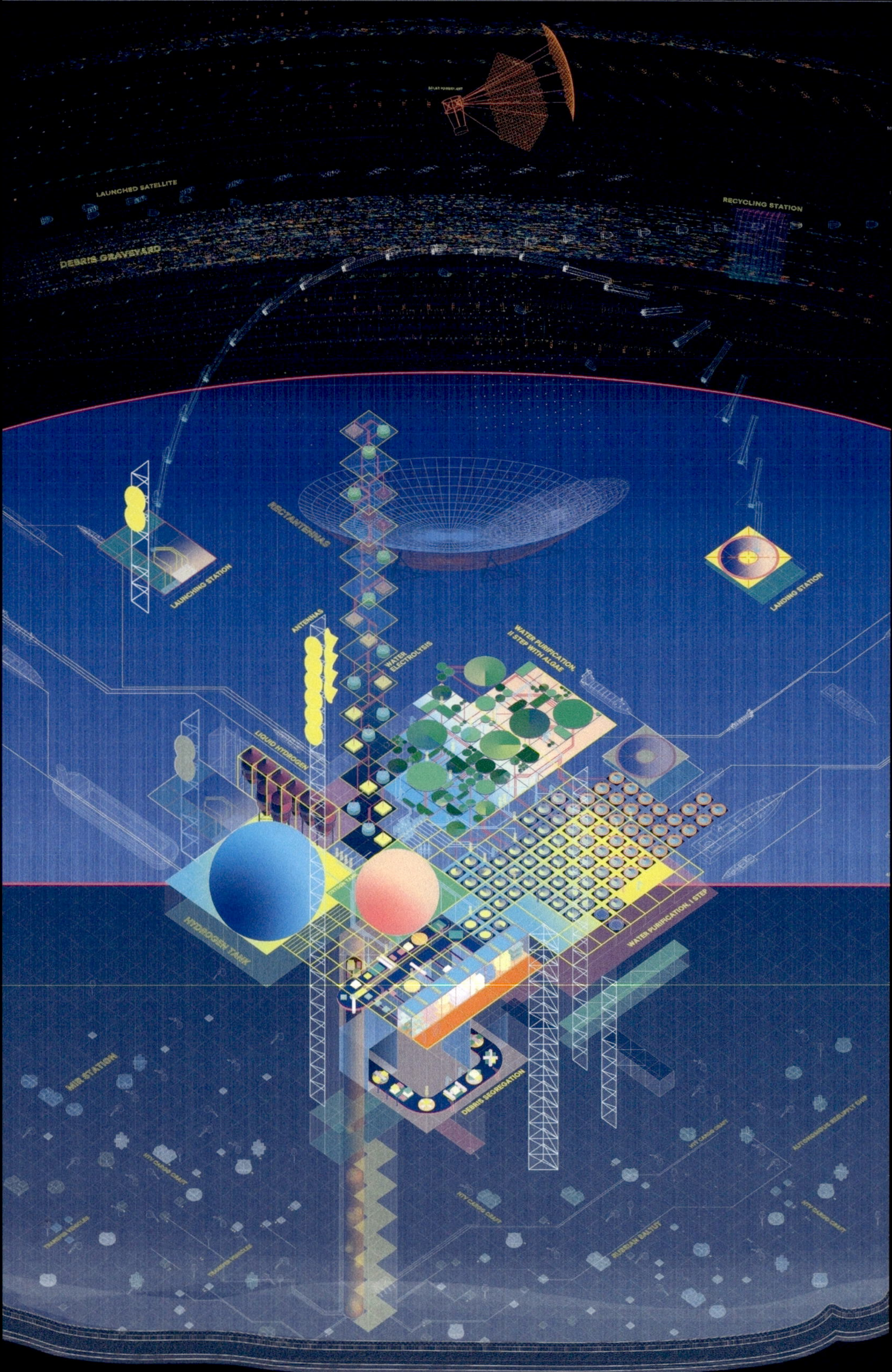
LAUNCHED SATELLITE
RECYCLING STATION
DEBRIS GRAVEYARD
RECTANTENNAS
LAUNCHING STATION
LANDING STATION
ANTENNAS
WATER ELECTROLYSIS
WATER PURIFICATION, II STEP WITH ALGAE
LIQUID HYDROGEN
HYDROGEN TANK
WATER PURIFICATION, I STEP
DEBRIS SEGREGATION
MIR STATION

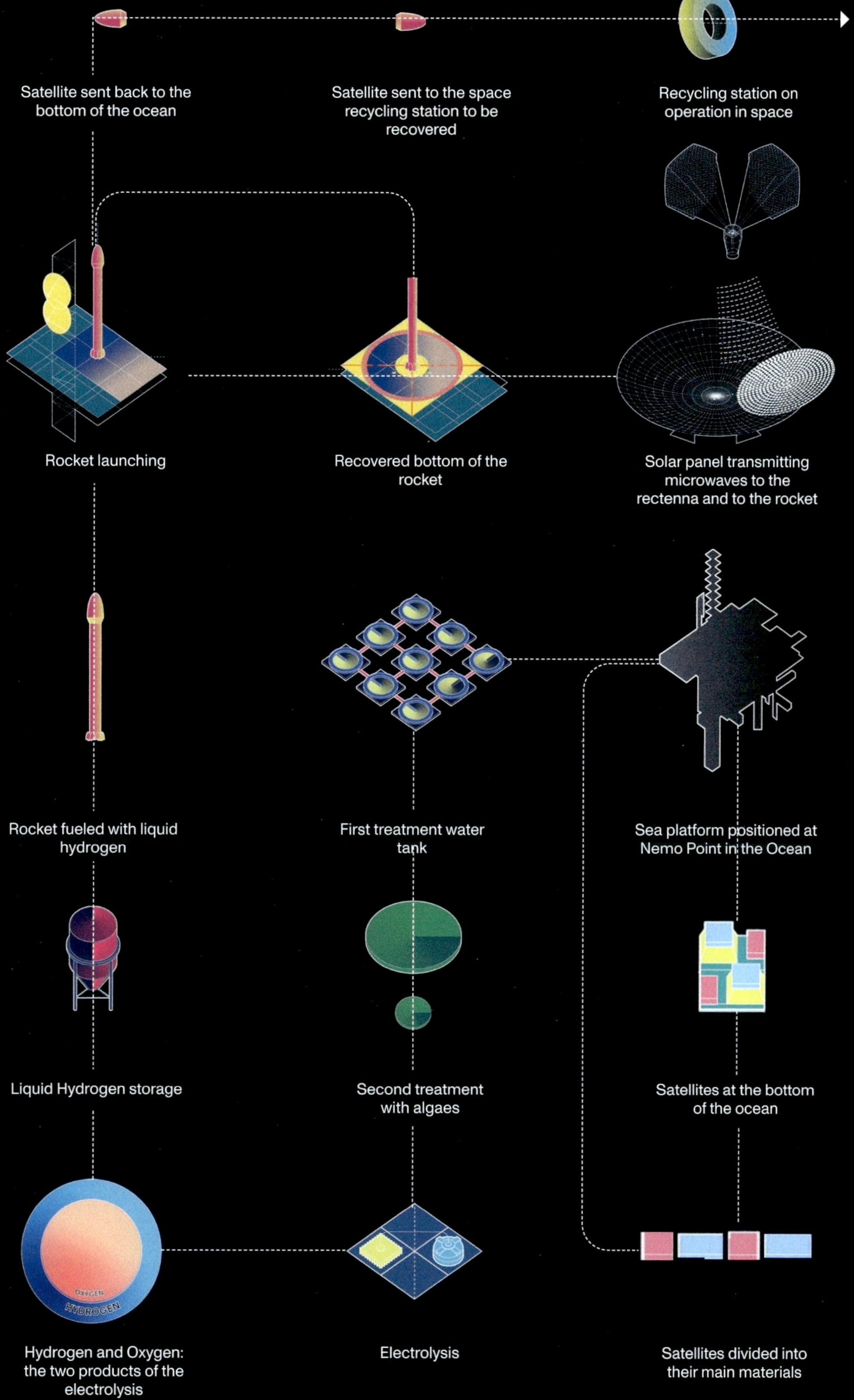


Fig. 6 System diagram showing the relations between designed elements across the Ocean, the Earth and Outer Space.

# CON(TRO)VERSING WITH MARS

Emma Chris Avrămiea, Martah Lundheim

Artificial objects in space are losing their purpose, either because they are breaking down or becoming redundant. In our quest for knowledge, fame and natural resources, we are leaving behind a trail of man-made objects, abandoning them when they are no longer useful. Instead of an obsession with the new, and a discarding of the old, patterns characterizing our capitalist society, what if we could extend a circular ecology to the energy-intensive endeavour of space research?

While space exploration is marketed as a global effort, statistics reveal a claim to space that is restricted to a selected few actors with the necessary knowledge, funds and infrastructure to explore space. Access to space relies on collaborating, and being allowed to collaborate, with the main players of the field. Thus, space exploration closely mirrors the present geopolitical climate on Earth.

Our position is to not strictly regulate and limit our knowledge of the universe, nor to limit the use of infrastructure to one or a couple of Space Agencies. On the contrary, we believe in extending our knowledge of space, while at the same time building a more sustainable infrastructure. Our proposed communication hub is a robotically operated base in space, orbiting around the planet Mars. Its main tasks are retrofitting and maintaining existing satellites, increasing the amount of data transferred between Mars and Earth, and managing an increasingly complex network of artificial celestial bodies.



In the short term, the hub will facilitate a fully robotic exploration of space. In this scenario, our position is that a human-focused mission on the surface of another planet would be too risky and counterproductive as of now. Current knowledge of the behaviour of the human body in space is insufficient to ensure a safe manned Mars mission. Furthermore, a human would contaminate the planet with trillions of Earth microorganisms. If these organisms were to infect the surface of Mars, it would be impossible to tell if Mars ever supported life on its own, and future scientific efforts would be hampered. When space research has reached a definitive conclusion in regard to life on Mars, and safe technology has been invented, we could re-evaluate using the hub to also support a human settlement.

The ISS has proven that a structure in space can be operated through shared consent and separately owned modules. However, the ISS is owned and shared by major space companies, with little room for recently established parties. An important challenge would be to ensure that governing power lies with the users of the hub, and that new actors are protected by a more neutral, scientifically focused support. To protect and enable new actors' access to space, it is likely that a new legal system, and a new infrastructure system, will be required to democratize space. The hub will enable the latter, and then it is up to the Earth's inhabitants to enable the former.

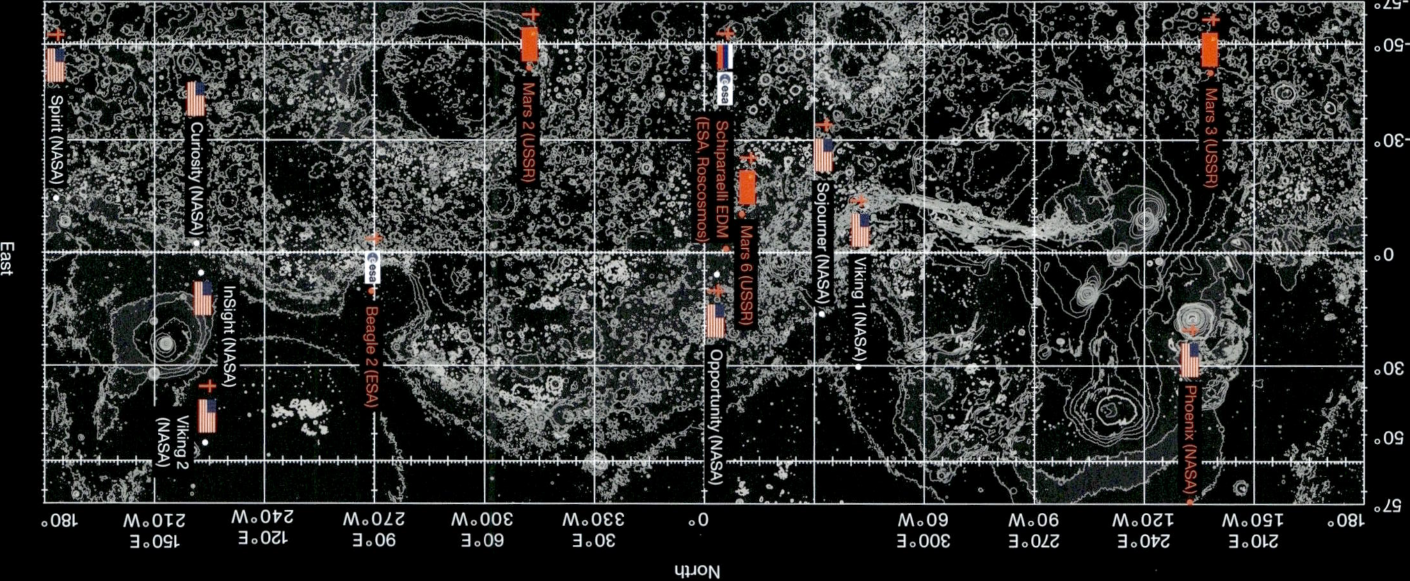
North
East
Phoenix (NASA)
Mars 3 (USSR)
Viking 1 (NASA)
Sojourner (NASA)
Mars 6 (USSR)
Schiparaelli EDM
(ESA, Roscosmos)
Opportunity (NASA)
Mars 2 (USSR)
Beagle 2 (ESA)
InSight (NASA)
Viking 2
(NASA)
Curiosity (NASA)
Spirit (NASA)
180°
210°E
150°W
240°E
120°W
270°E
90°W
300°E
60°W
0°
30°E
330°W
60°E
300°W
90°E
270°W
120°E
240°W
150°E
210°W
57°
50°
30°
-30°
-50°
-57°

Fig. 2 Missions to Mars.

Zond 2 – USSR
1964
Phobos 1, 2 – USSR
1988
Mars 5 – USSR
1974
Mars 4 – USSR
1973
Nozomi – ISAS
1998
Mars Climate
Orbiter – NASA
1998
Mariner 9 – NASA
1971
Mars 6 – USSR
1974
Mariner 4 – NASA
1964
Mars 2 – USSR
1971
Mariner 6&7-NASA
1969
Mars Polar Lander -
NASA
1999
2001 Mars
Odyssey – NASA
2001
Mars 7 – USSR
1974
MAVEN – NASA
2013
Curiosity – NASA
2011
MarsExplorationRoverSpirit –
NASA
2004
Viking 1, 2 – NASA
1975
Mars Express ESA
2003
ExoMars – ESA
2016
Reconnaissance
Orbiter – NASA
2005
Mars Orbiter
ISRO
2013
Mars Global
Surveyor – NASA
1996
Mars Pathfinder -
NASA
1996
MarsExplorationRoverOpportunity –
NASA
2004
Phoenix – NASA
2007
InSight – NASA
2018

= 3.125.000.000 microbes

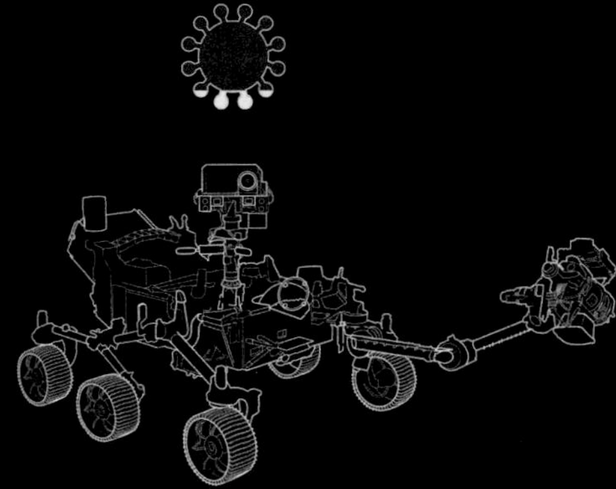

300.000

VS

100.000.000.000.000

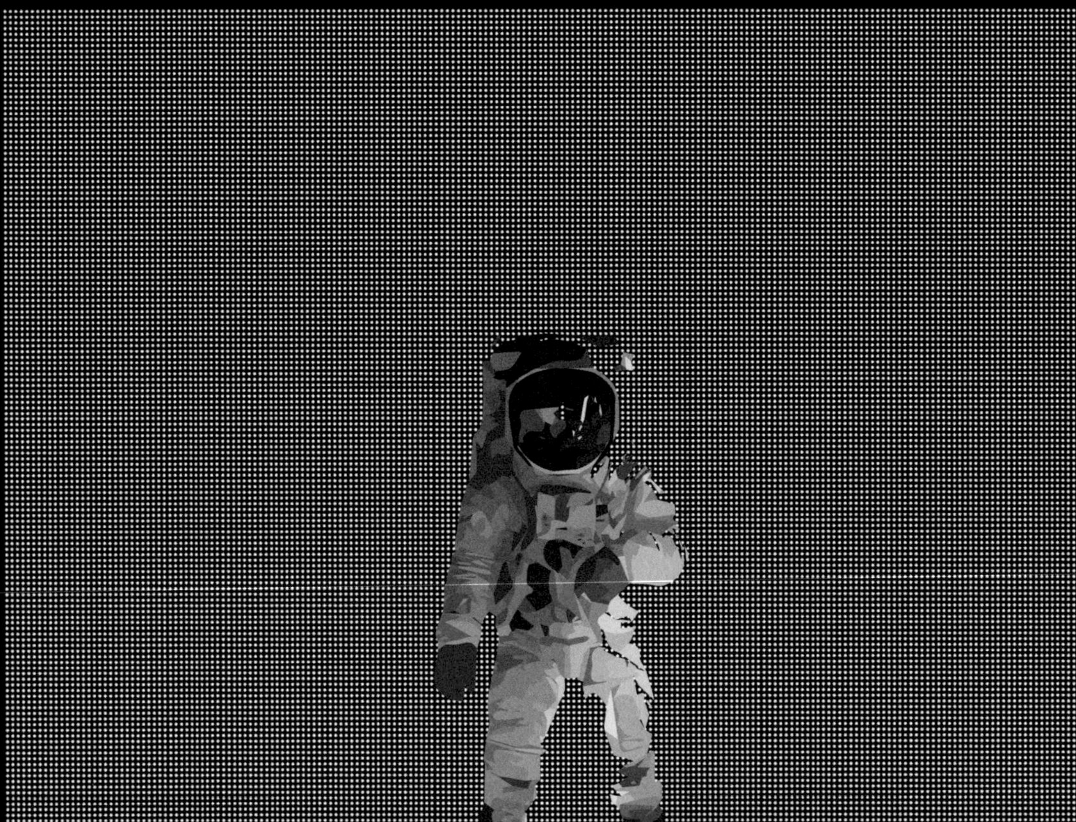

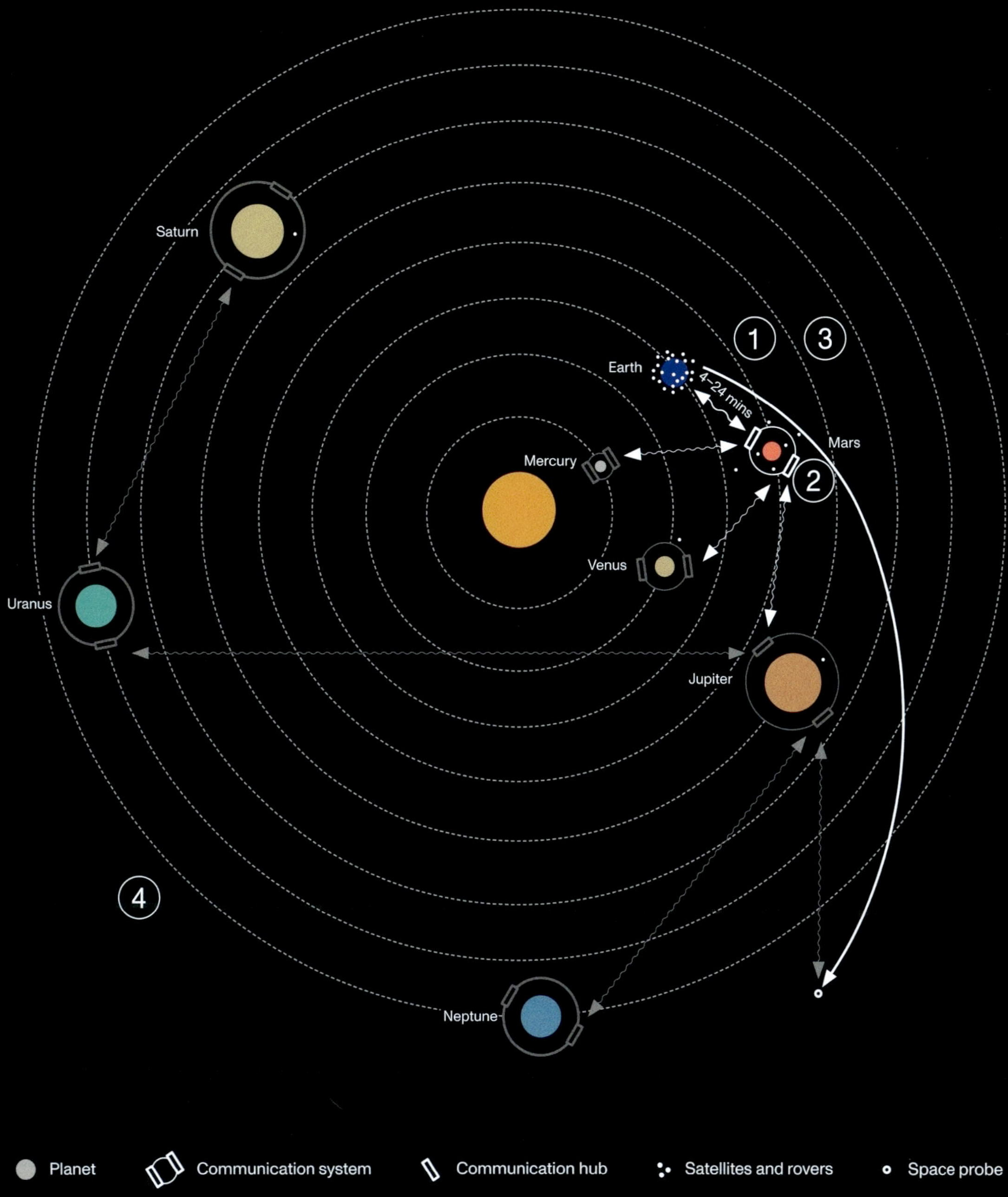


Fig. 4 Links of an interplanetary internet :

1 Most communication from and to Earth is sent via satellites to various strategic locations with powerful parabolic antennas.
2 Infrastructure is established on Mars to create the Martian internet. It is connected with Earth through a communication hub, similar to how the transatlantic cable connects the continents on Earth.
3 Mars is an important relay point and transfers data from other planets in the solar system as well as from its own planet.
4 As human expands further through the solar system the communication network is expanded to the other planets and it will be connected through a series of relay points to reach Earth. All interplanetary communication is done through optical laser.

REFUELING STATION
ASSEMBLY
COMM NODE STORAGE
SATELLITE COMPONENTS KIT
STORAGE
SATELLITE SCANNING

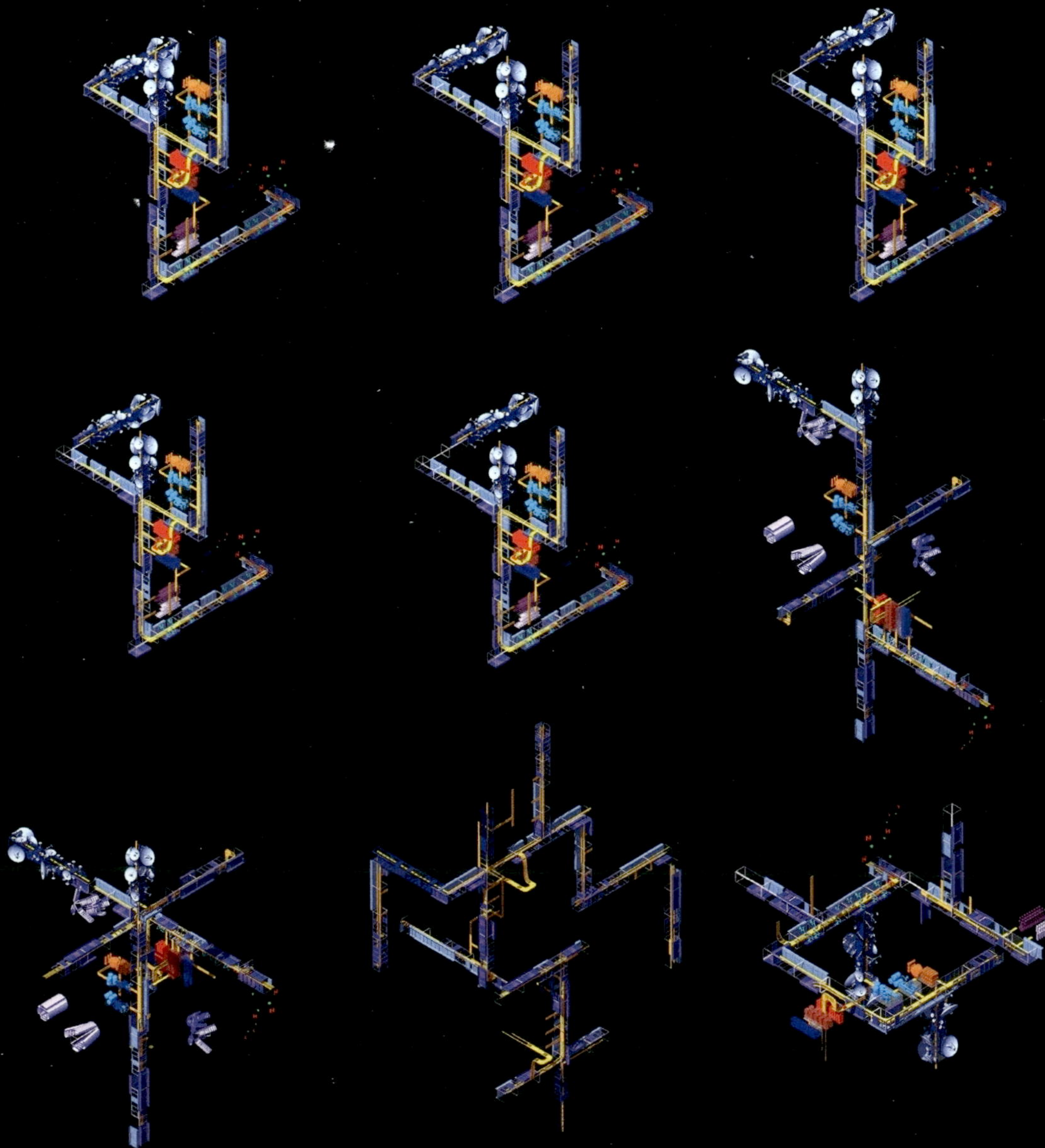

Fig. 5 A modular hub: Enabling communication across the Milky Way. The hub is constructed from identical modules that can be reconfigured to suit the communication or satellite refurbishment activities.

Sandra Baggerman, Marina Bonet

Over time, humankind has become increasingly dependent on modern healthcare and medicine. In its development, the idea of healthcare has become more and more institutionalized and entangled with modern technology, while the notion of care has changed course and meaning.

As the definition of care evolved, so did the definition of what a 'patient' is. One of the early providers of healthcare was the church. Due to a lack of medical knowledge, 'giving care' was defined as giving the grace of God and forgiveness. Mental care was the focus rather than providing a physical cure or remorse. From 1700 on, clinical knowledge increased and evolved because of studies and autopsies. The body was seen as a study object. From the middle of the nineteenth century, technology evolved and with it clinical healthcare. Curing the body was now looked at from a purely clinical point of view. The body was seen as a machine that, like a car, could be fixed by tuning it and adjusting its settings. Care therefore transitioned into the treatment of illnesses. Throughout the twentieth century, clinical knowledge evolved, and the patient was seen more and more as a machine. Healthcare in the twenty-first century is characterized by the heavy use of technology. This has led to the objectification of the body, as the patient is part of a factory-like healthcare system. At the same time, people are focusing more and more on their health—they perceive their body as a temple, but the healthcare system still functions like an inhumane factory.

The human body has been dehumanized by the system and is seen as a device that can be fixed with the proper intervention. Healthcare no longer targets healing, but instead promotes procedures that exhaust the body extensively, ultimately leading to a higher and higher consumption of health care. The expropriation of healthcare thus maintains a self-sustaining system where the emphasis lies on the institution and its treatment, rather than on providing the patient with sustainable care and good health. Humankind is becoming increasingly dependent on the consumption of modern medicine.

To break free from this nemesis, we need to regain autonomy and responsibility for our bodies. We need to reconsider the architecture of the healthcare system by placing the patient at the centre of care.

Implementation of healthcare 4.0, the gathering of body performance data and the incorporation of wearables can offer a certain level of autonomy. The suggested vision assembles a non-linear relationship between various parts of the organization of healthcare, including different flows and diverse operational parts, insurance companies, waste production, energy consumption and other factors.

Putting the patients at the centre of care and the implementation of healthcare 4.0 form the base of the future's decentralized healthcare typology, where the modern non-normative patient receives tailor-made care. A flexible, efficient and sustainable architecture is the result of the thorough implementation of technology and data across all levels of the system.

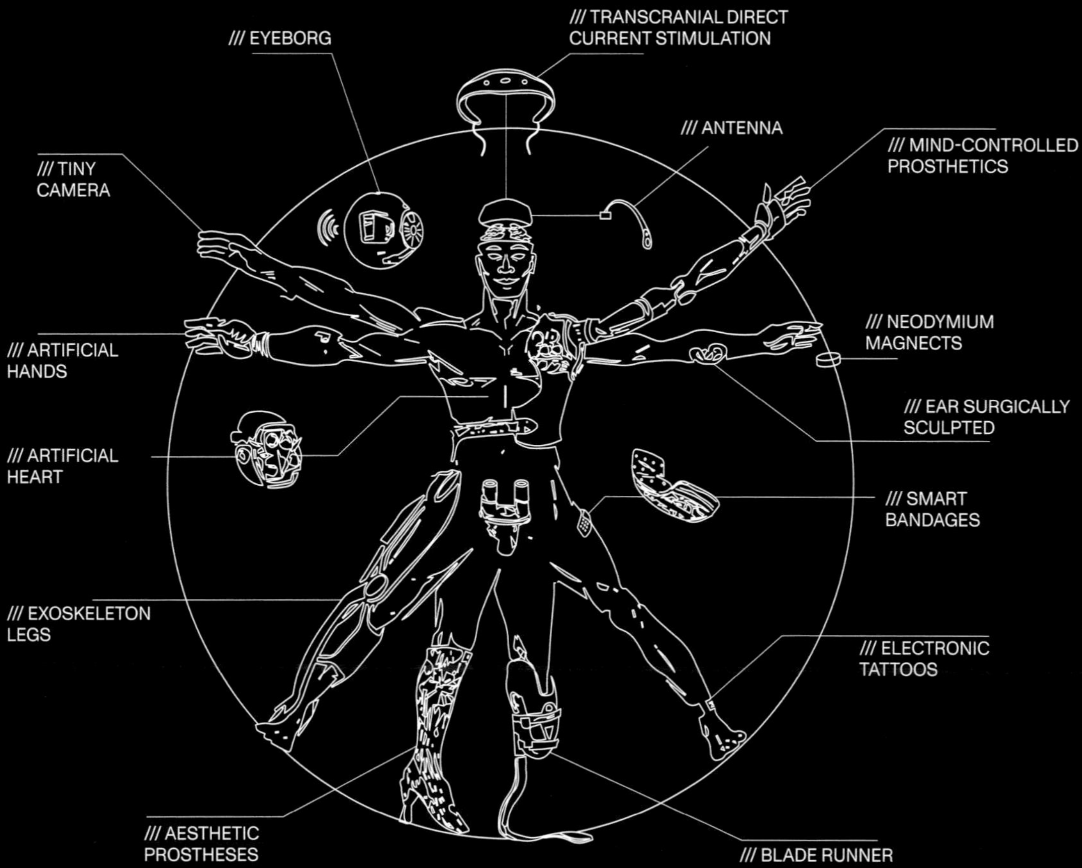


Fig. 1 Future human cyborg. Drawing by Jeremy Li Ho Kong and Alice Chau Ka Yee.

    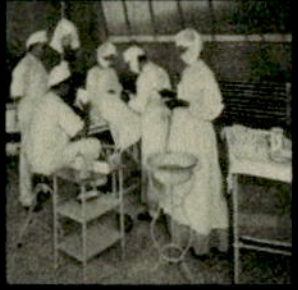 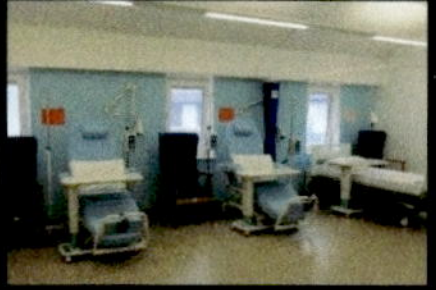

| | HOTEL – DIEU<br>PARIS, FR (650) | GREAT HOSPITAL<br>NORWICH, UK (1249) | HOTEL – DIEU<br>PARIS, FR (1787) | HÔPITAL LARIBOISIERE<br>PARIS, FR (1850) | MT. SINAI HOSPITAL<br>NEW YORK, USA (1904) | MODBURRY HOSPITAL<br>MODBURRY, AUS (1990) |
|---|---|---|---|---|---|---|
| TYPE | *"INN OF GOD"* | *CLOISTER TYPE* | *SYMMETRIC 'GREEK CROSS' PLAN* | *PAVILION TYPE 'NIGHTINGALE GALE' PRINCIPLE* | *MONOBLOCK 'STACKED PAVILION' TYPE* | *MEGA HOSPITAL 'DEEP PLAN'* |
| PLAN | *church*<br>*ward*<br>*research*<br>*support functions*<br>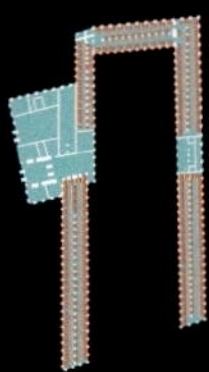<br>*5 levels* | <br>*2 levels* | 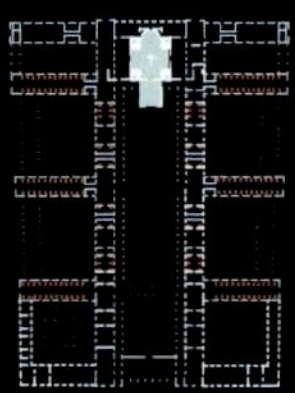<br>*4 levels* | 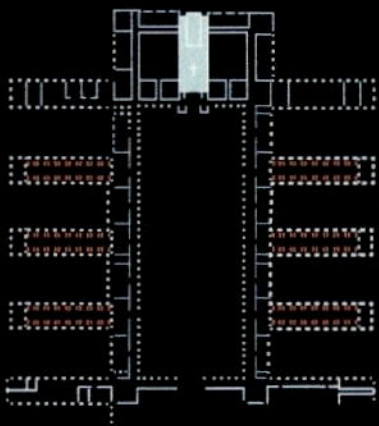<br>*6 levels* | <br>*5 levels* | <br>*6 levels* |
| PROGRAM | 15% | 40% | 20% | 24% | 44% | 40% |
| FUNCTIONS | • grace of God<br>• for sick/dying<br>• take the disease from the street | • grace of God<br>• for sick/dying<br>• give shelter to travelers | • clinical education<br>• cure the sick | • clinical education<br>• cure the sick<br>• research space | • clinical education<br>• cure the sick<br>• research space<br>• institution<br>• nursing school | • cure the sick<br>• private health care |
| SERVICE | | | | | | |
| TECHNICAL INNOVATION | • grace of God<br>• for sick/dying<br>• take the disease from the street | • grace of God<br>• for sick/dying<br>• give shelter to travelers | • gass pipes<br>• oil lamps | • washing facilities<br>• sewage<br>• surgery spaces<br>• germless ventilation<br>• modern heating | • elevator<br>• mechanical ventilation | • air conditioning<br>• artificial light<br>• electric bed<br>• central sterile supply<br>• automatic X-ray<br>• auto analyzers in lab<br>• plastic blood bags<br>• identification bracelets<br>• pneumatic tube system<br>• power plants |

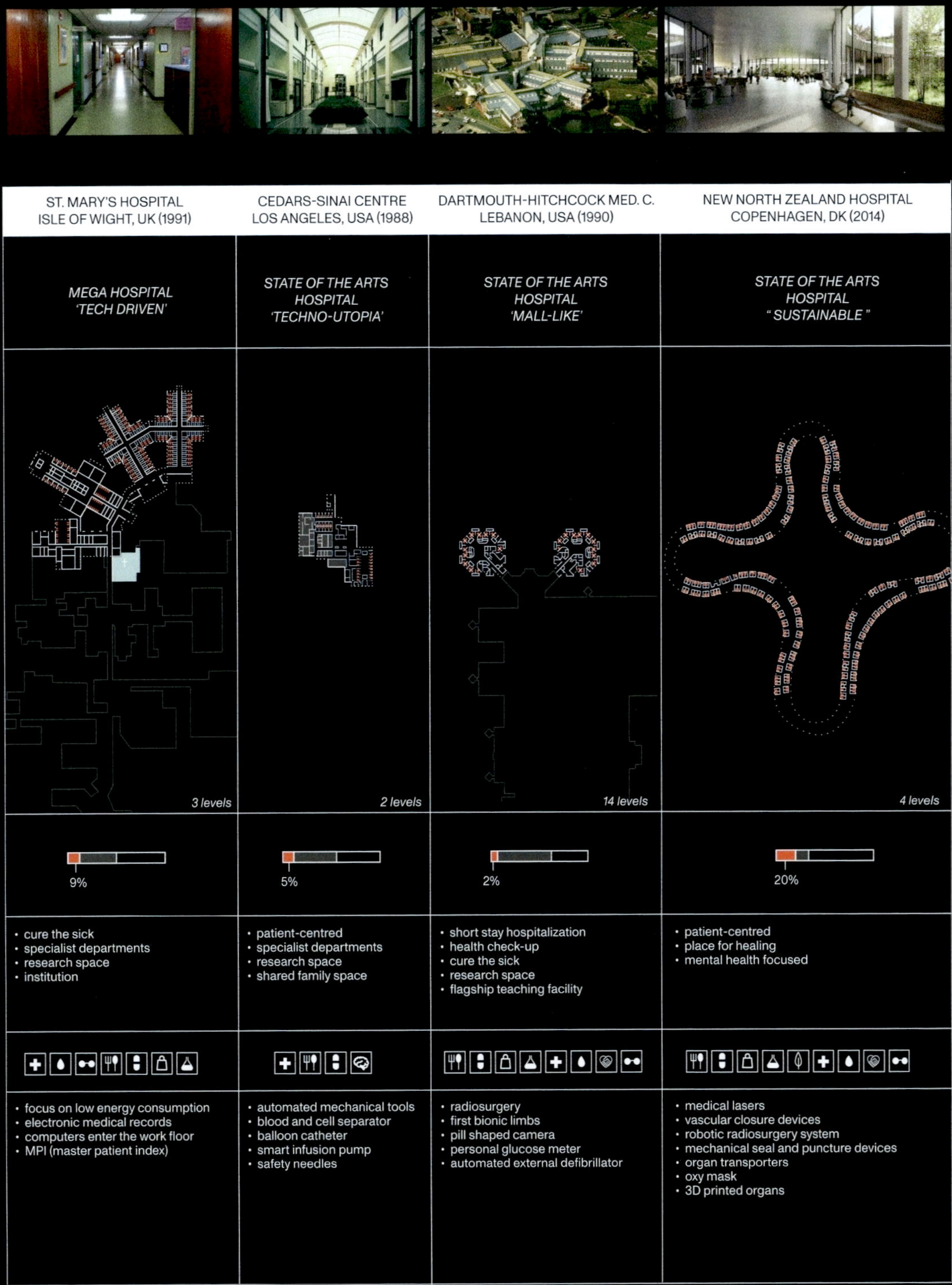


Fig. 2 The architecture of healthcare.

LEVEL0:BODY-TECH
LEVEL1:TECH-TECH
LEVEL2:SYSTEM-SYSTEM
LEVEL2:SYSTEM-DATACENTER
ANGIOGRAPHY
CARDIAC CATHETERIZATION LAB
AI AUTOMATED SURGERY
[INTERVENTION]
VIRTUAL DISPLAY
3D PRINTER
SCANNER
ULTRASOUNDS
MRI SCAN
HEMODIALISIS STATION
MONITORING SYSTEM
[DIAGNOSIS]
FULL BODY SCANNER
MENTAL STIMULATING VR
FITBIT
SMART PHONE HEATLH APPS
SOMNOX
[MONITORING]
SMART SURFACE
POWER SUPPLY
BIO FEEDBACK
ELECTRICAL SAFETY ANALYSER
[CONTROL STATION]
PATIENT RIBBON
HALO LIGHT BOX
TOUCH DISPLAY
VR
PATIENT COMPANION
EQUIPMENT
MOBIL LAB ROBOT
NO-ISOLATION ROBOT
PHARMA DELIVERY AI
[CARE]

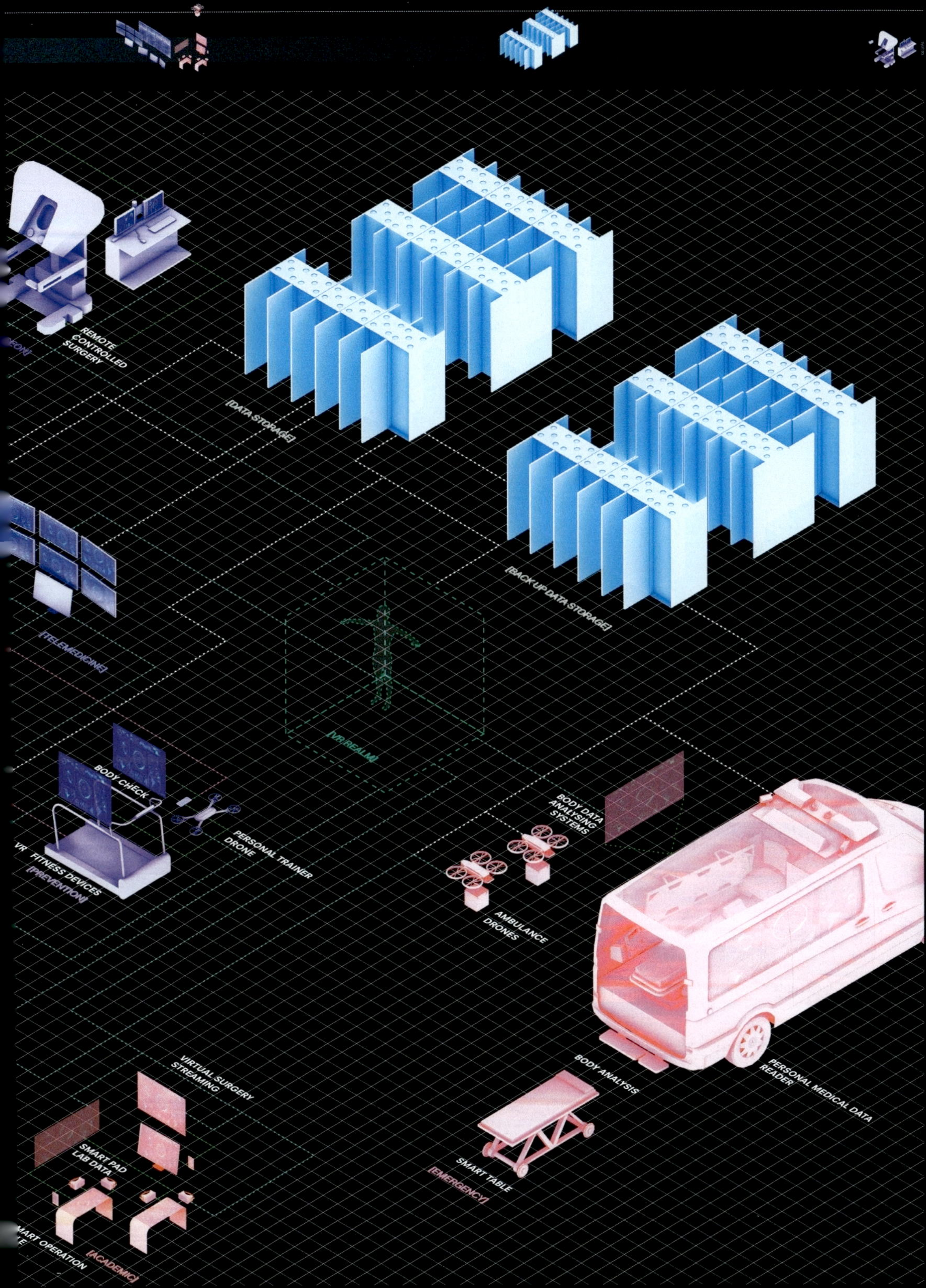


Fig. 3 De-centralized person-focused healthcare system.

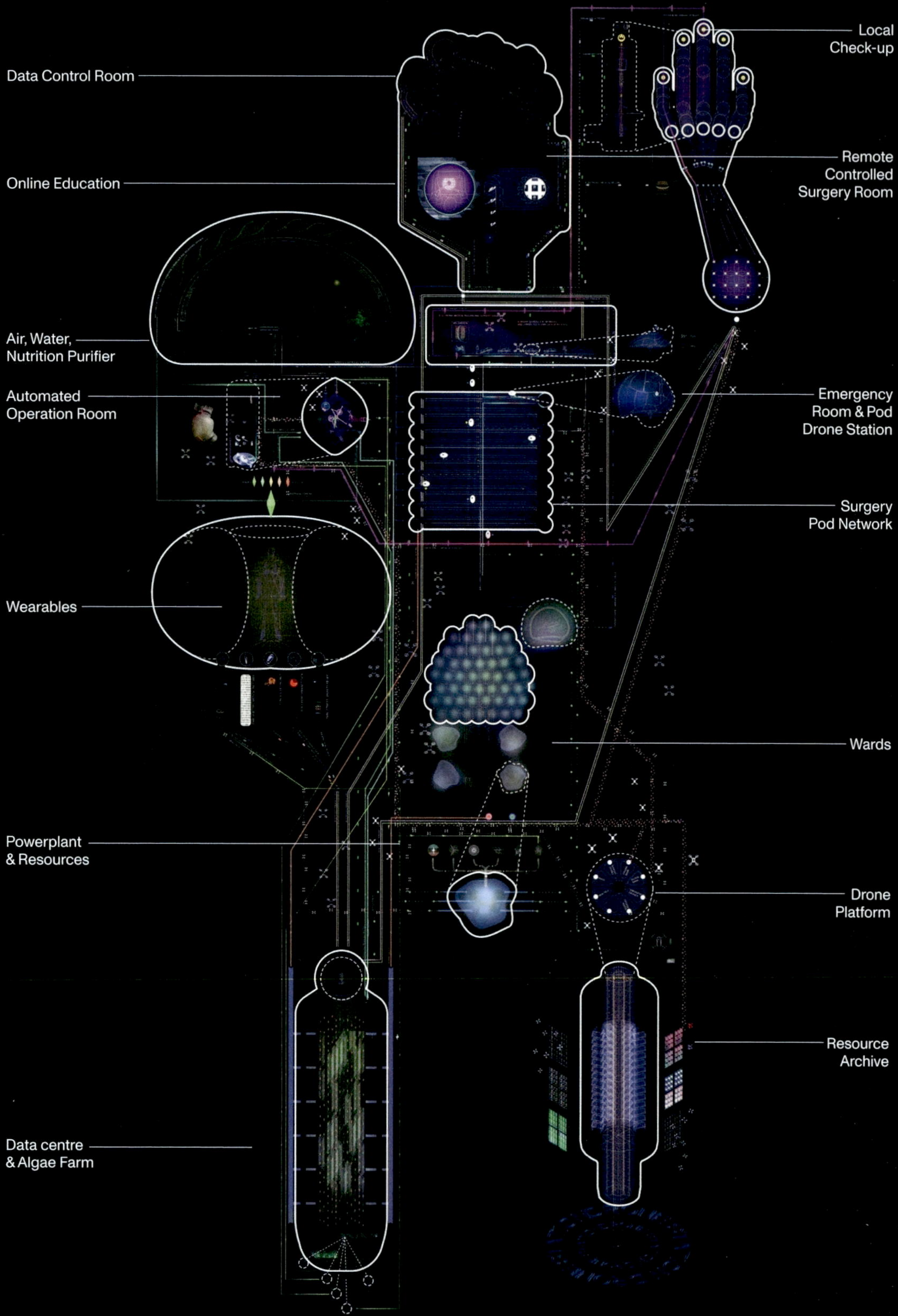


Fig. 4 The future architecture of health care by Sandra Baggerman.

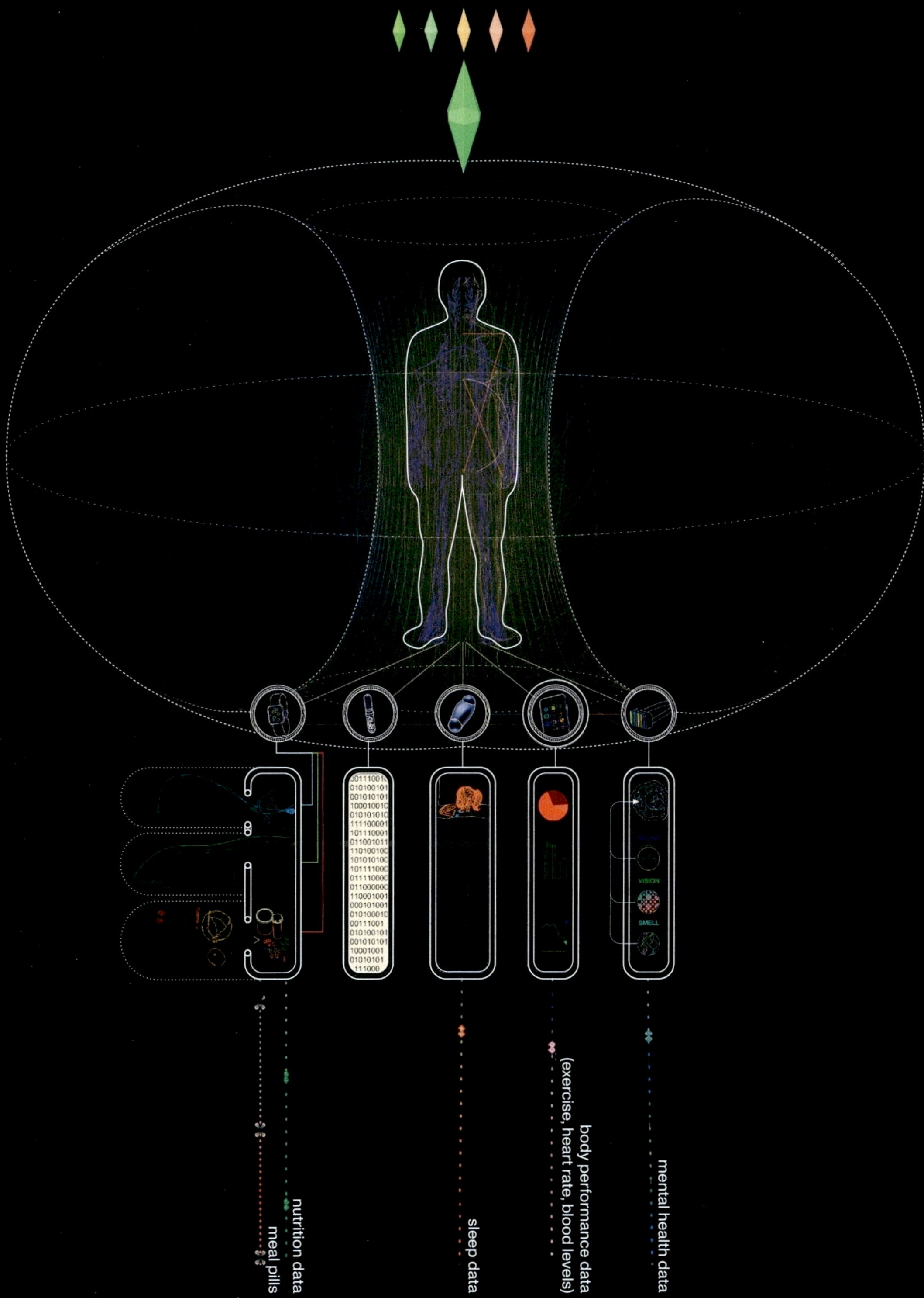


Fig. 5 Zoom in on 'The future architecture of health care' chapter 'Wearables'.

Wearables allow for a vicious circle of care. Body performance data is collected by the devices and shared with the human to improve the body's performance. Next to collecting and sharing data, Wearables allow for precision medicine, telehealth, and AI to operate under the most efficient circumstances. Wearables help humans maintain their' physical and mental health. The devices track the body's performance. The data is shared with AI and specialists to advise, update and guide a human to stay healthy.

## A NETWORK FOR DEMOCRACY

I-Hsuan Lee, Susanne van de Pol, Robert Bichlmaier

The way urban space is created and secured has undoubtedly changed since the terrorist attacks at the beginning of the twenty-first century. Security and surveillance became a frequent subject of discussion. Some elements in our everyday life, like barricades blocking entrances to public events as protection from terrorist attacks (such as the one that happened on a Christmas market in Germany in 2016), are easily recognizable as barriers in the space of conflict. The intrusion of surveillance into public and private life has certainly affected the way people move and act in the urban environment.

Taking the act of protesting as an example, the urban environment becomes the stage of conflict. Opposing ideologies and political positions physically come face to face in the public space. Often in the form of citizens expressing their dissatisfaction with the current government or political system. In recent years, news about political protests around the world has become more frequent, with the pro-democracy movement in Hong Kong and Black Lives Matter demonstrations as recent and ongoing examples.

To understand the impact of technology on our cities, as spaces of conflict in political protest, we investigated the 2019/2020 Hong Kong protests, specifically looking for the technological means used by the government and protestors alike and how those technological means led to a new way of protesting, organizing bodies, tactics and collectives.

Increasing surveillance capabilities, paired with facial recognition programs enabled by artificial intelligence, give the government a strong authority over the information concerning ongoing events in the city. It is possible to track nearly every corner of the public space from the comfort of a control room. Protecting their identity by wearing masks and disabling/attacking CCTV cameras are actions that the protestors employed. At first glance it seems that there is a strong imbalance in authority over the information gathered in the urban landscape, but the protestors managed to communicate efficiently by using peer to peer mesh networks, which enabled them to adapt to situations quickly. This efficient way of organization without the need of a designated leader is the main difference when comparing the current protests in Hong Kong with previous events like the 2014 Occupy movement. Following a quote by one of Hong Kong's most famous citizens, Bruce Lee, 'Be like water', the protestors were always on the move, quickly adapting to every new situation, while blocking parts of the city by building barricades fast, 'on the fly'.

With our design, we want to enable the tactics even further by proposing a city-wide Bluetooth mesh network. This network can be expanded by everyone who wants to participate. They can add devices like Bluetooth repeaters, storage or sensors to the network and thus enhance it in numbers and range. The effectiveness of the network is increased as more people participate, no matter if they're physically disabled or not, making it a democratic tool for the pro-democracy movement! The authority of information is no longer tied to the city's infrastructure of CCTV networks and data centres. Anyone can contribute, giving the authority over information and narrative back to the protestors, thus effectively formulating a counterproject to the official surveillance system by surveying the actions of the government itself.

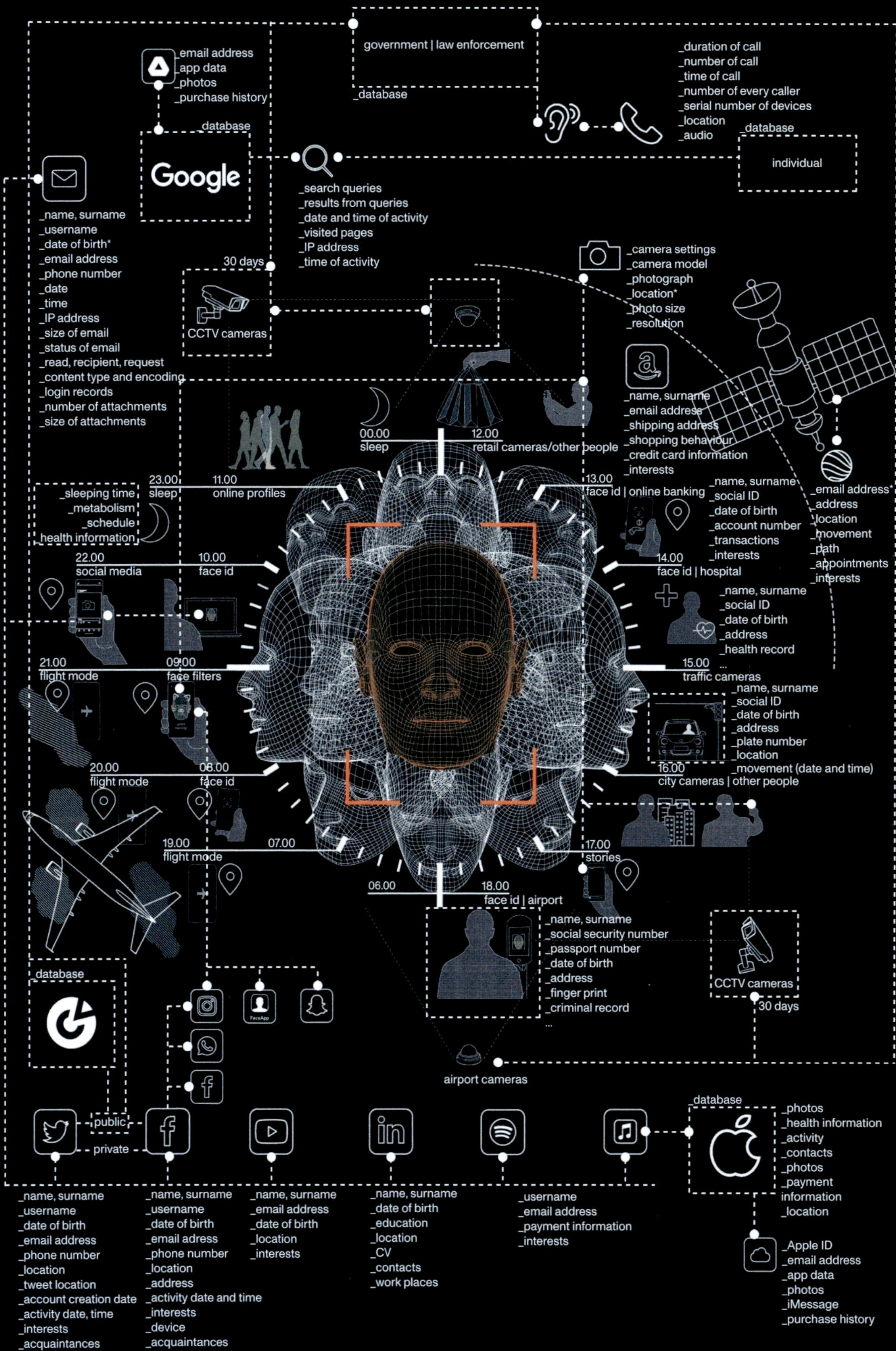


Fig. 1 Interlinked network of online personal data and everyday face recognition surveillance encounter. Drawing by Ali Fatih Gebeci and Dilara Turgut.

CCTV Image
Status: Connected
Queen's Rd E, Wan Chai, Hongkong
MEDIA PLANNING EXPERT
TEL:2988 8348
ID: 69j5nr
AGE: 43
SEX: M
ID: 74o0ln
AGE: 35
SEX: M
ID: 6v4h62
AGE: 27
SEX: M
ID: w28fx9
AGE: 26
SEX: M
ID: 2z5549
AGE: 43
SEX: F
ID: vz096v
AGE: 28
SEX: M
ID: 669k0l
AGE: 24
SEX: F
ID: au8fx9
AGE: 36
SEX: F
ID: n3n37f
AGE: 56
SEX: M
ID: 53cs38
AGE: 32
SEX: M
ID: f921z3
AGE: 25
SEX: M
WV-SF 138
15:34:49 - 2019 - June 09

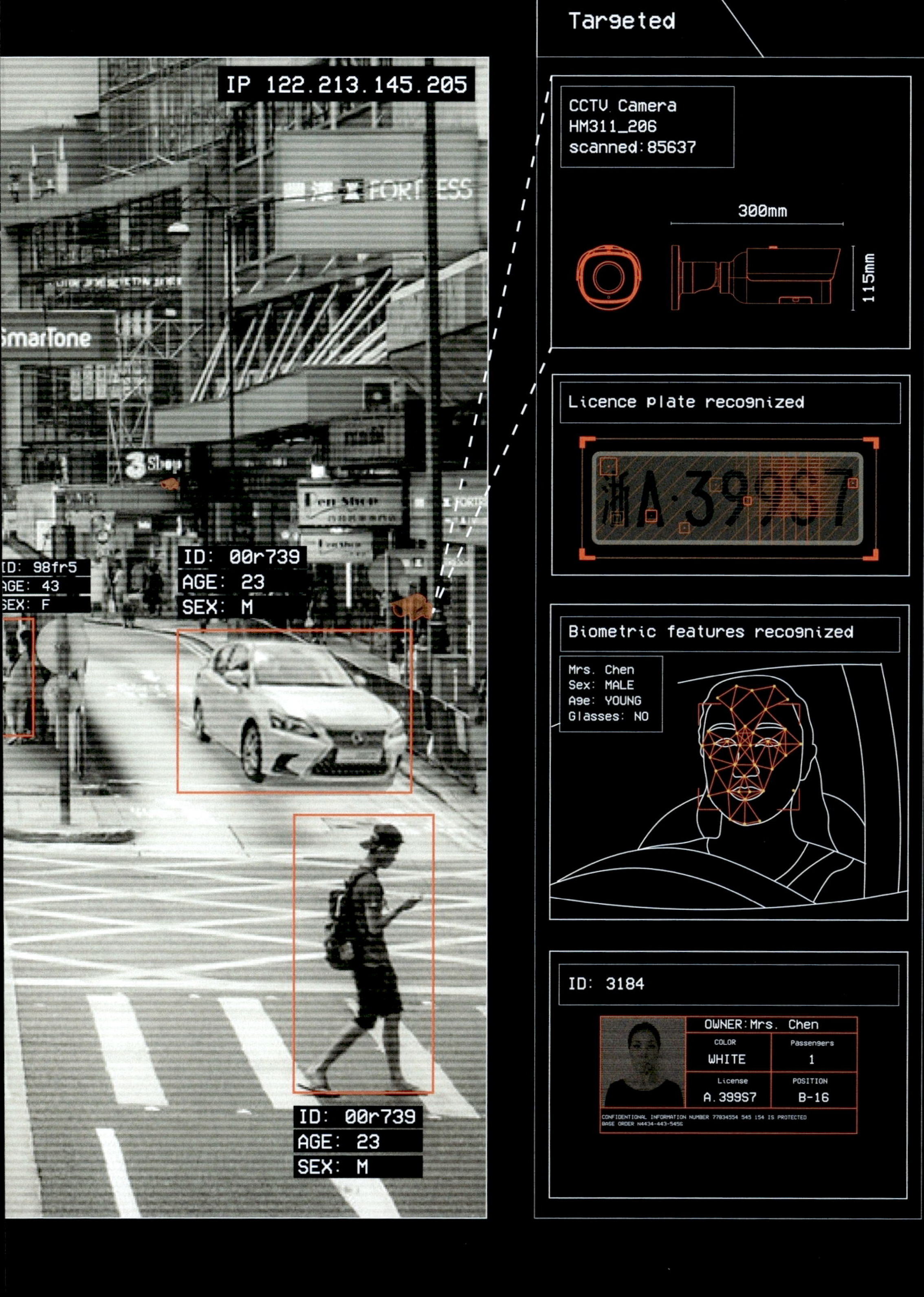


Fig. 2 Surveillance diagram, 'if you have nothing to hide, you have nothing to fear'.

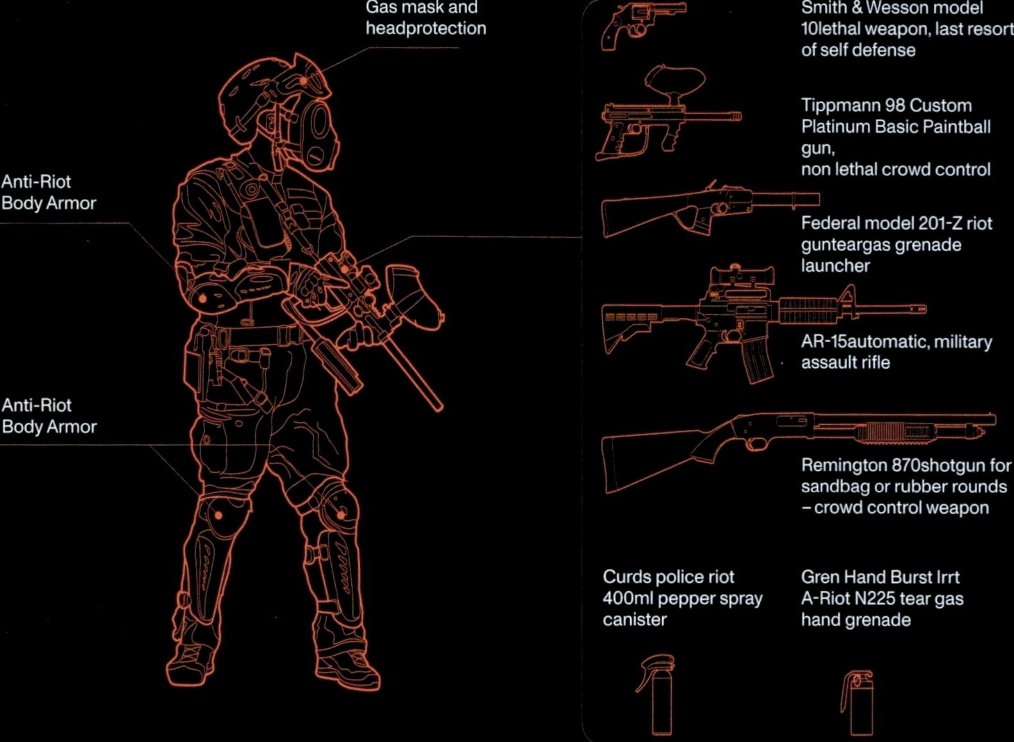


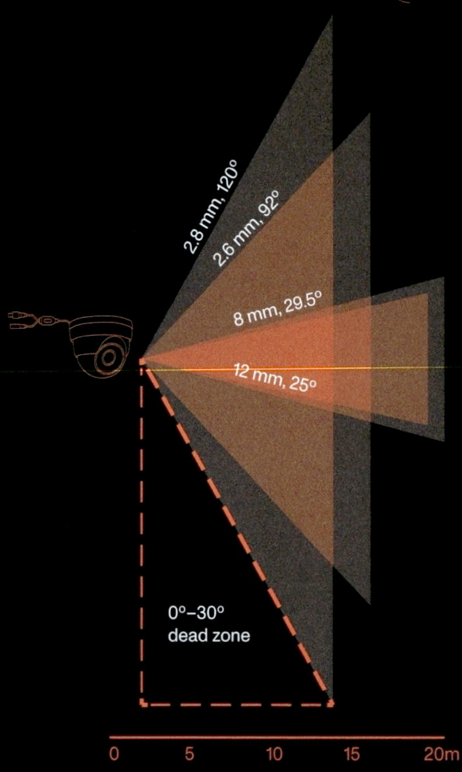


↑↑ Policeman with accessible equipment. ↑ CCTV recognition zone.

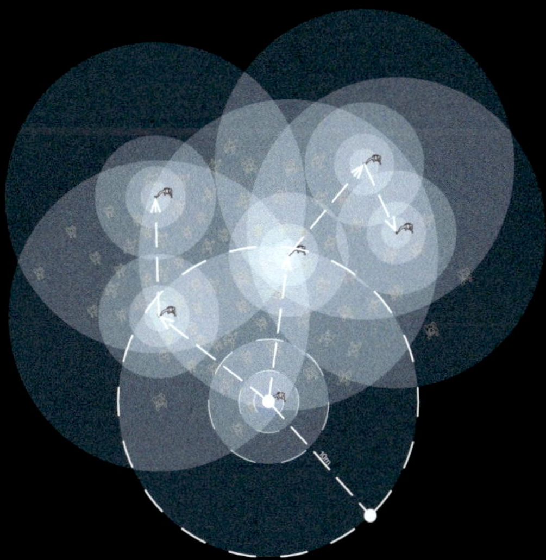

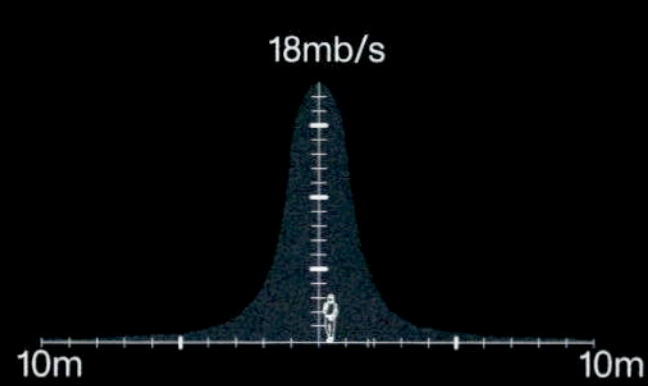


Fig. 4 ↑↑ Protester with accessible equipment. ↑ Bluetooth area range. Decentralized Network using Apples Airdrop technology Searching for close devices (10m) with Bluetooth. , directly connecting after request is accepted with Wifi for fast connections.

Protest crowd_10-02
FREE
DEMOCRACY
HKSAR Headquarters_20-01
光復香港
時代革命
香港獨立
HONG KONG INDEPENDENCE
WARNING
TEAR SMOKE

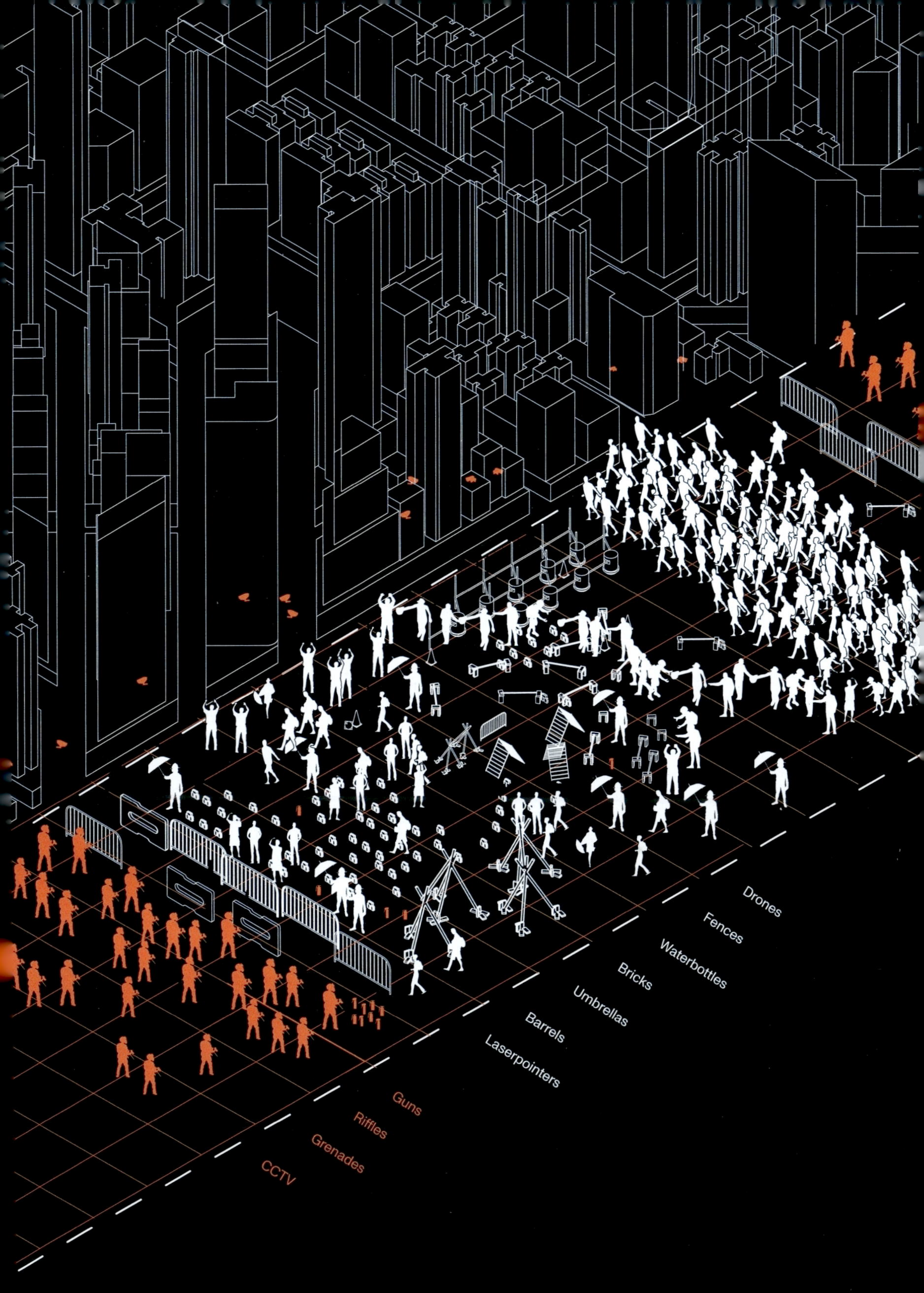


Fig. 6 Layering protest in Hongkong streets: city scape, battlefield, in conflict.

## META-OFFICE

Lauritz Bohne, Lea Scherer, Edward Zammit

What we think of as artificial intelligence is often manufactured by the routine clicks of countless workers. Hidden behind their screens they operate the digital production line of a rising labeling industry focused on data description services—services that seem to be automated but are in fact deeply human. Amazon employs workers from all over the planet that are paid cents per minute per task to label, categorise and validate vast arrays of raw data, providing the important meta-data. The so-called Amazon Mechanical Turks are the unseen workforce behind digital automation.

By using Amazon's crowdsourcing platform to enter into a dialogue with its workers, Meta Office exposes their decentralized 'office' spaces. While the workers are confronted with the mechanisms of the platform, they allow us a voyeuristic gaze into the most intimate interiors, where individual needs bend to those of work.

Photo by photo, the workers share with us which objects are most meaningful to them, revealing a repetitive pattern: desktops, mice and keyboards, office chairs, tables and trays, headphones, extension cables and desk fans meet bedrooms, kitchens and bathrooms, beds, pillows, and refrigerators.

Click by click, the workers provide information about square metres, geographical location or earnings, expanding our understanding of the photographs with their metadata.

Meta Office merges the physical imprint of that digital production line, whose common ground is the transformation of space itself: an improvised making of a workspace with all the means at its disposal. An office space that imposes itself, transforming its original surroundings into a working environment whose appearance does not resemble a corporate design, but underpins the influence of a globally operating corporation. An individualised and continuously expanding office space whose existence, just like that of its workers, remains concealed behind countless generic User-IDs.



Meta Office is what it is described as.

The meta office is self-descriptive. The task of its employees is the recording of their offices, being domestic interior spaces. The description of their environment defines it at the same time: Description becomes construction, heavily dependent on the subjective perception at this very moment. As such, the project applies the method of crowdsourcing, and pushes its boundaries further, becoming a method itself.

Meta Office is pure visualization.

The design challenges architecture as pure visualization applying different tools of spatial mapping. The visual reproduction of the explored decentralized spaces is seen as an act of centralization that allows us to understand their physical implications as a whole.

Meta Office is perceived from within.

The mapped (office) interiors are qualified by their objects. Once all these objects were interior objects of living rooms, sleeping chambers, kitchens, or bathrooms. Domestic objects became objects of work. The workers employ them to serve them in order to build up their office. Their transmutable interrelations define the program of the space.

Meta Office is meta-data.

It is statistics, labels, numbers, categories and quantities. It is size, location, effort and time—m2, coordinates, $ and sec. As the effort of its workers completely lies in the creation of these data, the meta office is never complete. As the meta office is constant description, it can never be described fully.

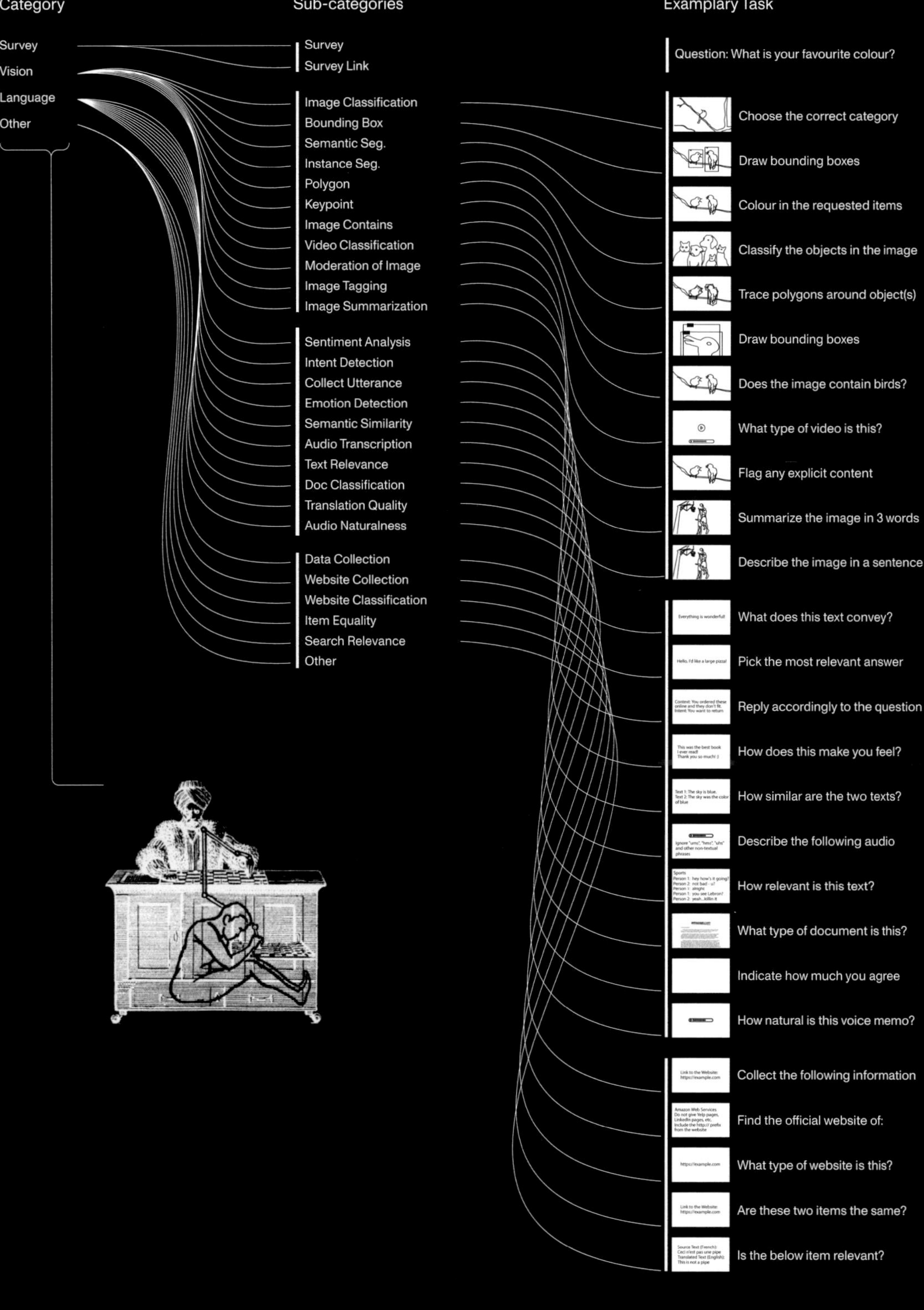


Fig.1 Interfaces of the labelling industry.
Index of various tasks that can be performed by the Mechanical Turk.

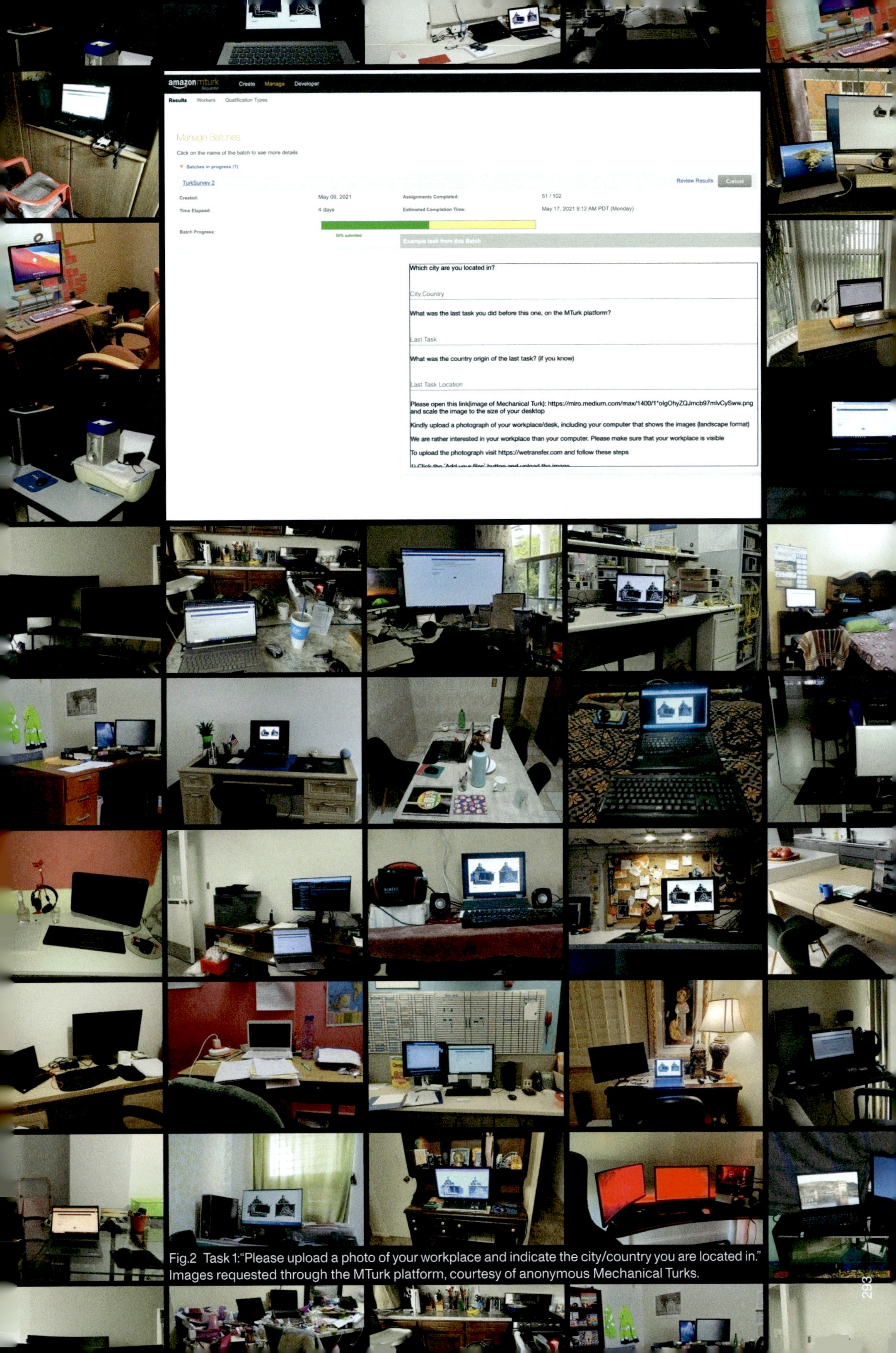


Fig.2 Task 1:"Please upload a photo of your workplace and indicate the city/country you are located in." Images requested through the MTurk platform, courtesy of anonymous Mechanical Turks.

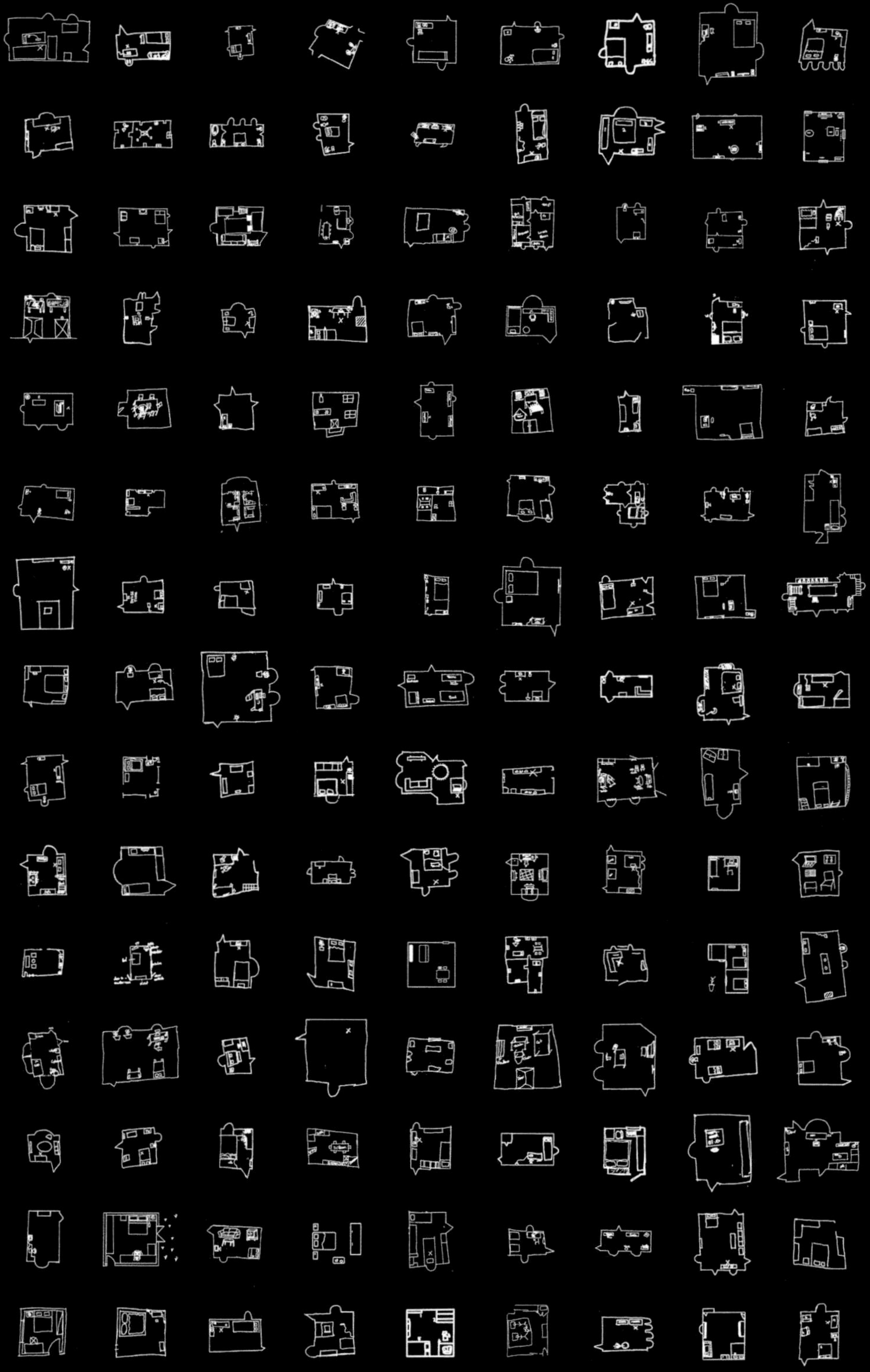

Fig.3 Database of the Mechanical Turks' office floor plans, hand-drawn by anonymous Mechanical Turks. The drawings were collected and compiled by authors and they were requested through www.mturk.com.

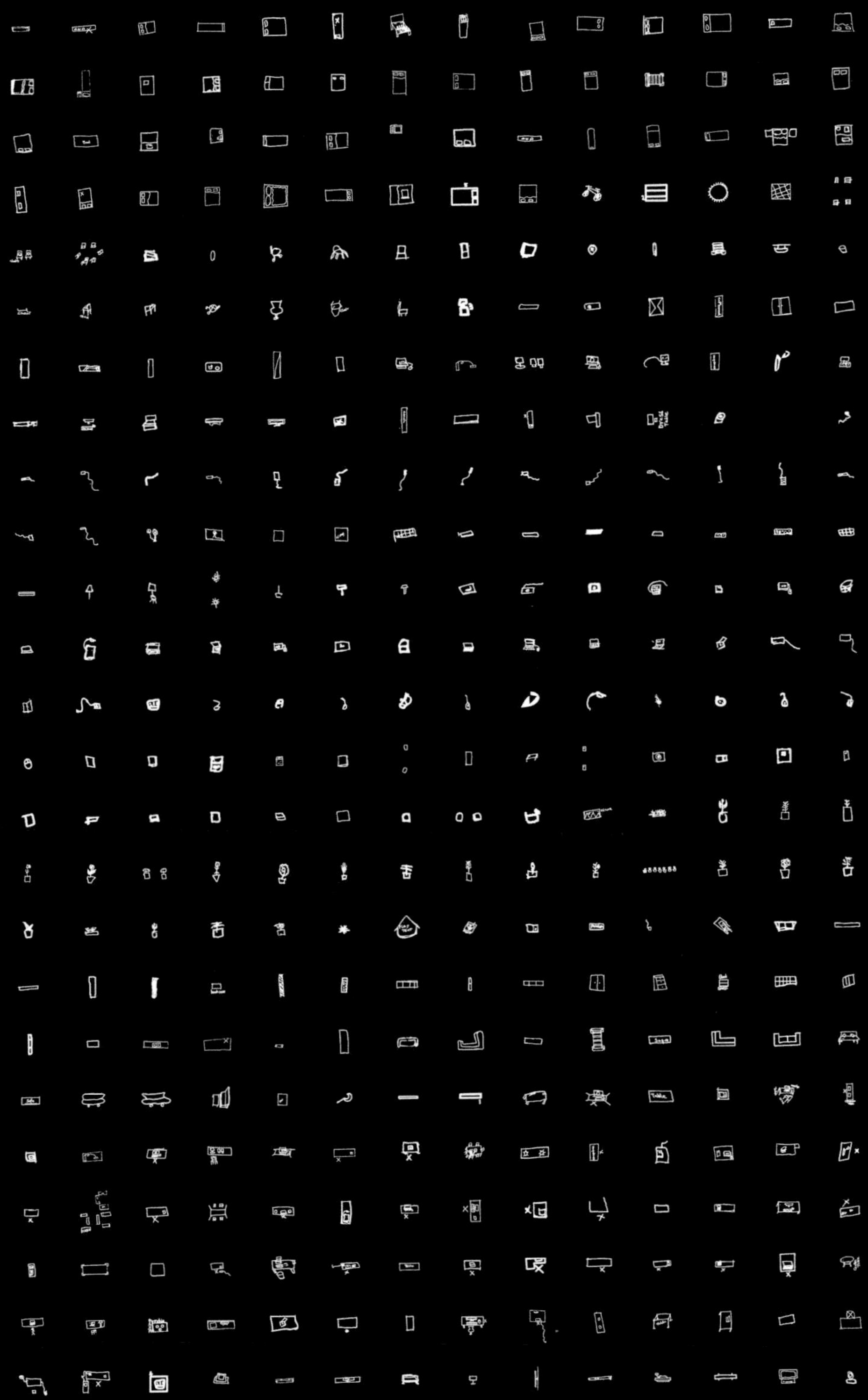

Fig.4 MTurks' office tools.

NL, Roermont
desk
$0.25
855 sec

USA, Dallas
television
$0.25
1876 sec

France, Perpignan
vacuum cleaner
$0.25
1449 sec

France, Grenoble
plant
$0.25
1679 sec

Spain, Madrid
storage
$0.25
618 sec

Brazil, Salvador
sidetable
$0.25
855 sec

Italy, Naples
sofa
$0.25
2269 sec

France, Grenoble
extension cable
$0.25
834 sec

Venezuela, Merida
bed
$0.25
1233 sec

India, Hyderabad
chair
$0.25
705 sec

USA, New York
laptop
$0.25
2638 sec

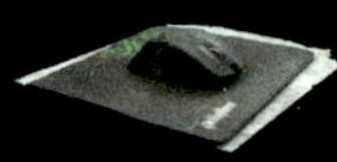

India, Chandigarh
mouse
$0.25
2112 sec

Z1
Brazil, Guarulhos
fan
$0.25
855 sec

Y3
Brazil, Sao Paolo
water
$0.25
801 sec

Z2
India, Chandigarh
electric heater
$0.25
947 sec

Fig.5 ↑ Database of the Mechanical Turks' chosen objects. A series of photos of each object were requested from each MTurk.Through photogrammetry, each object was converted into a 3d model and catalogued by the authors.
Fig.6 → Task 5: 'Meta Office is all of you in one building. How does the building look like?'

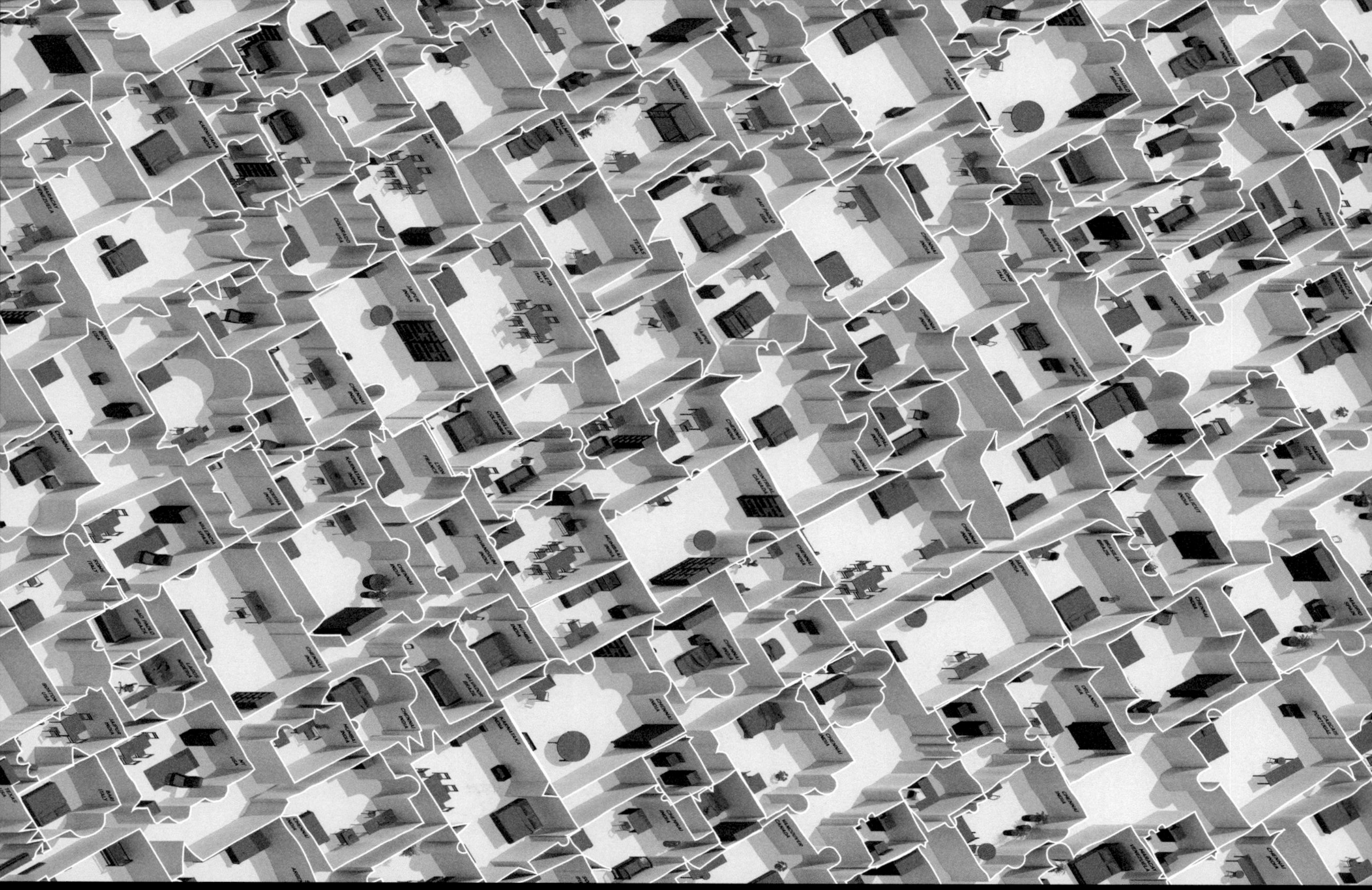

# ORGANICA: NOETIC HORIZONS OF DATA

Rafał Tarczyński, Ragna Nordström

Organica explores the possibilities of existing with technicity while reconnecting to our inherent spirituality. A spirituality that has been enabled through the evolution of information and communication and specifically the evolution of the digital. The project we are proposing is not just an organic data centre, it extends into the surrounding territories, potentially constructing healing datascapes. These datascapes in turn offer their inhabitants a monistic way of coexisting, technicity and nature reunited, reconnected, re-ligare.

The purpose of this research is to analyse transformations of religious thought throughout history and more specifically systems of belief related to the rapid spread of information and communication technologies in contemporary society. We are interested in the digital, in humans and in the effect of technology on humans. How does the use of technology and the authority of Big Data affect our lives? How does it change our relationship to the world and our relationship to other humans? What can we learn from our religious heritage in terms of ethical and moral thinking?

The question of authority and free will has been contemplated throughout every period of Western history and philosophy. The term 'free will' emerged as the common concept of designation of control over one's actions. Subsequent questions that emerged in regard to this topic concern the nature and existence of this same control as well as questions surrounding the significance of this control—for what reasons it might be of importance and whether it is essential for moral and ethical reasoning. The idea of a global biotechnical network is elaborated by Howard Bloom in The Global Brain. 1 Since the birth of the personal computer, scientists have been working on this global brain composed of computer networking. This radical human transformation began by attaching billions of minds of humanity together into a single system. Nearly 20 years ago, Bloom predicted a collective consciousness where we all come together in a global network that has learned our ways of thinking and can provide us with the knowledge we need before we even know we want it. A network that will turn the planet into a single spiritual super-being, a massive 'collective consciousness'.

By recultivating the land with organic infrastructure, biodiversity and habitats can recover. This is a way for humans to coexist with nature without losing the accessibility of information and technicity. The organic data centre works as the focal point of this society. It is the main source and creator of the landscapes and the society. It functions so that it can attach to any data flow: undersea cables, underground networks and to external power grids, thus allowing for floating, digging and flying. Its main task is to translate as much binary data as it possibly can into organic form and then return it to the land, thereby 'pollinating' the Earth with new organic data. The programme of the structure responds to the simple unity: one part belowground, which reacts to the intake of binary data, and one part aboveground, which responds to an output of organic information through the cultivation of plants, bacteria and minerals. The intersection of technicity and nature can also be observed when the data is retrieved. How would such an interaction be visualized? The Apostles of Data are provided with a starter pack consisting of portable devices to decode data from datascapes.

The starter pack comes with a backpack with integrated solar panels for charging the devices. The device for plants is equipped with a DNA sequencer that analyses an inserted sample and decodes text and audio files. While lying and resting in the picturesque landscape, you can listen to the sound of the flowers.

1 Howard K. Bloom, Global Brain: The Evolution of Mass Mind from the Big Bang to the 21st Century (New York: Wiley, 2000).

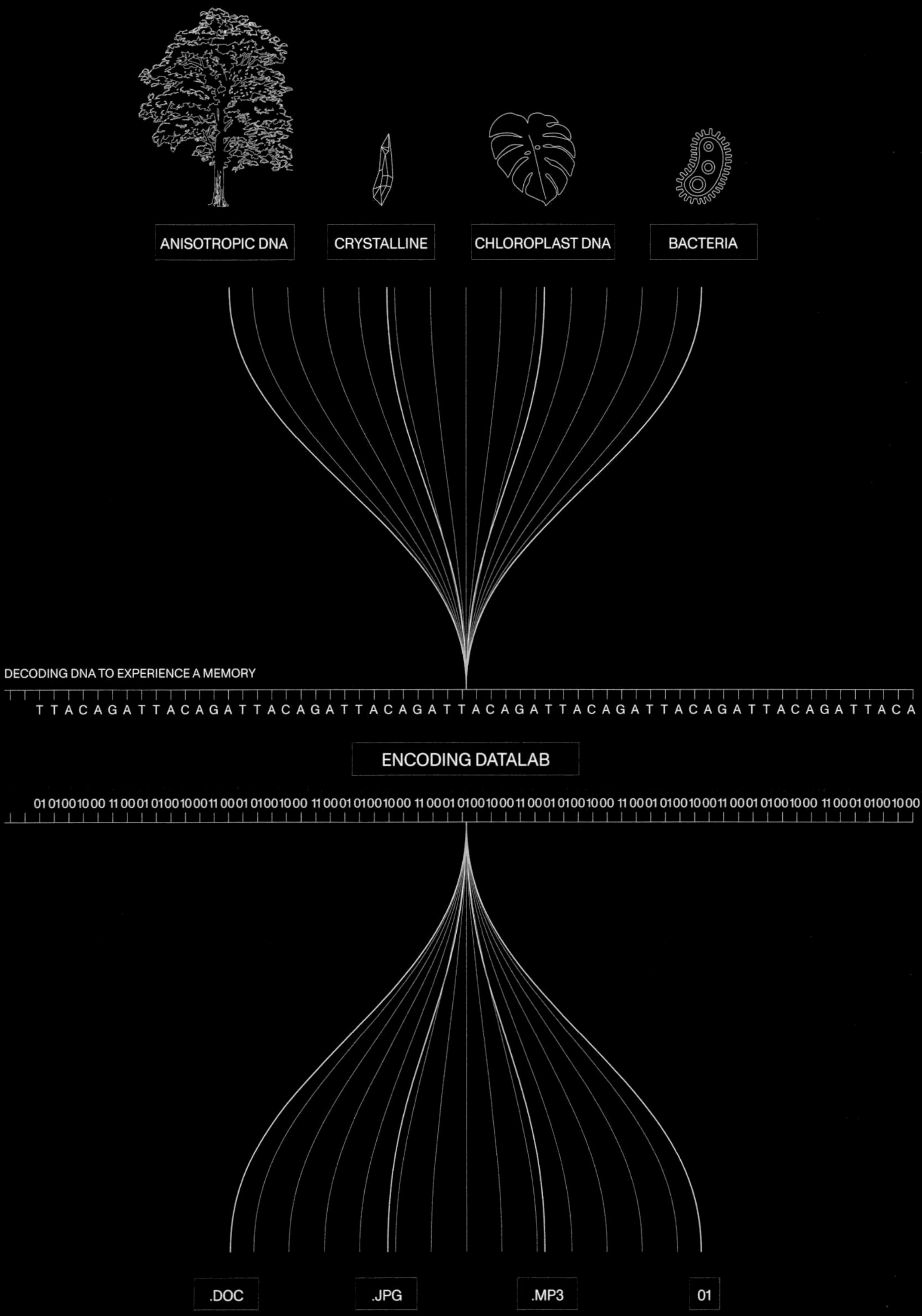


Fig. 1 Any binary file can be translated into autotroph DNA, bacteria's DNA and crystalline structures.

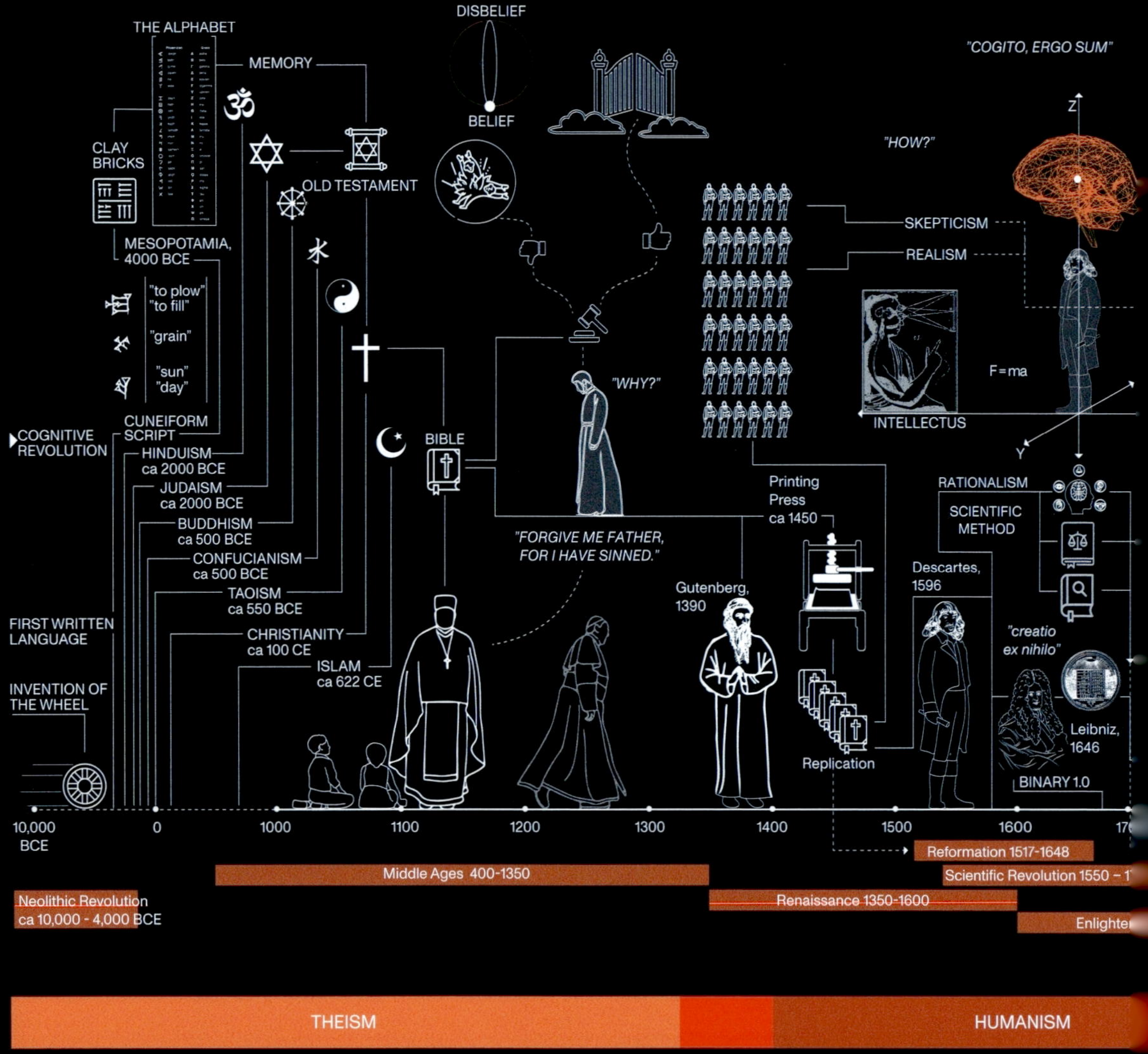


Fig. 2 The timeline shows an alternative future of data, where data is transferred back to the Earth.

Source: Yuval N. Harari, Homo Deus: A Brief History of Tomorrow (London: Harvill Secker, 2016); Yuval N. Harari, 21 lessons for the 21st century (New York: Random House Audio, 2018); Shawn DuBravac, Digital Destiny (Washington DC: Regnery Publishing a Salem Communications Company, 2015); Kevin Kelly, What Technology Wants (New York: Viking, 2010); Jonna Bornemark, The Renaissance of the Immeasurable (New York: Meridian Books, 2018); "History of Writing System", Britannica, accessed 2021-05-12; "Big History Project", Khan Academy, accessed 2021-05-12.

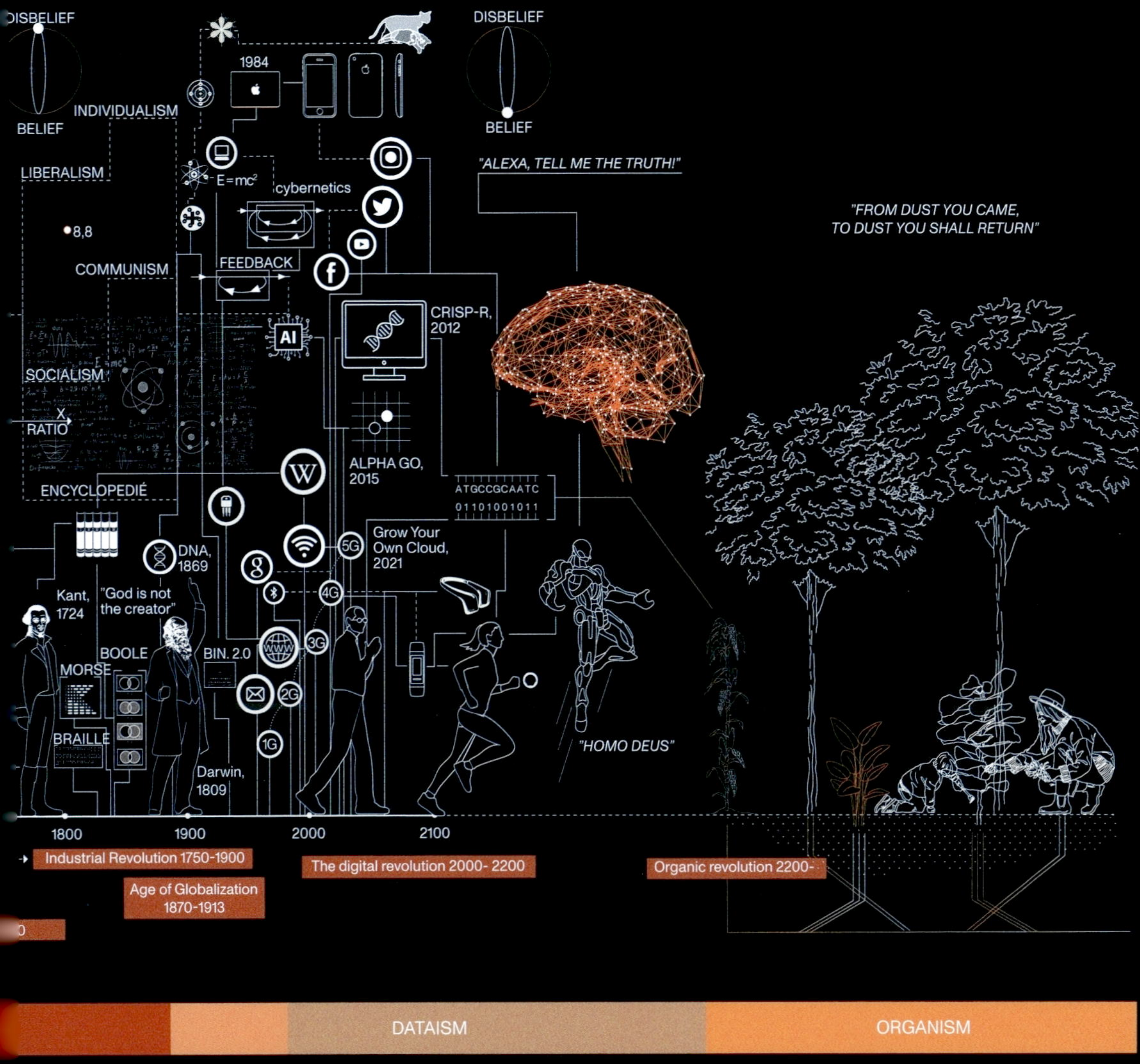
DISBELIEF
BELIEF
INDIVIDUALISM
LIBERALISM
8,8
COMMUNISM
SOCIALISM
X
RATIO
ENCYCLOPEDIÉ
Kant,
1724
"God is not
the creator"
BOOLE
MORSE
BRAILLE
DNA,
1869
BIN. 2.0
Darwin,
1809
1984
E=mc²
cybernetics
FEEDBACK
AI
CRISP-R,
2012
ALPHA GO,
2015
ATGCCGCAATC
01101001011
Grow Your
Own Cloud,
2021
5G
4G
3G
2G
1G
DISBELIEF
BELIEF
"ALEXA, TELL ME THE TRUTH!"
"FROM DUST YOU CAME,
TO DUST YOU SHALL RETURN"
"HOMO DEUS"
1800
1900
2000
2100
Industrial Revolution 1750-1900
Age of Globalization
1870-1913
The digital revolution 2000- 2200
Organic revolution 2200-
DATAISM
ORGANISM

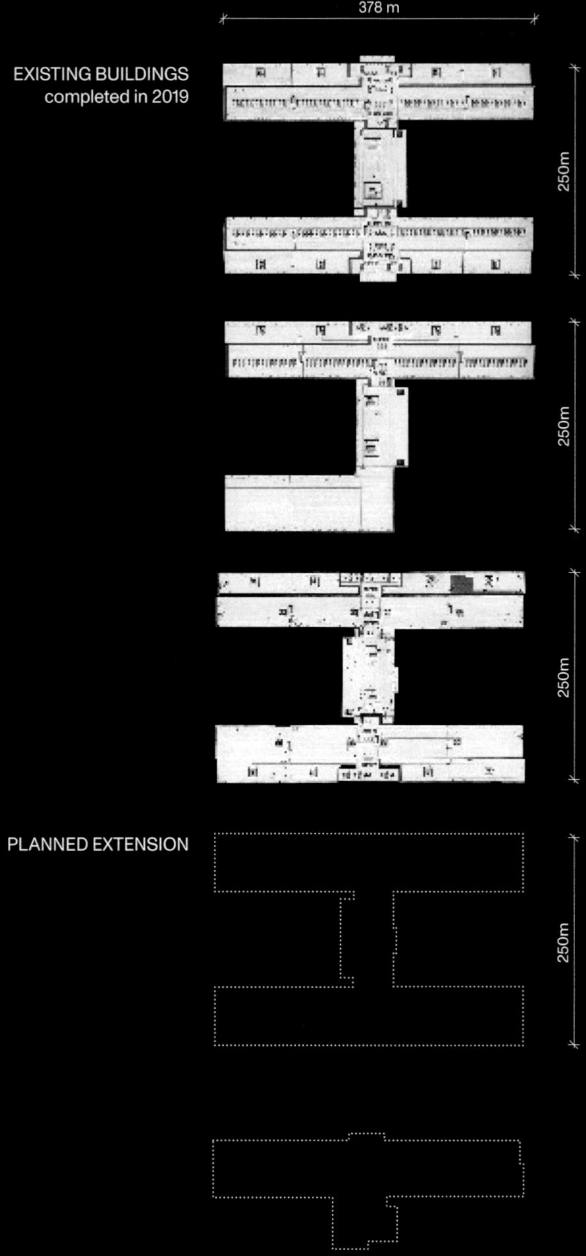


SURFACE OF THE DATA CENTER: 371.600 m2
ENERGY REQUIRED TO OPERATE THE DATA CENTER: 310 MW
LIFESPAN OF A SERVER: 3–5 years

of META's latest Sarpy Data Centre in Sarpy County, Nebraska (USA) & DNA Data
Source: DNA has a 521 year half-life. Matt Kaplan, Nature (2012).

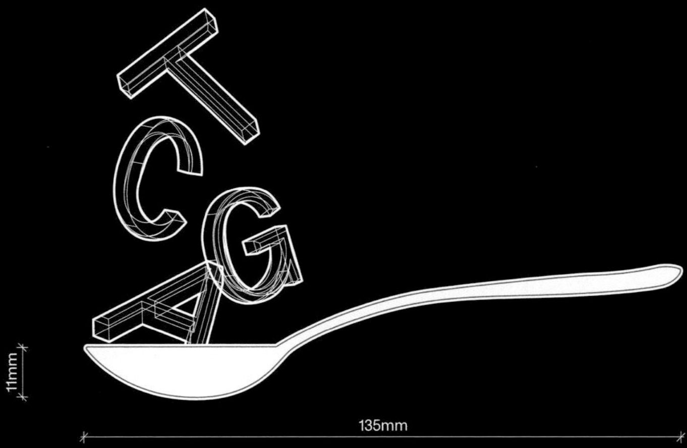


ESTIMATED VOLUME OF TODAY'S WORLD DIGITAL DATA: 64.2 zetabytes
NECESSARY AMOUNT OF DNA TO STORE IT: 20g
LIFESPAN OF DNA STORAGE (AT IDEAL PRESERVATION TEMPERATURE): ~1.5 million years

Fig. 4 Envisioned landscape of Orgánica: a Noetic Horizons of Data.

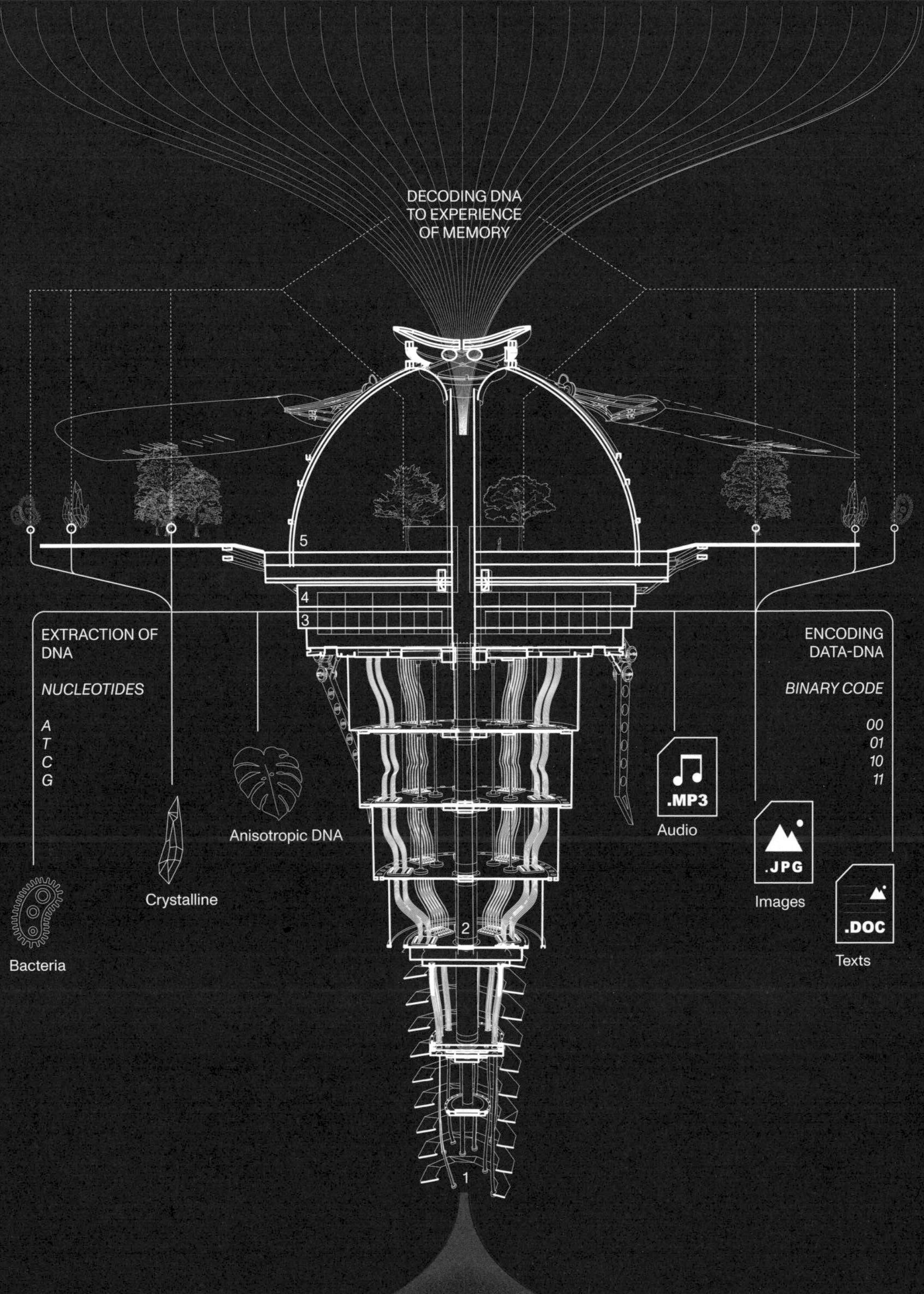


Fig. 5 Section of the organic data center:

1 Binary Data intake from underwater cables
2 Water cooling system
3 Data servers for temporary storage of binary data before encoding
4 Bio-data lab for DNA encoding
5 Greenhouse for cultivation of plants before re-planted outside

Ali Fard is a researcher, designer and educator, currently an assistant professor at the University of Virginia School of Architecture. Fard's work operates at the intersection of design and urban processes. He is particularly interested in the spatial imprints of technology and the urban disposition of infrastructure, and how design research can help ground the operational complexity, spatial hybridity and territorial scales of large technical systems. Though his design research practice AF/DR, Fard has been involved in a range of award-winning projects globally that reinforce design's critical and multiscalar role in the territorial dynamics of contemporary urbanization.

Georg Vrachliotis is Full Professor and Head of Chair for Theory of Architecture and Digital Culture, Department of Architecture, TU Delft. From 2016 Georg was dean of the Karlsruher Institut für Technologie (KIT) Faculty of Architecture and Chair for Architecture Theory (2014–2020). Previously he conducted research at the Institute for the History and Theory of Architecture at ETH Zürich. He studied architecture at the Berlin University of the Arts and did his PhD at the ETH Zürich. Besides that Georg was a visiting researcher and lecturer at the UC Berkeley Department of Architecture in California and the TU Wien in Vienna. Georg Vrachliotis is member of the advisory board of the magazine ARCH+, external examiner at Bartlett School of Architecture, UCL London, and academic advisory board member of the *Jaap Bakema Study Centre* at *The Het Nieuwe Instituut*, Museum for Architecture, Design and Digital Culture in Rotterdam. Georg Vrachliotis is the author and co-editor of several books on the history and theory of digital culture in architecture, such as *Complexity. Design Strategy and World View* (2008), *Simulation. Presentation Technique and Cognitive Tool* (2008), *Patterns. Ornament, Structure and Behaviour* (2009), *Code. Between Operation and Narration* (2010), and *Datatopia* (ARCH+ special issue, 2019). He recently published *The New Technological Condition: Architecture and Design in the Age of Cybernetics* (2022). Georg Vrachliotis curated the exhibition "Fritz Haller. Architect and Researcher" at the SAM Swiss Architecture Museum in 2014 in collaboration with the Institute for History and Theory (gta) at ETH Zurich (catalogue published by gta Publisher in 2014, with Laurent Stalder), the exhibition "Sleeping Beauty. Reinventing Frei Ottos Multihalle" (catalogue published by Spector Books in 2018) on the occasion of the 16th International Architecture Exhibition of the Venice Biennale in 2018. He recently curated the exhibition "Models, Media and Methods. Frei Otto's Architectural Research" at the School of Architecture at Yale University (2020).

Gökçe Önal is a doctoral candidate at the Faculty of Architecture at Delft University of Technology, with the research group Borders & Territories. She taught at METU and at Delft University of Technology and contributed to TU Delft's Massive Open Online educational portal as a course developer. She is currently an assistant editor for 4S Backchannels. Her doctoral research investigates the agency of remote sensing practices in the production of spatial knowledge.

Jeroen van Loon is a Dutch artist researching and visualising digital culture through the lens of ephemerality and permanence. His work has been exhibited internationally at venues such as Centre Pompidou, Transmediale, IDFA Doclab, Aksioma, Centraal Museum, HMKV | Dortmunder U, IMPAKT festival, Verbeke Foundation, SPRING performing arts festival, Nemo Biennale, V2_, KIKK Festival and MU Artspace. He received the K.F. Hein Art grant, the European Youth Award, was nominated for the New Technological Art Award (NTAA) 2022 and is part of the Brave New World Speaking Agency as a speaker on art and technology. Van Loon (b. 1985 in 's Hertogenbosch, The Netherlands, lives and works in Utrecht, NL) received a bachelor in Digital Media Design and a European Media Master of Arts from the HKU University of the Arts Utrecht.

Joost Grootens is a graphic designer, educator and researcher. Based in Amsterdam (NL) and Biel/Bienne (CH), his studio SJG designs books, maps, typefaces, spatial installations and digital information environments for publishers, including Lars Müller Publishers, nai010 publishers, Park Books and Phaidon press; for educational and research institutes such as ETH Zurich, Future Cities Laboratory Singapore, Technical University of Munich and Delft University of Technology; and for museums like Astrup Fearnly Museet Oslo, Serpentine Galleries London and Van Abbemuseum Eindhoven. Grootens is Professor of Artistic Research in Visual Design at the Royal Danish Academy in Copenhagen and formerly the head of the master programme Information Design at Design Academy Eindhoven and university lecturer at Leiden University. Grootens studied architectural design at the Gerrit Rietveld Academy Amsterdam and obtained his doctorate from Leiden University. His research addresses the transformation of the fields and practices of graphic design and mapmaking resulting from technological changes in tools to record, create, edit, produce and distribute visual information.

Kate Crawford is a leading international scholar of the social implications of artificial intelligence. She is a Research Professor at USC Annenberg, a Senior Principal Researcher at MSR, an Honorary Professor at the University of Sydney, and the inaugural Visiting Chair for AI and Justice at the École Normale Supérieure in Paris. Her latest book, *Atlas of AI* (Yale, 2021) won the Sally Hacker Prize from the Society for the History of Technology, the ASSI&T Best Information Science Book Award, and was named one of the best books in 2021 by *New Scientist* and the *Financial Times*. Over her twenty-year research career, she has also produced groundbreaking creative collaborations and visual investigations. Her project *Anatomy of an AI System* with Vladan Joler is in the permanent collection of MoMA in NYC and the V&A in London, and was awarded with the Design of the Year Award in 2019 and included in the Design of the Decades by the Design Museum of London. Her collaboration with the artist Trevor Paglen, *Excavating AI*, won the Ayrton Prize from the British Society for the History of Science. She has advised policy makers in the United Nations, the White House, and the European Parliament, and she currently leads the Knowing Machines Project, an international research collaboration that investigates the foundations of machine learning.

Kees Kaan founded KAAN Architecten in 2014. Kaan graduated in architecture at Delft University of Technology in 1987. Kaan is Chair of the department of Complex Projects at the faculty of Architecture at Delft University of Technology since 2006. In 2019 Kaan was appointed Chairman of the Architecture Department. Moreover, he has been Principal Investigator at Amsterdam Institute for Advanced Metropolitan Solutions (AMS) since 2016. Kaan is an international lecturer Numerous books and exhibitions have been dedicated to his body of work.

Marija Marić is an architect and researcher, working as a postdoctoral research associate at the Master in Architecture programme, University of Luxembourg, where she also teaches. Marija has been appointed as a co-curator of the Luxembourgish Pavilion at the 2023 Architecture Biennale in Venice, with the project that critically unpacks space mining through the perspective of resources. In 2020, she obtained her doctoral degree from the Institute for the History and Theory of Architecture (gta), ETH Zurich. Her dissertation, titled *Real Estate Fiction: Branding Industries and the Construction of Global Imaginaries*, looked into the role of communication strategists in the mediation, design and globalization of the built environment. Marija's research is organized around the questions of architecture, real estate, media and technology, and the production of the built environment and urban imaginaries in the context of global capitalism and the global flow of information.

Marina Otero Verzier is the head of the Social Design Masters at Design Academy Eindhoven. The programme focuses on new roles for designers attuned to contemporary ecological and social challenges. From 2015 to 2022, she was the director of research at Het Nieuwe Instituut (HNI), the Dutch Institute for Architecture, Design and Digital Culture. At HNI, she led initiatives focused on labour, extraction and mental health from an architectural and post-anthropocentric perspective, including 'Automated Landscapes', 'BURN OUT: Exhaustion on a Planetary Scale' and 'Lithium'. Previously, she was the director of Global Network Programming at Studio-X, Columbia University GSAPP, in New York. Otero was a co-curator at the Shanghai Art Biennial 2021, curator of the Dutch Pavilion at the Venice Architecture Biennale in 2018 and chief curator of the 2016 Oslo Architecture Triennale. She has co-edited *Lithium: States of Exhaustion* (2021), *More-than-Human* (2020), *Unmanned: Architecture and Security Series* (2016–2020), *Architecture of Appropriation* (2019), *Work, Body, Leisure* (2018) and *After Belonging* (2016), among others. Otero studied at

Delft University of Technology, ETSA Madrid and Columbia University GSAPP. In 2016, she received her PhD from ETSA Madrid.

Natalie P. Koerner (1987) is assistant professor at the Royal Danish Academy of Architecture (Copenhagen) and at the Centre for Privacy Studies at the University of Copenhagen, a Centre of Excellence funded by the Danish National Research Foundation (DNRF). Koerner combines research with teaching at the master's programme Spatial Design where she is responsible for running the first semester. In her practice, she builds and works on artistic research projects. Koerner has worked for several firms, including the artist Studio Olafur Eliasson. She studied architecture at the University of Cambridge (2005–2008), at the ETH Zurich (2009–2011), and received her PhD from the Royal Danish Academy (2015–2019). Her thesis *Towards the Meteorological: The Architecture of Data Centres and the Cloud* linked notions of a planetary imaginary to digital archival practices.

Negar Sanaan Bensi is an architect, educator, and researcher. She is a lecturer at the Architcture faculty, TU Delft , the chair of Borders & Territories. She holds a Ph.D. in History and Theory of Architecture from TU Delft and she is an UKNA fellow. She received an honourable mention in the National Archiprix 2011 in the Netherlands for her master graduation project. Her research focuses on the relation between architecture and territory, infrastructure, and inhabitation, specifically in the context of the Iranian Plateau and the Middle East. She has taught at TU Delft, Tilburg Fontys and Rotterdam Academy, Syracuse University NYC, and lectured at the University of Antwerp and IIAS in Leiden. She is part of the editorial board of the Footprint Journal and online Modi Operandi publication series (B&T). Some of her recent publications are: the upcoming book *An Inhabitable Infrastructure: Rethinking the Architecture of the bazaar* (Routledge), "The Qanat System: A Reflection on the Heritage of the Extraction of Hidden Waters" in *Adaptive Strategies for Water Heritage* edited by Carola Hein and *Footprint #23: The Architecture of Logistics* as the (co)editor. She is part of the research group "Borders & Territories" in the Architecture Department at TU Delft. She has experience in practice individually and in collaboration with several offices such as ZUS [Zone Urbaine Sensible], Import. Export Architecture, and GFC.

Paul Cournet is a French architect based in Rotterdam, The Netherlands. In 2022, Paul founded CLOUD, a creative agency that focuses on architecture, research and design. He is also the deputy editor of Capsule—the International Review of Radical Design & Desire Theory—a new publication delving into the world of design and its broader definition. Previously, Paul worked at OMA*AMO as a Senior architect between 2010 and 2021. At OMA, Paul worked on several projects currently under construction or built such as the future headquarters of SNCB*NMBS in Brussels and Lafayette Anticipations, the cultural foundation for Galeries Lafayette in Paris. Paul also worked within AMO, the think-tank branch of OMA. Projects Paul worked on for AMO include the Pierre Paulin Program exhibition in Bordeaux (2019), the Palermo Atlas—an urban study for the art biennale Manifesta 12 (2018), exhibitions and set designs for the fashion house Prada, and Monditalia—an exhibition for the 14th architecture Venice Biennale titled Fundamentals (2014).

Paul Swagerman is graduated in Visual Communication from the AKI, academy for Art and Industry. He has a broad interest in visual communication, cartography, infographics, photography, visual identity and multimedia design. Paul has the talent to bring together project content and representation in a readable and insightful way. Paul worked for, amongst others, ZUS [Zones Urbaines Sensibles] Architects, Artist Melle Smets (Department of Search), Bureau B+B Landscape Architecture, Kraaijvanger Architects, Maas Architects, LOER Architects, ECHO Architects, Musician Joris Voorn, Villa Zebra, art museum for children, Exhibition designers Koehorst in 't Veld, website builder Fotomat, and Ydoc Foundation, Platform for Visual Storytelling.

Ruha Benjamin is a professor of African American studies at Princeton University, founding director of the ida B.Wells Just Data Lab and a faculty associate in the centre for Infromation Technology Policy, Programme on History of Science, Centre for Health and Wellbeing, and Programme on Gender and Sexuality Studies. She is the author of three books, *Viral Justrice* (2022), *Race After Technology* (2019), *People's Science* (2013), and editor of *Captivating Technology* (2019). She is currently working on her fifth book, *Imagination: A Manifesto*. She writes and speaks widely about the relationship between innovation and inequity, knowledge and power, race and citizenship, health and justice.

Sabrina Chou is a Taiwanese-American artist and writer from Los Angeles, USA. She received her doctorate in practice-led art at the University of Oxford. Currently she is a lecturer in Critical Studies Fine Art at Central Saint Martins in London and a research affiliate of the Research Network for Philosophy and Technology. Sabrina holds an MFA from the Piet Zwart Institute in Rotterdam and a BA in Visual and Environmental Studies from Harvard University in Cambridge. She has presented her work internationally in exhibitions and performances, at, for instance, Human Resources Los Angeles, the 6th Moscow International Biennale for Young Art and Het Nieuwe Instituut. She has been an artist-in-residence at Centre International d'art et du paysage Île de Vassiviere and Sitterwerk Foundation. Sabrina writes on theory and art, including in *Thresholds 48: Kin* (MIT Press, 2020) and *Ringgummimatte* (Spector Books, 2021).

Shannon Mattern is a Professor of Anthropology at The New School for Social Research. Her writing and teaching focus on archives, libraries, and other media spaces; media infrastructures; spatial epistemologies; and mediated sensation and exhibition. She is the author of *The New Downtown Library: Designing with Communities*, *Deep Mapping the Media City*, and *Code and Clay, Data and Dirt: Five Thousand Years of Urban Media*, all published by University of Minnesota Press. In addition to writing dozens of articles and book chapters, she also contributes a regular long-form column about urban data and mediated infrastructures to *Places*, a journal focusing on architecture, urbanism, and landscape, and she collaborates on public design and interactive projects and exhibitions.

Shumon Basar is a writer, editor and curator. He is co-author of *The Age of Earthquakes: A Guide to the Extreme Present* with Douglas Coupland and Hans Ulrich Obrist. His edited books include *Translated By*, *Cities from Zero* and *Hans Ulrich Obrist Interviews Volume 2,The World of Madelon Vriesendorp, Did Someone Say Participate?, With/Without*. He is the Commissioner of the Global Art Forum in Dubai; Editor-at-Large of *Tank* magazine and Contributing Editor at *Bidoun* magazine; Director of the Format programme at the AA School; a member of Fondazione Prada's Thought Council and Art Jameel's Curatorial Council. Basar studied at Cambridge University, the Architectural Association, and Goldsmiths, University of London.

Datapolis Studio alumni
Aleksandra Milanowska, Alex Da Costa Gomez, Ali Fatih Gebeci, Alice Chau Ka Yee, Allard Meijer, Andrea Bautil, Arda Yildiz Data, Callum Birchall, Casper Aussems, Chris Aerts, Coen de Vries, Cyriac Levet, Damian Lachtara, Daniela Míková, Deniz Tichelaar, Dilara Turgut, Dimitrije Milić, Dominik Stoschek, Dorsa Ghaemi, Edward Zammit, Emma Chris Avrămiea, Fabio Sala, Farian Anwar, Federica Longoni, Filip Romaniuk, Fola-Sade Victoria Oshinusi, Francesca Monteleone, Franek Morka, Hanna Rudner, Hannah Zhu, Haruka Maeda, Li Ho Kong, iahui Yan, I-Hsuan Lee, Jacopo Zani, Jaka Korla, Jasmijn van der Harst, Joana Bessa e Silva Machado Vilaça, Jose Pedro Candido Morandeira, Julianne Guevara, Julie Moraca, Ka Yee Chau, Kaja Monika Wiewióra, Kelly Lai, Kianu Goedemond, Kim Sinnige, Lauritz Bohne, Léa Alapini, Lea Scherer, Luca Fontana, Luca Parlangeli, Marcello Carpino, Marina Bonet, Martah Lundheim, Matteo Gumirato, Max McGrath, Mine Çelik, Niklas Bahn, Patrycja Raszka, Pedro Pantaleone, Phat Ho, Qi Jing, Rafal Tarczyński, Ragna Nordström, Robert Bichlmaier, Rogier Tamminga, Rohan Shenoy, Romy Weeteling, Sandra Baggerman, Sandra Jasionytė, Scott Spoon, Sua Kim, Susanne van de Pol, Thanat Prathnadi, Theodor Reinhardt, Thierry Syriani, Tijmen Dijk, Vittoria Poletto, Weiming Yin, Xinyuan Zhang, Yanchen Wang, Yiyin Yu, Wu Yongyi , Yosua Yosua Adisapta, Pranata Andoko.

Students' assistants (Datapolis Studio)
Federica Longoni, Léa Alapini, Patrycja Raszka

# ACKNOWLEDGEMENTS

First, we would like to thank all the incredible students of the Datapolis studio at TU Delft. This book would have never existed without their collective force.

We are thankful to Kees Kaan, the head of the architecture department at the Technical University of Delft and the chair of Complex Projects who believed in this project since the beginning. We also thank our colleagues in the chair of Borders & Territories for their support and feedback during the process.

We are grateful to the publishing house nai010 publishers who has supported the book from its inception. Special thanks goes to Marcel Witvoet and Marja Jager who made this project a reality.

The book has benefited enormously from the photographer Paul Swagerman, the graphic design studio of Joost Grootens and the copy editor D'laine Camp. Without their input, the book would have never reached to this point.

We would like to thank the Creative Industries Fund NL Stimuleringsfonds, the Climate Institute as well as the chairs of Complex Projects and Theory of Architecture and Design Culture from TU Delft and KAAN Architecten who provided funding for the research and the production of this book.

We are thankful to all the intelligent minds that contributed to the seminar series during the Datapolis studio at TU Delft including Stavros Kousoulas, Tanja Herdt, Andrej Radman, Arman Noroozian, Tobias Borsdorff, Gökçe Önal, Georg Vrachliotis, Guillaume Pitron, Victor Munoz Sanz, Minke van Wingerden, Peter van Winderden, Lucas de Ruiter, Niels Schrader, Roel Backaert, Stijn Grove, Beatriz Colomina and Marina Otero Verzier.

We are also grateful to all the devoted collaborators who allowed us to develop our photographic documentation and open the doors of some exceptional sites including: Martine Vriens, Nanny van Hese, Anouk De Bont, Rebecca McFedries, Eva Wittebrood, Laurence Tu Mai, Tommaso Ghidini, Eric van Scherpenzeel, Marko van der Puil, Minke Van Wingerden, Peter Van Wingerden, Klaas Wierda, Peter Benedick.

Our Thank goes to E-flux and Nick Axel who allowed us to reach incredible thinkers to share their research with our project.

We are Thankful to the team behind the event Deep City 2021: Climate Crisis, Democracy and the Digital organized by EPFL in Lausanne in collaboration with SUTD Singapore University of Technology and Design and Hong Kong University, who through their productive feedback and insights allowed us to push our research further.

Thanks to Indira van 't Klooster and ARCAM who gave us the opportunity to share our knowledge during the event "Architecture and the Computer—what is the physical impact of data on our built environment?" at Pakhuis de Zwijger in Amsterdam in 2020.

To all our friends and collaborators who provided continuous support and reflections on the ongoing work during these years.

*The book is dedicated to the memory of Niklas Bahn, a Datapolis studio alum.*

# The Cloud is

Pages from *The Extreme Self: Age of You*. Douglas Coupland, Hans-Ulrich Obrist and Shumon Basar, design by Daly & Lyon. Published by Verlag der Buchhandlung Walther und Franz König, 2021.

CREDITS

Editors
Paul Cournet, Negar Sanaan Bensi

Contributors
Kate Crawford, Ruha Benjamin, Shannon Mattern, Marina Otero Verzier, Georg Vrachliotis, Kees Kaan, Sabrina Chou, Ali Fard, Joost Grootens, Marija Marić, Gökçe Önal, Natalie P. Koerner, Jeroen van Loon, Shumon Basar

Copy editing
D'Laine Camp

Graphic Design
SJG / Joost Grootens, Dimitri Jeannottat, Julie da Silva

Photography
Paul Swagerman

Lithography and Printing
NPN Drukkers, Breda

Paper
Maxioffset 90 g/m² and Heaven42 115 g/m²

Publisher
Marcel Witvoet, nai010 publishers, Rotterdam

Production
Marja Jager, nai010 publishers, Rotterdam



nai010 publishers is an internationally orientated publisher specialized in developing, producing and distributing books in the fields of architecture, urbanism, art and design. www.nai010.com

nai010 books are available internationally at selected bookstores and from the following distribution partners:

North, Central and South America
Artbook | D.A.P., New York, USA, dap@dapinc.com

Rest of the world
Idea Books, Amsterdam, the Netherlands, idea@ideabooks.nl

For general questions, please contact nai010 publishers directly at sales@nai010.com or visit our website www.nai010.com for further information.

Printed and bound in the Netherlands.

ISBN 978-94-6208-719-4
NUR 648
BISAC ARC010000

Datapolis is also available as e-book:
ISBN 978-94-6208-751-4 (e-book)

This publication was made possible by financial support from Creative Industries Fund NL/Stimuleringsfonds, The Climate Institute, TU Delft, Complex Projects, TU Delft, Theory of Architecture and Digital Culture, TU Delft, KAAN Architecten

JOURNEY OF AN IPHONE

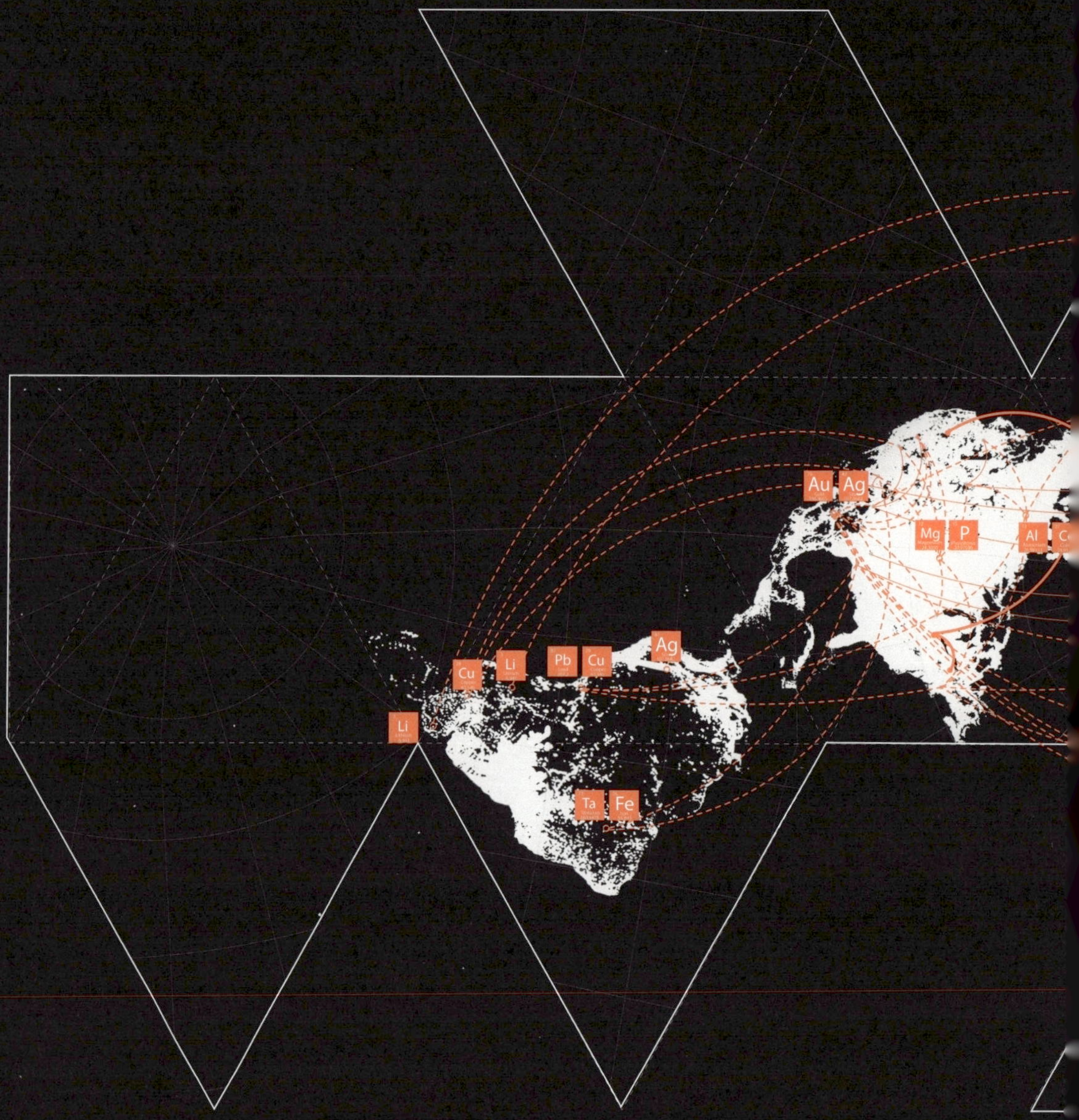


Raw Material Flow
Component Flow
Product Flow
Mine
Component Manufacturer
Distribution Centre
Apple Selling Point

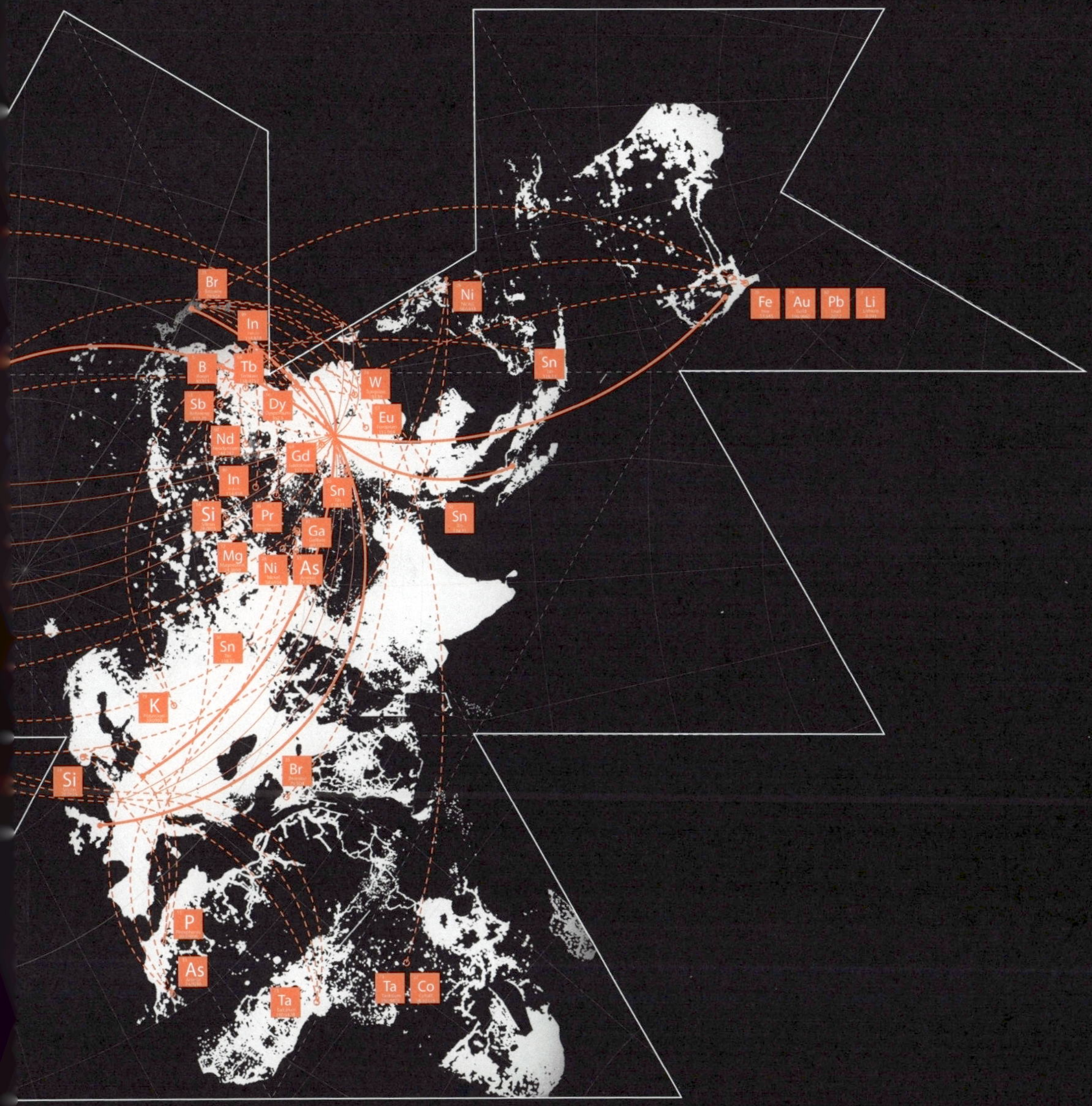


Starting from raw material extractions, component refining, and manufacturing process, to the distribution in worldwide's stores, your iPhone travels 800,000 km to reach your pocket. A smartphone contains around more than 34 raw materials, where seven of the 34 were classified as 'critical raw materials' or rare earth elements by the EU commission in 2014. Drawing by Jeremy Li Ho Kong and Alice Chau Ka Yee. Source: Jay Greene/CNET.